Power Systems

More information about this series at http://www.springer.com/series/4622

Lei Wang · Man-Chung Wong
Chi-Seng Lam

Adaptive Hybrid Active Power Filters

Springer

Lei Wang
Department of Electrical and Computer Engineering, Faculty of Science and Technology
University of Macau
Macau, China

Man-Chung Wong
Department of Electrical and Computer Engineering, Faculty of Science and Technology
University of Macau
Macau, China

Chi-Seng Lam
State Key Laboratory of Analog and Mixed Signal VLSI
University of Macau
Macau, China

ISSN 1612-1287 ISSN 1860-4676 (electronic)
Power Systems
ISBN 978-981-10-8826-1 ISBN 978-981-10-8827-8 (eBook)
https://doi.org/10.1007/978-981-10-8827-8

Library of Congress Control Number: 2018948613

Printed on acid-free paper

This Springer imprint is published by the registered company Springer Nature Singapore Pte Ltd.
The registered company address is: 152 Beach Road, #21-01/04 Gateway East, Singapore 189721, Singapore

To My Parents: Jun-Feng Wang and Xin Tao
To My Beloved Wife: Liang Dong
To My Brother: Yu Wang

From Lei Wang

To My Parents: Kok Kay and Yuk Ying
To My Beloved Wife: Ebbie Wong
To My Daughter: Oriana Wong

From Man-Chung Wong

To My Parents: Sio-Hong Lam and In-Fong Lei
To My Beloved One
To My Brothers: Chi-Kit Lam and Chi-Man Lam

From Chi-Seng Lam

Preface

The smart grid, regarded as the next-generation power grid, is considered as a promising solution for energy crisis. Since smart grid is interconnected with different grids, the development of smart grid brings many new challenges for power quality. Power quality compensators are one of the key technologies for power quality improvement in smart grids. Undoubtedly, lower cost, lower loss, wider compensation range, and better performance will be the development trends and goals for power quality compensators in the coming decades. Among different current quality compensators, the state-of-the-art hybrid active power filters (HAPFs) possess high potential to be further developed due to the desired characteristics of lower cost, lower loss, higher reliability, and better performance. As a result, gaining the knowledge of state-of-the-art design and control techniques of HAPFs is important and useful for students, researchers, and engineers.

In this book, the different state-of-the-art power quality compensators are mainly focused on. The content starts by comprehensively reviewing the power quality issues in Chap. 1. Then, Chap. 2 provides the comprehensive study of characteristics, cost, reliability, power loss, and tracking performance of different power quality compensators. The mathematical analysis, modeling, design, control, and applications of a state-of-the-art adaptive thyristor-controlled LC-coupling HAPF (TCLC-HAPF) are investigated in Chaps. 3–8 as the perspectives solution for power quality compensations. The design and implementation of a three-phase three-wire TCLC-HAPF experimental hardware prototype and the experimental results are provided in Chap. 9. Finally, Chap. 10 draws the conclusions and perspectives for future works. In addition, the system analysis, design, and control techniques presented in this book can be extendable to other power quality compensators, such as static var compensators (SVCs), active power filters (APFs), static synchronous compensators (STATCOMs), etc.

This book is intended for undergraduates, postgraduates, doctorates, or professionals with power electronics background. The book can also be useful to fellow researchers and electrical power engineers who are specializing in power quality issues, power electronics, power filters, and control techniques. After reading this book, the readers can gain understanding of the state-of-the-art techniques of inverter design, pulse width modulation (PWM), compensation methods, control methods, hardware design, etc.

Macau, China

Lei Wang
Man-Chung Wong
Chi-Seng Lam

Acknowledgements

The research works presented in this book were carried out in the Power Electronics Laboratory/Electric Power Engineering Laboratory, Department of Electrical and Computer Engineering (ECE), Faculty of Science and Technology (FST), and State Key Laboratory of Analog and Mixed-Signal VLSI, University of Macau (UM), Macao, China. We would like to express our gratitude to the Science and Technology Development Fund, Macao SAR (FDCT) with project codes: 109/2013/A3, 015/2008/A1, 025/2017/A1, and Research Committee (RC) of University of Macau (UM) with project codes: MRG012/WMC/2015/FST, MYRG2015-00009-FST, MYRG2017-00038-FST, MYRG2015-00030-AMSV, and MYRG2017-00090-AMSV.

We would like to express our hearty and profound gratitude to Prof. Han Ying-Duo from Tsinghua University, China for having opened our vision. We also appreciate him for the immensely inspirational guidance and precious advices.

Last, but certainly not least, we would like to send our heartfelt appreciation to our families, who endured our dedication to this book.

Contents

1 Introduction 1
1.1 Power Quality Issues and Its Market 1
1.1.1 Low Power Factor 2
1.1.2 Current Harmonic Pollution 2
1.1.3 Current Unbalanced Problem 3
1.1.4 Power Quality Compensation Market 4
1.2 Development of Power Quality Compensators 5
1.3 Inductive-, Capacitive-and Adaptive-Coupling Power Quality Compensators and Their Control Issues 6
1.3.1 The Reference Signals Determination Methods 6
1.3.2 Pulse Width Modulations 8
1.4 TCLC-HAPF and Its Potential Advantages 8
1.5 Research Challenges and Goals 9
1.6 Book Organization 11
1.7 Appendix: Voltage and Current standards 12
References 13

2 Comparisons Among Thyristor Controlled LC-Coupling Hybrid Active Power Filter (TCLC-HAPF) and Other Different Power Quality Filters 17
2.1 Introduction 17
2.2 Structures of APF, HAPF and TCLC-HAPF 19
2.3 V-I Characteristics of APF, HAPF and TCLC-HAPF 20
2.3.1 V-I Characteristics of APF, HAPF and TCLC-HAPF 21
2.3.2 Simulation Case Studies 23
2.4 Cost Comparison Among APF, HAPF and TCLC-HAPF 26
2.4.1 Cost Comparison Among APF, HAPF and TCLC-HAPF Under Low Voltage Levels 26
2.4.2 Cost Comparisons Among APF, HAPF and TCLC-HAPF Under High Voltage Levels 28

2.5 Reliability Comparison Among APF, HAPF and TCLC-HAPF 29
2.6 Power Loss Comparison Among APF, HAPF and TCLC-HAPF 31
2.7 Tracking Performance Comparison Among APF, HAPF and TCLC-HAPF 33
2.8 Experimental Results 35
2.8.1 Experimental Results of APF 35
2.8.2 Experimental Results of HAPF 35
2.8.3 Experimental Results of TCLC-HAPF 37
2.9 Summary 37
2.10 Appendix: Calculation of Failure Rate for Different Components 40
References 44

3 Mitigation of the Harmonic Injection in TCLC Part and Nonlinear Hysteresis PWM Control in Active Inverter Part of Thyristor Controlled LC-Coupling Hybrid Active Power Filter (TCLC-HAPF) 47
3.1 Introduction 48
3.2 Circuit Configuration and Modeling of TCLC Part and TCLC-HAPF 49
3.3 Mitigation of the Harmonic Injection in TCLC Part 49
3.3.1 Mitigation of the Harmonic Injection in TCLC Part 49
3.3.2 The Selection of n_1 and n_2 Through the Design of L_c 52
3.3.3 Simulation and Experimental Verifications of the Mitigation of the Harmonic Injection in TCLC Part 53
3.4 Nonlinear Hysteresis PWM Control in Active Inverter Part of TCLC-HAPF 59
3.4.1 Compensating Current Characteristics of TCLC-HAPF 60
3.4.2 Non-linear Adaptive Hysteresis Band PWM Controller for TCLC-HAPF 63
3.5 Summary 71
References 72

4 Modeling and Parameter Design Method of Thyristor Controlled LC-Coupling Hybrid Active Power Filter (TCLC-HAPF) for Balanced/Unbalanced Loading Compensation 75
4.1 Introduction 75
4.2 Circuit Configuration and Modeling of TCLC-HAPF 77
4.3 Proposed TCLC-HAPF Parameter Design Method Design for Balanced Loads 79

4.4 Proposed TCLC-HAPF Parameter Design Method Design for Unbalanced Loads . . . 82
4.4.1 Design of V_{DCf}, C_{PF} and L_{PF} Based on Power Flow Analysis Under Fundamental Frequency Consideration . . . 82
4.4.2 Design of V_{DCh} Based on Harmonic Frequency Analysis . . . 86
4.4.3 Design of L_c for Current Ripple Filtering . . . 88
4.4.4 Summary of TCLC-HAPF Parameter Design . . . 88
4.5 Simulation Case Studies . . . 89
4.6 Experimental Results . . . 93
4.7 Summary . . . 96
References . . . 101

5 Proposed Unbalanced Control Strategy for Thyristor Controlled LC-Coupling Hybrid Active Power Filter (TCLC-HAPF) . . . 103
5.1 Introduction . . . 103
5.2 Circuit Configuration of Three-Phase Three-Wire TCLC-HAPF . . . 105
5.3 Proposed Unbalanced Control Strategy for TCLC-HAPF . . . 106
5.3.1 TCLC Part Control Strategy . . . 107
5.3.2 The Active Inverter Part Control Strategy . . . 112
5.3.3 The Proposed Hybrid Controller for TCLC-HAPF . . . 114
5.4 Simulation and Experimental Results . . . 115
5.5 Summary . . . 122
5.6 Appendix: Balancing Three-phase Fundamental Active Power by Reactive power compensation . . . 122
References . . . 126

6 Minimizing Inverter Capacity Design and Comparative Performance Evaluation of Static Var Compensator Coupling Hybrid Active Power Filters (SVC-HAPFs) . . . 129
6.1 Introduction . . . 129
6.2 Circuit Configuration and Modeling of SVC-HAPFs . . . 131
6.3 Ratio of Phase Active Inverter Rating and SVC Part Rating and Required Minimum DC-Link Voltage . . . 132
6.3.1 The Parameter Design and Characteristics of FC-TCR and TCLC . . . 134
6.3.2 Fundamental Frequency Analysis of R_{tot1} and V_{DC_tot1} . . . 136
6.3.3 Harmonic Frequency Analysis of R_{totn} and V_{DC_totn} . . . 137
6.3.4 The Minimizing Inverter Capacity Design of Total R_{tot} and V_{DC_Tot} . . . 141

6.3.5 Comparison of SVC-HAPFs 144
6.4 Simulation Results 144
6.5 Experimental Results 146
6.6 Summary 148
References 148

7 Adaptive DC-Link Voltage Control of Thyristor Controlled LC-Coupling Hybrid Active Power Filter (TCLC-HAPF) 151
7.1 Introduction 151
7.2 Circuit Configuration of Three-Phase Three-Wire TCLC-HAPF 154
7.3 Proposed Simplified Minimum DC-Link Voltage Calculation Method 156
7.3.1 Deduction of DC-Link Voltage (V_{DCxf}) at Fundamental Frequency 156
7.3.2 Deduction of DC-Link Voltage (V_{DCxh}) at Harmonic Frequency 158
7.3.3 Comparison Between Conventional and Proposed Minimum V_{DC} Calculation Methods 159
7.4 Control Block of the Proposed Adaptive DC-Link Voltage Controlled TCLC-HAPF 160
7.4.1 TCLC Control Block 160
7.4.2 Active VSI Control Block 162
7.4.3 Adaptive DC-Link Voltage Control Block 162
7.5 Simulation Case Studies 163
7.5.1 Under Compensation by Adaptive V_{DC} Controlled TCLC-HAPF 164
7.5.2 Over Compensation by Adaptive V_{DC} Controlled TCLC-HAPF 169
7.6 Experimental Results 171
7.6.1 Dynamic Performance of Adaptive V_{DC} Controlled TCLC-HAPF to Load Variation 172
7.6.2 Comparison with Fixed V_{DC} Controlled TCLC-HAPF 172
7.7 Summary 176
References 177

8 Selective Compensation of Distortion, Unbalanced and Reactive Power of a Thyristor Controlled LC-Coupling Hybrid Active Power Filter (TCLC-HAPF) 181
8.1 Introduction 181
8.2 Circuit Configuration of the TCLC-HAPF 183
8.3 Power Analysis of the Proposed Selective Compensation 184

8.4 Proposed Selective Compensation Control Strategy of TCLC-HAPF 185
8.4.1 Active Inverter Part Control 186
8.4.2 TCLC Part Control 188
8.4.3 Compensation Priority Selection Among k_Q, k_U and k_H 189
8.4.4 Control Block of TCLC-HAPF 191
8.5 Simulation and Experimental Verifications 193
8.5.1 PSCAD Simulations 194
8.5.2 Experimental Results 196
8.6 Summary 201
References 201

9 Implementation of a 110 V-5 kVA Three-Phase Three-Wire of Thyristor Controlled LC-Coupling Hybrid Active Power Filter (TCLC-HAPF) Experimental Prototype 205
9.1 Introduction 205
9.2 Circuit Configuration of the TCLC-HAPF Experimental Prototype 206
9.3 Hardware Design of TCLC-HAPF Experimental Prototype 207
9.3.1 Power Semiconductor Switching Devices and Their Drivers 208
9.3.2 Transducer with Signal Conditioning Boards 211
9.4 Software Design of TCLC-HAPF Experimental Prototype 212
9.5 Experimental Results 214
9.5.1 Experimental Results of TCLC-HAPF for Dynamic Inductive and Capacitive Reactive Power Compensations 214
9.5.2 Experimental Results of TCLC-HAPF for Reactive Power and Harmonic Compensations 217
9.5.3 Experimental Results of TCLC-HAPF for Unbalanced Loading Compensation 217
9.5.4 Experimental Results of TCLC-HAPF Compensation During Voltage Dip 218
9.5.5 Experimental Results of TCLC-HAPF Compensation Under Voltage Fault 221
9.6 Summary 224
References 224

10 Conclusions and Prospective for Further Work 225
10.1 Conclusions 225
10.2 Perspectives for Future Works 227

Biography of Authors 229

Abbreviations

A	Amperes
AC	Alternating current
A/D	Analog-to-digital
ADC	Analog-to-digital converter
APF	Active power filter
ASD	Adjustable speed drive
CB	Capacitor bank
CT	Coupling transformer
CGCI	Capacitive-coupling grid-connected inverter
DC	Direct current
DPF	Displacement power factor
DSP	Digital signal processor
DVR	Dynamic voltage restorer
FFT	Fast Fourier Transform
FC-TCR	Fixed capacitors and thyristor-controlled reactor
HAPF	Hybrid active power filter
HPF	High-pass filter
HVDC	High-voltage dc transmission
HGCI	Hybrid grid-connected inverter
IEEE	Institute of electrical and electronics engineering
IGBT	Insulated gate bipolar transistor
IPM	Intelligent power module
IGCI	Inductive-coupling grid-connected inverter
L	Inductor
LC	Inductor and capacitor
LPF	Low-pass filter
P	Proportional
PCB	Printed circuit board
PF	Power factor
PI	Proportional and integral

P.M.	Phase margin
PPF	Passive power filter
PSCAD/EMTDC	Power system computer-aided design/electromagnetic transient in dc system
PWM	Pulse width modulation
RMS	Root mean square
S/H	Sample and hold
STATCOM	Static synchronous compensator
SVC	Static var compensator
SVC-HAPF	Static var compensator-coupling hybrid active power filter
TCLC	Thyristor-controlled LC
TCLC-HAPF	Thyristor-controlled LC-coupling hybrid active power filter
TDD	Total demand distortion
THD	Total harmonic distortion
UPQC	Unified power quality compensator
UPS	Uninterruptible power supplies
V	Volts
VSI	Voltage source inverter

Chapter 1
Introduction

Abstract This chapter presents a summary and overview of the major power quality issues (low power factor, harmonic pollution and unbalanced problem) by taking the power quality measurement data in Macau and China as examples. And, the potential market for reactive power and harmonic compensation in China is investigated. In order to solve the above major power quality issues, the historical review of different power quality compensator topologies and control methods are provided. Finally, the organization of this book is introduced at the end of this chapter.

Keywords Active power filter · Power quality compensation · Power factor
Passive power filter · Pulse width modulation · Hybrid active power filter
Harmonic distortion · Inverter

1.1 Power Quality Issues and Its Market

With the proliferation of power electronics and motor loading in power system, the loading becomes dynamic, fast time varying and nonlinear, which can cause many power quality issues such as: low power factor, harmonic pollution, unbalanced problem [1–4], etc. Thus, it is becoming more different for electricity utility to supply the customers with the voltage and current within the standards [5, 6], Taking the measurement data in Macau as example, some of the major power quality problems are low power factor, current harmonic pollution and unbalanced current. The aforementioned power quality problems are discussed in the following with the power quality measurement data recorded by Electric Power Engineering Laboratory, University of Macau, Macao, China [7, 8].

L. Wang et al., *Adaptive Hybrid Active Power Filters*, Power Systems,
https://doi.org/10.1007/978-981-10-8827-8_1

1.1.1 Low Power Factor

The application of induction motor loads can cause a phase shift between the voltage and current in the power network. This results in the lowering of the power factor of the loads. Loads with low power factor draw more reactive current than those with high power factor. And larger reactive power can increase the system losses and lower the power network reliability. Besides, electricity utilities usually charge the industrial and commercial customers a higher electricity fee in low power factor situation. According to the policy of the electricity company of Macau, Companhia de Electricidade de Macau (CEM), when the power factor of Tariff Group B or C is lower than 0.857, the customers are required to pay extra charges for the reactive power. Table 1.1 gives the power factor measurement data for 7 different buildings in Macau [7–9]. In Table 1.1, the shaded area means that the corresponding building is required to pay extra charge for the reactive power consumption.

1.1.2 Current Harmonic Pollution

The widespread applications of power electronic devices can increase the distortion and disturbances on the current in the power network, as the power electronic devices draw harmonic current from the power utility. Harmonic current distortion causes various problems in power network and consumer products, such as mal-operation of control devices, transformer overheating, the defection of accuracy in power meters, etc. There are many standards related to the power quality issues. Among them, the IEEE standard 519-2014 [5] is one of the most widely used standards, and the detailed descriptions related to voltage and current standards are given in Appendix. With reference to the IEEE standard 519-2014 [5], the

Table 1.1 Power factor measurement data for 7 buildings in Macau

Type of building			Commercial building	Hotel	Middle school	Indoor sport center	Public admin. building A	Public admin. building B	Public admin. building C
Power factor	Phase A	Max.	0.953	0.997	0.953	0.686	0.963	0.885	0.959
		Min.	0.808	0.910	0.761	0.634	0.792	0.735	0.724
	Phase B	Max.	0.950	0.996	0.996	0.809	0.972	0.879	0.967
		Min.	0.825	0.911	0.750	0.748	0.791	0.727	0.724
	Phase C	Max.	0.915	0.993	0.969	0.901	0.962	0.922	0.937
		Min.	0.758	0.887	0.627	0.854	0.807	0.758	0.714
	Three-phase	Max.	0.940	0.994	0.967	0.771	0.957	0.887	0.951
		Min.	0.801	0.905	0.771	0.723	0.797	0.742	0.720

Note The shaded area means undesirable measurement data

Table 1.2 Current *THDi* measurement data for 7 buildings in Macau [7–9]

Type of building			Commercial building	Hotel	Middle school	Indoor sport center	Public admin. building A	Public admin. building B	Public admin. building C
Total current harmonic distortion (THD_i %)	Phase A	Max.	37.04	33.21	46.83	13.75	28.09	19.27	25.61
		Min.	9.08	6.04	3.01	10.89	3.92	2.82	2.54
	Phase B	Max.	39.96	30.80	64.13	17.89	50.75	20.06	28.84
		Min.	9.61	7.14	3.54	14.40	3.78	2.42	3.10
	Phase C	Max.	45.41	28.66	63.73	24.47	47.65	27.12	27.01
		Min.	9.67	6.47	2.89	18.79	5.60	3.56	2.37

Note The shaded area means unsatisfactory data

acceptable Total Demand Distortion (*TDDi*) is $\leq 15\%$ with I_{SC}/I_L in the 100 < 1000 scale. The nominal rate current is assumed to be equal to the fundamental load current in the worst case analysis, which results in $THD = TDD \leq 15\%$. Therefore, this book evaluates the current harmonics performance by setting an acceptable standard with $THD \leq 15\%$. However, nearly all of the measured *THDi* of phase current for the buildings in Macau exceed the IEEE standards 519-2014 [5] as shown in the shaded area in Table 1.2 [7–9].

1.1.3 Current Unbalanced Problem

When unequal loads are connected to the three phase utility distribution system, unbalanced problem could occur. The unbalanced problem leads to the additional heating and loss in the stator windings, damage on the over-loaded phase power cable, reduction of transmission capability, increase in transmission loss, etc. [10–13]. Figure 1.1 shows the phase and neutral current RMS measurement data for

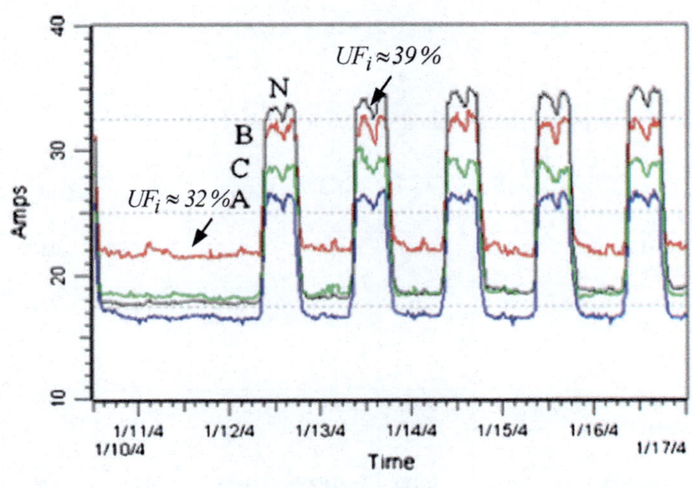

Fig. 1.1 Phase and neutral current RMS measurement data for one building in Macau

one building in Macau [7–9], in which the current unbalanced factor (UF_i) is always larger than 30% during office hours.

1.1.4 Power Quality Compensation Market

Concerning the above three major power quality issues, there is a large power quality compensation market. Taking China as an example, the reactive power and harmonic compensation market is shown in Tables 1.3 and 1.4.

From Tables 1.3 and 1.4, it can be seen that the proportions of reactive power and harmonic compensations in China were just 13.0 and 13.5% until the year of 2012. Nevertheless, according to the China National Bureau of Statistic's release, by 2020, China's related market will amount to 15 billion RMB. Therefore, there is a large potential market for the power quality compensators in China.

Concerning the above three major power quality issues and the large potential market, researchers have proposed many different power quality compensators to

Table 1.3 Reactive power compensation market in China [11, 12]

Year	2006	2007	2008	2009	2010	2011	2012
Reactive power compensation (× 10^8 kvar)	0.41	0.37	0.39	0.42	0.45	0.49	0.53
Reactive power compensation market (× 10^9 RMB)	26.81	33.50	24.40	25.97	28.57	32.33	37.10
The proportion of reactive power compensation (%)	1.5	2.5	3.5	3.7	6.0	9.0	13.0
Actual reactive power compensation (×10^9 RMB)	0.41	0.59	0.85	0.96	1.71	2.91	4.82

Table 1.4 Harmonic power compensation market in China [11, 12]

Year	2006	2007	2008	2009	2010	2011	2012
The national electricity load (×10^9 kWH)	2836.7	3245.8	3440.0	3645.0	3936.6	4251.5	4591.7
Harmonic power (15% of loads) (×10^9 kvar)	0.049	0.056	0.059	0.062	0.067	0.073	0.079
Harmonic power compensation market with 1200 RMB/kvar (×10^9 RMB)	1.06	1.58	2.18	3.00	4.85	7.86	12.7
The proportion of Harmonic compensation (%)	3.0	3.2	4.4	5.0	5.5	7.5	10.0
Harmonic compensation market (×10^9 RMB)	0.032	0.050	0.095	0.15	0.269	5.90	12.74

address those problems. In the following part, the development of power quality compensators will be introduced.

1.2 Development of Power Quality Compensators

Installation of the current quality compensators is one of the solutions for the low power factor, harmonic current pollution and unbalanced problem. Different power quality compensators and their parameter design methods are summarized in Table 1.5 and compared in historical order in the following:

Shunt capacitor banks (CBs) were firstly applied in power systems in around 1900s for power-factor correction and feeder voltage control due to its advantages of low cost and flexibility of installation. However, CBs can easily get burnt if the current harmonics level is high. To compensate the current harmonics, the passive power filters (PPFs) were proposed in 1940s. Unfortunately, the PPFs have many disadvantages such as low dynamic performance, resonance problem and lack of unbalanced compensation ability [14]. The thyristor controlled static var compensators (SVCs) were firstly proposed in 1960s [15]. And the SVCs are popularly for dynamic reactive power [16–19] and unbalanced power compensations [20–22]. However, the SVCs suffer from drawbacks such as resonance problem, harmonic current injection and poor harmonic compensation ability. To overcome the drawbacks and improve the performances of SVCs simultaneously, the inverter based controlled active power filters (APFs) were proposed in the year of 1976 [23, 24]. Unfortunately, APFs still cannot have large scale development in the power quality markets due to the high initial and operational costs. Afterwards, the *LC*-coupling hybrid active filters (HAPFs) were proposed in the year of 2003. And, HAPFs have lower rating of active inverter part than the APFs. Since the rating of active inverter part is proportional to the cost of compensators, the HAPFs are more cost-effective than the APFs [25–30]. However, the HAPFs have a quite narrow compensation range, which limits its compensation ability. When the loading reactive power is outside its designed range, it loses its low-inverter rating advantages [30]. In the year of 2014, the topology of thyristor controlled *LC*-

Table 1.5 Characteristics of different active current quality compensators and their corresponding parameter design methods

Year	1900s	1940s	1960s	1976	2003	2014
Compensators	CBs	PPFs	SVCs	APFs	HAPFs	TCLC-HAPF
Resonance prevention	Yes	Yes	Yes	No	No	No (*)
Compensation Range	Fixed	Fixed	Large	Large	Narrow	Large (*)
Harmonics Compensation	No	Yes	No	Yes	Yes	Yes
Unbalance problem	No	No	Yes	Yes (High V_{DC})	No	Yes (Lower V_{DC})(*)

Notes The shaded area means undesirable characteristic, and (*) means the potential characteristic which will be discussed in this book coupling hybrid active power filter (TCLC-HAPF) is proposed in [31], in which this state-of-the-art TCLC-HAPF has the characteristics of a wider compensation range than HAPFs and lower dc-link voltage than APFs for power quality compensation. Until now, the complete studies of characteristics, design techniques and applications of the TCLC-HAPF are still lacking, and hopefully this book can fill up this gap.

Based on the above discussions and Table 1.5, the traditional power quality compensators like CBs, PPFs and SVCs have relatively poor performance and suffer from many inherent problems. On the other hand, the APFs, HAPFs and TCLC-HAPF have much better performance than the CB, PPFs and SVCs. The existing control issues of APFs, HAPFs and TCLC-HAPF will be given and compared in Sect. 1.3. And the detailed comparison of APFs, HAPFs and TCLC-HAPF will be given in Chap. 2.

1.3 Inductive-, Capacitive-and Adaptive-Coupling Power Quality Compensators and Their Control Issues

To control the inductive-coupling power quality compensators (like APFs) and capacitive-coupling power quality compensators (like HAPFs) and adaptive-coupling power quality compensator (like TCLC-HAPF), two steps are most necessary, namely: (a) the reference signals detection (discussed in Sect. 1.3.1) and (b) the pulse width modulations (PWMs) (discussed in Sect. 1.3.2). In the following, the control issues of inductive-, capacitive- and adaptive-coupling power quality compensators will be discussed and compared for reactive power, harmonic current and unbalanced power compensation.

1.3.1 The Reference Signals Determination Methods

The reference current or voltage is used for power compensators to compensate the power quality issues, which include the reactive, harmonic, and unbalance power components. To determine the reference current (or voltage), different control methods have been proposed for APFs, HAPFs and TCLC-HAPF.

Akagi et al. [32] firstly proposed instantaneous p-q control method in order to eliminate the reactive, harmonic, and unbalanced power of the loading instantaneously. In order to adapt instantaneous p-q control method under different voltage conditions (distorted, unbalanced, etc.), many other control techniques were further developed, such as: d-q control method [24, 25, 33–36], p-q-r control method [37, 38], Lyapunov function-based control method [39], etc. However, those instantaneous power control methods [24, 25, 32–39] are dedicated to inverter/converter-based structures. And, their corresponding performances are highly dependent on the computation speed and the switching frequency of the digital controllers and the switching devices. Another popular control method for APFs and HAPFs is to balance the system by compensating the negative- and zero-sequence current components in unbalanced loading situation, as the oscillating power/voltage/current can be analytically expressed as positive-, negative- and zero-sequence components (+, − and 0 sequences) [40]. The major drawback of this control method is that the sequence components introduced by harmonics are not taken into consideration. To solve this problem, the authors in [41–44] combine the above instantaneous control methods with the +, − and 0 sequence control method, but the computation steps increase a lot, thus significantly increasing the control complexity. Recently, Leszek et al. [45, 46] proposed a power analysis control method based on the theory of current' physical component (CPC) to compensate the reactive power in unbalanced three-phase four-wire system. However, after this power analysis method compensation, the active power remains unbalanced, which means the unbalance power cannot be completely eliminated.

With all the above control methods, both APFs [24, 32–40, 45, 46] and HAPFs [25–29, 41–44] can effectively compensate the reactive power and harmonic current under unbalanced loading compensation (Table 1.6). In 2014, Rahmani et al. [31] proposed a state-of-the-art control method for TCLC-HAPF to reduce the steady-state error of the TCLC part and improve the performance of current tracking and voltage regulation of the active inverter part. However, the control method proposed in [31] was designed based on the assumption of balanced loading condition. If this control method is applied to TCLC-HAPF for unbalanced loading compensation, it either fails to perform acceptable current quality compensation or requires a high rating active inverter part for compensation, which results in increasing system initial cost, switching loss and switching noise. Therefore, this book (in Chap. 5) aims to propose an unbalanced control method for the TCLC-HAPF, which can balance source side active power and compensate the reactive power and harmonic current with small rating of the active inverter part.

Table 1.6 Characteristics of different compensators and their control methods

	APFs [24, 32–40, 45, 46]	HAPFs [25–29, 41–44]	TCLC-HAPF [31]
Balance/ Unbalanced compensation methods	(1) Instantaneous power methods [24, 32–35] (2) +, − and 0 sequence methods [36–40] (3) Power analysis method [45, 46]	(1) Instantaneous power methods [25–29, 41–43] (2) +, − and 0 sequence methods [44]	Hybrid method [31] (balance control) Lack of unbalanced control study (*Proposed in Chap.* 5)

1.3.2 Pulse Width Modulations

After determining the reference current (or voltage), the reference signals are sent to the pulse width modulations (PWMs) control. Different PWMs, from hysteresis to space vector modulations (SVM), have been proposed for controlling APFs [47]. For example, PWM methods such as ramp comparison, ripple prediction [48, 49] and improved asymmetric space vector modulation [50] were developed to obtain faster dynamics and greater robustness. Wei et al. [51] investigated an algorithm that doubles the switching frequency in the area where the ripple current is larger. This algorithm could reduce switching loss with the same current ripple requirements. However, all of the above PWMs are dedicated to APFs.

Akagi in [25] proposed an HAPF with a low dc-link voltage to reduce the cost of APFs. Today, two main strategies have been used to control the HAPFs:

- equivalent resistor transfer function [25, 31, 44, 51–55]; and
- hysteresis PWM [26–29, 56, 57].

The equivalent resistor transfer function can be classified as an indirect current control (multiplying the reference current by a gain) to calculate the voltage reference, which is then inserted back into the equivalent circuit as a system transfer function to characterize the total system response. In contrast, the hysteresis band defines the allowable error in the current that turns the power electronic switches on or off. The nonlinear current characteristics of the hysteresis PWM approach were not studied by Rahmani in 2009 [58]. Until 2012, the non-linear current characteristics were studied and classified into non-linear, quasi-linear and linear regions in [26]. In [26], the non-linear current was linearised under the criteria defined in [26] for power electronic switching so that HAPFs could be equivalent to linear systems, such as active filters operations. However, the-state-of-the art hysteresis PWM approach still suffers from relatively high switching loss. Therefore, in this book (in Chap. 3), based on the consideration of the users' requirements on total harmonics distortion (*THD*), a non-linear adaptive hysteresis band controller will be discussed and proposed to reduce the switching loss.

1.4 TCLC-HAPF and Its Potential Advantages

In this section, the proposed compensator: TCLC-HAPF [59–64] and its possible advantages are introduced. The system topology of a three-phase three-wire TCLC-HAPF is provided in Fig. 1.2. v_{sx}, v_x and v_{invx} are the source voltage, load voltage and inverter output voltage, respectively. And i_{sx}, i_{Lx} and i_{cx} are source current, load current and compensating current, respectively, where the subscript 'x' denotes phase $x = a, b, c$.

The TCLC-HAPF consists of a TCLC part and an active inverter part. The TCLC part is composed of a coupling inductor L_c, a parallel capacitive

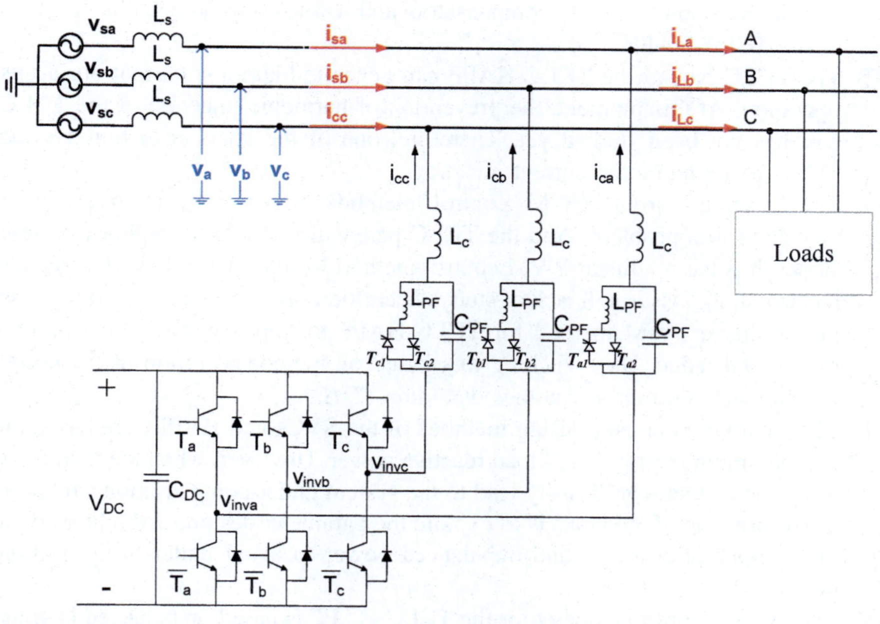

Fig. 1.2 System topologies of three-phase three-wire TCLC-HAPF

C_{PF} and a thyristor controlled reactor (TCR) with an inductor L_{PF}. And the active inverter part of the TCLC-HAPF is a three-phase voltage source inverter (VSI) with a dc-link capacitor C_{DC}.

The TCLC part provides a wide reactive power compensation range and a large voltage drop between system voltage and inverter voltage, so that the active invert part can keep a low dc-link voltage level. A small rating active inverter part is used to improve the performance of the TCLC part by absorbing its harmonic current injection, avoiding mistuning of the firing angles and preventing the resonance problem.

1.5 Research Challenges and Goals

The objective of this research aims to develop a cost-effective three-phase three-wire TCLC-HAPF with low dc-link voltage and wide compensation range for reactive power, unbalanced power and harmonic current compensation. The challenges and goals of this research work can be summarized as follows:

(1) The comparison among the different power quality compensators has not been studied yet in terms of V-I characteristic (compensation range and required active inverter rating), cost, reliability, power loss and tracking performance. Therefore, one of the goals is to provide a comprehensive comparison among

the different power quality compensators and to show the potential advantages of the TCLC-HAPF.

(2) The TCLC part of the TCLC-HAPF can generate harmonic current during its operation. At this moment, the prevention of harmonic injection of the TCLC part has not been studied yet. Therefore, one of the goals is to mitigate the TCLC part's harmonic current injection.
(3) The existing current PWM control methods suffer from relatively high switching loss problem. And the TCLC part will generate a nonlinear current slope, thus the nonlinear PWM control method for the TCLC-HAPF to reduce the switching loss is still lack of study. Therefore, one of the goals is to propose the nonlinear PWM method for TCLC-HAPF to improve the system performance and reduce the switching loss based on the consideration of the users' requirements on total harmonics distortion (*THD*).
(4) The existing parameter design methods of the SVCs and HAPFs are based on the maximum pre-measured load reactive power. However, when the loading is unbalanced, this would easily lead to the system under-compensation situation. Therefore, one of the goals is to explore the parameter design method based on the compensation range and unbalanced power to avoid under-compensation problem.
(5) The existing control strategy for the TCLC-HAPF is based on balanced loading assumption, while the control strategy for unbalanced loading compensation has not been developed. Therefore, one of the goals is to investigate the control strategy for TCLC-HAPF to compensate reactive power, harmonic current and balance active power under unbalanced loading.
(6) The minimizing inverter capacity design of different SVC-HAPFs (FC-TCR-HAPF and TCLC-HAPFs) structures is lack of study. Therefore, one of the goals is to propose the minimizing inverter capacity design of different SVC-HAPFs to obtain a structure with low inverter capacity and acceptable compensation performance.
(7) The DC-link voltage for the TCLC-HAPFs in reactive power and current harmonics compensation can be adaptively controlled to reduce the switching frequency and switching loss. Therefore, one of the goals is to propose the adaptive DC-link voltage control technique for TCLC-HAPFs.
(8) When the load generated harmonic, unbalanced and reactive power is beyond the limited capacity of the TCLC-HAPF, the conventional control methods cannot provide satisfactory compensation performance. In this book, a selective compensation control method among harmonic distortion, unbalanced and reactive power of the TCLC-HAPF will be proposed.
(9) Last but not least, one of the goals is to develop a three-phase three-wire 110V-5kVA TCLC-HAPF hardware experimental prototype. The reactive power, current harmonics and unbalanced power can be compensated simultaneously by the prototype under different current and voltage conditions like current unbalanced, voltage fault, voltage dip, etc. The compensated power quality should meet the requirements of the international standards.

1.6 Book Organization

In this book, the introduction is provided in this chapter. Then, the comparisons of different (hybrid) active power filters are given in Chap. 2. After that, the characteristics, design and control of the SVC part and active inverter part of the TCLC-HAPF are selectively provided in Chap. 3. Then, under the unbalanced loads condition, the unbalanced modeling and parameter design method of the TCLC-HAPF is provided in Chap. 4 and the corresponding unbalanced control strategy is given in Chap. 5. Following that, the minimizing inverter capacity design and comparative performance evaluation of TCLC-HAPF is given in Chap. 6. The adaptive DC-link voltage control strategy for the TCLC-HAPF is proposed in Chap. 7 and the selective compensation control strategy during over compensator's capacity situation is given in Chap. 8. In Chap. 9, the hardware and software design of an 110V-5kVA three-phase three-wire experimental prototype is presented and discussed. Finally, the conclusion and the prospective of further works are given in Chap. 10.

The detailed chapter descriptions are shown in below:

This chapter presents an overview and summary of the major power quality issues (low power factor, harmonic pollution and unbalanced problem) by taking consideration of the power quality measurement data in Macau as an example. And, the potential market for reactive power and harmonic compensation in China is introduced, which illustrates a large potential market in China for power quality compensation. In order to solve the above major power quality issues, the historical review of different power quality compensator topologies and control methods are introduced.

Chapter 2 compares TCLC-HAPF with other different power quality compensators in terms of V-I characteristic, cost, reliability, power loss and tracking performance.

Chapter 3 investigates the characteristic, design and control for the TCLC part and the active inverter part of the TCLC-HAPF. Specifically, the solutions for the TCLC part to prevent harmonic current injection and nonlinear PWM method for the active inverter part to reduce the switching loss are proposed.

Chapter 4 proposes the unbalanced modeling and parameter design of the TCLC-HAPF based on its compensation range and the unbalanced loading power to avoid under-compensation problem.

Chapter 5 proposes an unbalanced control method for the TCLC-HAPF to compensate reactive power, harmonic current and balance the active power with low DC-link voltage.

Chapter 6 discusses the minimizing inverter capacity design and comparative performance evaluation of different SVC-HAPFs (FC-TCR-HAPF and TCLC-HAPFs). And, the purpose of this chapter is to obtain the superior structure with acceptable compensation performance and low inverter capacity.

Chapter 7 provides the adaptive DC-link voltage control technique for the TCLC-HAPF to reduce the switching loss.

Chapter 8 proposes a selective compensation control method of harmonic distortion, unbalanced and reactive power with the consideration of the limited capacity of the TCLC-HAPF.
Chapter 9 presents the implementation of an 110V-5kVA three-phase three-wire TCLC-HAPF experimental prototype. Experimental results will be provided to show its low DC-voltage and wide compensation range characteristics under different voltage and current conditions such as unbalanced current, voltage fault, voltage dip, etc.
Chapter 10 draws the conclusion, gives the study limitations of this book and proposes prospective future research works.

1.7 Appendix: Voltage and Current Standards

Power quality standards have been issued by the standard organizations, which provide some guidelines or indexes to limit the power quality pollution in the electrical power network, thus maintaining the health of the power grid. Both voltage *THDv* standards and current *THDi* standards are defined in *IEEE Standard 519:2014IEEE Recommended Practices and Requirements for Harmonic Control in Electrical Power Systems* [5]. The details of the voltage *THDv* standards and current *THDi* standards are summarized in Tables 1.7 and 1.8.

Due to the practical power system capacity, when the short-circuit current at the point-of-coupling (PCC) is not high enough, harmonic current may cause voltage distortion. When TDD of the current (given in Table 1.8 for current waveform) cannot fulfill the IEEE standard, it can cause voltage distortion. The voltage waveform may be over the IEEE standard (defined in Table 1.7 for voltage waveform) due to harmonic current. Tables 1.7 and 1.8 imply that the required limits are getting tighter as the system capacity gets larger. The detailed discussion is given below, which can explain the reason for choosing the current harmonic standard with THD = 15% in this book. The basic information of the transformer used in experiments is summarized in Table 1.9.

According to the voltage THD_v standard (shaded area in Table 1.7), for the laboratory-scaled low voltage applications (<1 kV as shown in Table 1.9), the maximum acceptable voltage THD_v range based on IEEE Standard 519-2014 [5] is required to be lower than 8%. Therefore, this book evaluates the compensation performance by setting voltage THD_v< 8%.

Table 1.7 Voltage distortion limits for general distribution systems [5]

Bus voltage V at PCC	Individual harmonic (%)	Total harmonic distortion THD (%)
V ≤ 1.0 kV	5.0	8.0
1 kV < V ≤ 69 kV	3.0	5.0
69 kV < V ≤ 161 kV	1.5	2.5
161 kV < V	1.0	1.5[a]

[a]High-voltage systems can have up to 2.0% THD where the cause is an HVDC terminal whose effects will have attenuated at points in the network where future users may be connected

Table 1.8 Current distortion limits for general distribution systems (120–69,000 V) [5]

Maximum Harmonic Current Distortion in Percent of I_L						
Individual Harmonic Order (Odd Harmonics)						
I_{SC}/I_L	$h<11$	$11 \quad h<17$	$17 \quad h<23$	$23 \quad h<35$	$35 \quad h\leq 50$	*TDD*
<20[a]	4.0	2.0	1.5	0.6	0.3	5.0
20<50	7.0	3.5	2.5	1.0	0.5	8.0
50<100	10.0	4.5	4.0	1.5	0.7	12.0
100<1000	12.0	5.5	5.0	2.0	1.0	15.0
>1000	15.0	7.0	6.0	2.5	1.4	20.0

Even harmonic are limited to 25% of the odd harmonic limits above

Current distortions that result in a dc offset, e.g., half-wave converters, are not allowed

[a]All power generation equipment is limited to these values of current distortion regardless of actual I_{SC}/I_L

Where I_{SC} = maximum short-circuit current at PCC, I_L = maximum demand load current (fundamental frequency component) at PCC

Table 1.9 Datasheet of the transformer used in experiments

Model	Voltage (V)	Rate impedance (Z_T) (%)	Full load current (I_F) (A)	Maximum short circuit current at the PCC (I_{SC}) (A)
SE-10	110	3	26	867

Notes $I_{SC} = I_F/Z_T$

Based on current THD_i standard (shaded area in Table 1.9), the Total Demand Distortion (TDD) = 15% with I_{SC}/IL falling in 100 < 1000 scale. According to the *THD* and *TDD* definition given below, in the worst case analysis, the nominal rate current (I_{rated}) is assumed to be equal to the fundamental load current (I_1), which results in *THD* = *TDD* = 15%. Therefore, this book evaluates the compensation performance by setting current *THDi* < 15%.

$$THD : \sqrt{\sum_{n=2}^{\infty} I_n^2} / I_1 \quad TDD : \sqrt{\sum_{n=2}^{\infty} I_n^2} / I_{\text{rated}}$$

References

1. S.S. Williamson, A.K. Rathore, F. Musavi, Industrial electronics for electric transportation: current state-of-the-art and future challenges. IEEE Trans. Ind. Electron. **62**(5), 3021–3032 (2015)
2. A. Javadi, K. Al-Haddad, A single-phase active device for power quality improvement of electrified transportation. IEEE Trans. Ind. Electron. **62**(5), 3033–3041 (2015)
3. S.M. Mousavi, A. Tabakhpour, E. Fuchs, K. Al-Haddad, Power quality issues in railway electrification: a comprehensive perspective. IEEE Trans. Ind. Electron. **62**(5), 3081–3090 (2015)

4. P. Salmeron, S.-P. Litran, Improvement of the electric power quality using series active and shunt passive filters. IEEE Trans. Power Deliv. **25**, 1058–1067 (2010)
5. IEEE Recommended Practices and Requirements for Harmonic Control in Electrical Power Systems, 2014, IEEE Standard 519-2014
6. IEEE Recommended Practice on Monitoring Electric Power Quality, 1995, IEEE Standard 1159:1995
7. S.-U. Tai, M.-C. Wong, M.-C. Wong, Y.-D. Han, Some findings on harmonic measurement in Macao, in *Proceedings of 7th International Conference on Power Electronics and Drive Systems, PEDS 07* (2007), pp. 405–410
8. S.-U. Tai, Power quality study in Macau and virtual power analyzer, in *Master Book* (University of Macau, 2012)
9. Annual Reports, The Electricity Company in Macau (www.cem-macau.com)
10. B. Singh, K. Al-Haddad, A. Chandra, A review of active filters for power quality improvement. IEEE Trans. Ind. Electron. **46**(5), 960–971 (1999)
11. National Bureau of Statistics of the People's Republic of China (Dec 2013) [Online]. Available: http://www.stats.gov.cn/
12. Annual Report of Rongxin Power Electronic Co., Ltd. Dec 2013
13. H. Rudnick, J. Dixon, L. Moran, Delivering clean and pure power. IEEE Power Energy Mag. **1**(5), 32–40 (2003)
14. J.C. Das, Passive filters - potentialities and limitations. IEEE Trans. Ind. Appl. **40**(1), 232–241 (2004)
15. V. Trujillo, C.R. Fuerte-Esquivel, J.H. Tovar Hernandez, Advanced three-phase static VAr compensator models for power flow analysis. IEE Proc. Gener. Transm. Distrib. **150**(1), 119–126 (2003)
16. T. Baldwin, T. Hogans, S. Henry, F. Renovich, P. Latkovic, Reactive power compensation for voltage control at resistance welders. IEEE Trans. Ind. Appl. **41**(6), 1485–1492 (2005)
17. Y.C. Chang, Multi-objective optimal SVC installation for power system loading margin improvement. IEEE Trans. Power Syst. **27**(2), 984–992 (2012)
18. H. Ambriz-Perez, E. Acha, C.R. Fuerte-Esquivel, Advanced svc models for Newton-Raphson load flow and Newton optimal power flow studies. IEEE Trans. Power Syst. **15**(1), 129–136 (2000)
19. J.E.R. Alves, L.A.S. Pilotto, E.H. Watanabe, Thyristor-controlled reactors nonlinear and linear dynamic analytical models. IEEE Trans. Power Deliv. **23**(1), 338–346 (2008)
20. N. Daratha, B. Das, J. Sharma, Coordination between OLTC and SVC for voltage regulation in unbalanced distribution system distributed generation. IEEE Trans. Power Syst. **29**(1), 289–299 (2014)
21. F.R. Quintela, J.M.G. Arevalo, R.C. Redondo, Power analysis of static VAr compensators. Int. J. Electr. Power **30**, 376–382 (2008)
22. F.R. Quintela, J.M.G. Arevalo, R.C. Redondo, Single-phase power supply to balanced three-phase loads through SVAr compensators. Int. J. Electr. Power **33**, 715–720 (2011)
23. F.Z. Peng, H. Akagi, A. Nabae, A new approach to harmonic compensation in power systems-a combined system of shunt passive and series active filters. IEEE Trans. Ind. Appl. **26**, 983–990 (1990)
24. H. Hu, Y. Xing, Design considerations and fully digital implementation of 400-Hz active power filter for aircraft applications. IEEE Trans. Ind. Electron. **61**(8), 3823–3834 (2014)
25. S. Srianthumrong, H. Akagi, A medium-voltage transformerless ac/dc power conversion system consisting of a diode rectifier and a shunt hybrid filter. IEEE Trans. Ind. Appl **39**, 874–882 (2003)
26. C.S. Lam, M.C. Wong, Y.D. Han, Hysteresis current control of hybrid active power filters. IET Power Electron. **5**(7), 1175–1187 (2012)
27. C.S. Lam, X.X. Cui, W.H. Choi, M.C. Wong, Y.D. Han, Minimum inverter capacity design for three-phase four-wire LC-hybrid active power filters. IET, Power Electron. **5**(7), 956–968 (2012)

28. C.-S. Lam, W.-H. Choi, M.-C. Wong, Y.-D. Han, Adaptive dc-link voltage controlled hybrid active power filters for reactive power compensation. IEEE Trans. Power Electron. **27**(4), 1758–1772 (2012)
29. C.S. Lam, M.C. Wong, W.-H. Choi, X.-X. Cui, H.-M. Mei, J.-Z. Liu, Design and performance of an adaptive low-dc-voltage-controlled LC-Hybrid active power filter with a neutral inductor in three-phase four-wire power systems. IEEE Trans. Power Electron. **61**(6), 2635–2647 (2014)
30. S. Rahmani, N. Mendalek, K. Al-Haddad, Experimental design of a nonlinear control technique for three-phase shunt active power filter. IEEE Trans. Ind. Electron. **57**(10), 3364–3375 (2010)
31. S. Rahmani, A. Hamadi, K. Al-Haddad, A combination of shunt hybrid power filter and thyristor-controlled reactor for power quality. IEEE Trans. Ind. Electron. **61**(5), 2152–2164 (2014)
32. H. Akagi, Y. Kanazawa, A. Nabae, Instantaneous reactive power compensators comprising switching devices without energy storage components. IEEE Trans. Ind. Appl. **IA-20**(3), 625–630 (1984)
33. Y. Hu, Z. Zhu, K. Liu, Current control for dual three-phase permanent magnet synchronous motors accounting for current unbalance and harmonics. IEEE Trans. Emerg. Sel. Topics Power Electron. **2**(2), 272–284 (2014)
34. W.C. Lee, T.K. Lee, D.S. Hyun, A three-phase parallel active power filter operating with PCC voltage compensation with consideration for an unbalanced load. IEEE Trans. Power Electron. **17**(5), 807–814 (2002)
35. S. Senini, P.J. Wolfs, Hybrid active filter for harmonically unbalanced three phase three wire railway traction loads. IEEE Trans. Power Electron. **15**(4), 702–710 (2000)
36. S. Rahmani, K. Al-Haddad, F. Fnaiech, A three phase shunt hybrid power filter adopted a general algorithm to compensate harmonics, reactive power and unbalanced load under non ideal mains voltage, in *Proceedings of the IEEE International Conference on Industrial Technology, IEEE ICIT04* (2004), pp. 651–656
37. M. Aredes, H. Akagi, E.H. Watanabe, E. Vergara Salgado, L.F. Encarnacao, Comparisons between the p-q and p-q-r theories in three-phase four-wire systems. IEEE Trans. Power Electron. **24**(4), 924–933 (2009)
38. B. Wen, D. Boroyevich, R. Burgos, P. Mattavelli, Z. Shen, Analysis of D-Q small-signal impedance of grid-tied inverters. IEEE Trans. Power Electron. **31**(1), 675–687 (2016)
39. S. Rahmani, A. Hamadi, K. Al-Haddad, A Lyapunov-function-based control for a three-phase shunt hybrid active filter. IEEE Trans. Ind. Electron. **59**(3), 1418–1429 (2012)
40. L. Shaohua, W. Xiuli, Y. Zhiqing, L. Tai, P. Zhong, Circulating current suppressing strategy for MMC-HVDC based on non ideal proportional resonant controllers under unbalanced grid conditions. IEEE Trans. Power Electron. **30**(1), 387–397 (2015)
41. X. Guo, W. Liu, X. Zhang, X. Sun, Z. Lu, J.M. Guerrero, Flexible control strategy for grid-connected inverter under unbalanced grid faults without PLL. IEEE Trans. Power Electron. **30**(4), 1773–1778 (2015)
42. K. Ma, W. Chen, M. Liserre, F. Blaabjerg, Power controllability of a three-phase converter with an unbalanced AC source. IEEE Trans. Power Electron. **30**(3), 1591–1604 (2015)
43. M. Castilla, J. Miret, A. Camacho, L. Garcia de Vicuna, J. Matas, Modeling and design of voltage support control schemes for three-phase inverters operating under unbalanced grid conditions. IEEE Trans. Power Electron. **29**(11), 6139–6150 (2014)
44. P. Salmeron, S.P. Litran, A control strategy for hybrid power filter to compensate four-wires three-phase systems. IEEE Trans. Power Electron. **25**(7), 1923–1931 (2010)
45. L.S. Czarnecki, S.E. Pearce, Compensation objectives and current' physical components-based generation of reference signals for shunt switching compensator control. IET Power Electron. **2**(1), 33–41 (2009)
46. L.S Czarnecki, P.M. Haley, Unbalanced power in four-wire systems and its reactive compensation. IEEE Trans. Power Deliv. **30**(1), 53–63 (2015)

47. W. Jiang, W. Li, Z. Wu, Y. She, Z. Tao, Space-vector pulse-width modulation algorithm for multilevel voltage source inverters based on matrix transformation and including operation in the over-modulation region. IET Power Electron. **7**(12), 2925–2933 (2014)
48. X. Mao, R. Ayyanar, H. Krishnamurthy, Optimal variable switching frequency scheme for reducing switching loss in single-phase inverters based on time-domain ripple analysis. IEEE Trans. Power Electron. **24**(4), 991–1001 (2009)
49. D. Jiang, F. Wang, Variable switching frequency PWM for three-phase converters based on current ripple prediction. IEEE Trans. Power Electron. **28**(11), 4951–4961 (2013)
50. D. Zhang, F. Wang, S. El-Barbari, J. A. Sabate, D. Boroyevich, Improved asymmetric space vector modulation for voltage source converters with low carrier ratio. IEEE Trans. Power Electron. **27**(3), 1130–1140 (2012)
51. L. Wei, R.A. Lukaszewski, Pulse width modulation (PWM) rectifier with variable switching frequency. U.S. patent 7 190 143 132, Mar 2007
52. B. Angélico, L. Campanhol, S. da Silva, Proportional integral/proportional integral-derivative tuning procedure of a single-phase shunt active power filter using Bode diagram. IET Power Electron. **7**(10), 2647–2659 (2014)
53. M.S. Hamad, M.I. Masoud, B.W. Williams, S. Finney, Medium voltage 12-pulse converter: ac side compensation using a shunt active power filter in a novel front end transformer configuration. IET Power Electron. **5**(8), 1315–1323 (2012)
54. A.F. Zobaa, Optimal multiobjective design of hybrid active power filters considering a distorted environment. IEEE Trans. Ind. Electron. **61**(5), 107–114 (2014)
55. J.C. Wu, H.L. Jou, H.H. Hsaio, S.T. Xiao, A new hybrid power conditioner for suppressing harmonics and neutral-line current in three-phase four-wire distribution power systems. IEEE Trans. Power Deliv. **29**(4), 1525–1532 (2014)
56. S.K. Khadem, M. Basu, M.F. Conlon, Harmonic power compensation capacity of shunt active power filter and its relationship with design parameters. IET Power Electron. **7**(2), 418–430 (2014)
57. S.K. Chauhan, M.C. Shah, R.R. Tiwari, P.N. Tekwani, Analysis, design and digital implementation of a shunt active power filter with different schemes of reference current generation. IET Power Electron. **7**(3), 627–639 (2014)
58. S. Rahmani, A. Hamadi, N. Mendalek, K. Al-Haddad, A new control technique for three-phase shunt hybrid power filter. IEEE Trans. Ind. Electron. **56**(8), 2904–2915 (2009)
59. L. Wang, C.S. Lam, M.C. Wong, A SVC-HAPF with wide compensation range and low dc-link voltage. IEEE Trans. Ind. Electron. **63**(6), 3333–3343 (2016)
60. L. Wang, C.S. Lam, M.C. Wong, An unbalanced control strategy for a thyristor controlled LC-coupling hybrid active power filter (SVC-HAPF) in three-phase three-wire systems. IEEE Trans. Power Electron. **32**(2), 1056–1069 (2017)
61. L. Wang, C.S. Lam, M.C. Wong, Hardware and software design of a low dc-link voltage and wide compensation range thyristor controlled LC-coupling hybrid active power filter, in *TENCON 2015 IEEE Region 10 Conference proceedings*, Nov 2015
62. L. Wang, C.S. Lam, M.C. Wong, Modeling and parameter design of thyristor controlled LC-coupled hybrid active power filter (SVC-HAPF) for unbalanced compensation. IEEE Trans. Ind. Electron. **64**(3), 1827–1840 (2017)
63. C.S. Lam, L. Wang, S.I. Ho, M.C. Wong, Adaptive thyristor controlled LC-hybrid active power filter for reactive power and current harmonics compensation with switching loss reduction. IEEE Trans. Power Electron. **32**(10), 7577–7590 (2017)
64. L. Wang, C.S. Lam, M.C. Wong, Selective Compensation of distortion, unbalanced and reactive power of a thyristor controlled LC-coupling hybrid active power filter (SVC-HAPF). IEEE Trans. Power Electron. **32**(12), 9065–9077 (2017)

Chapter 2
Comparisons Among Thyristor Controlled LC-Coupling Hybrid Active Power Filter (TCLC-HAPF) and Other Different Power Quality Filters

Abstract The design of shunt active power filters (APF) and hybrid active power filters (HAPFs) characterized by better performance, high reliability, low cost and failure rates will be the major trend in the future. Over the past few decades, significant researches were focused on addressing the technical challenges associated with the parameter design, operation and control of the APF and different HAPFs. However, the comprehensive review of their V-I characteristics, cost, reliability, power loss and tracking performance has been rarely studied. Therefore, this chapter aims to provide a clear picture of the selection of APF and different HAPFs based on V-I characteristic, cost, reliability, power loss and tracking performance. The supreme one will be selected among APF and different HAPFs by considering the V-I characteristics, cost, reliability, power loss, and tracking performance. Finally, the experimental results of APF and different HAPFs will be provided to verify their compensation performances.

Keywords V-I characteristics · Cost · Reliability · Power loss
Tracking performance · Active power filter (APF) · Hybrid active power filters (HAPFs)

2.1 Introduction

Installation of the current quality compensators is one of the solutions for different power quality problems such as the low power factor, harmonic current pollution and unbalanced problem. The historical review of different power quality compensators are summarized in Table 2.1 and compared in the following.

To compensate the reactive power, current harmonics, the passive power filters (PPFs) were proposed in 1940s. Unfortunately, the PPFs have many disadvantages like poor dynamic performance, resonance problem, and fixed reactive power compensation [1, 2]. The static var compensators (SVCs) were firstly proposed in 1960s. And, the SVCs are popularly used for dynamic reactive power compensation [3]. However, the SVCs still suffer drawbacks such as resonance problem, harmonic

L. Wang et al., *Adaptive Hybrid Active Power Filters*, Power Systems,
https://doi.org/10.1007/978-981-10-8827-8_2

Table 2.1 Characteristics of different active current quality compensators

Year	1940s	1960s	1976	2003	2014
Compensators	PPFs	SVCs	APFs	HAPFs	TCLC-HAPFs
Compensation Range	Narrow	Wide	Wide	Narrow	Wide
Harmonics Compensation	Normal	Poor	Good	Good	Good
Cost	Lowest	Low	High	Normal	Normal
Reliability	High	High	Low	Normal	Normal
Switching loss	Low	Low	High	Normal	Normal
Tracking performance	Poor	Poor	Good	Poor	Good

Note The shaded area means the undesirable characteristic

current injection and poor harmonic compensation ability. To overcome those drawbacks of SVCs and provide better performances simultaneously, the active power filters (APFs) were proposed in the year of 1976 [4–6]. Unfortunately, APFs still cannot large scale development in the power quality markets due to the high initial and operational costs. Afterwards, the *LC*-coupling hybrid active filters (HAPFs) were proposed in the year of 2003 with low active inverter part rating [7–9]. Since the rating of the active inverter part is proportional to the cost of compensators, the HAPFs are more cost-effective than the APFs. However, the HAPFs have a quite narrow compensation range, which limits its compensation ability. When the load reactive power is outside its designed range, it loses its low-inverter rating advantages. From 2014 onwards, the thyristor controlled *LC*-coupling hybrid active power filters (TCLC-HAPFs) have been widely studied in [10–16], which have the desirable characteristics of a wider compensation range than HAPFs and lower dc-link voltage than APFs for power quality compensation.

Based on above discussions and Table 2.1, the traditional power quality compensators, PPFs and SVCs have inherent problems like resonance problem, slow response, poor harmonic compensation ability, etc. The above mentioned inherent problems can be solved if the active inverter part has been added to the topologies such as APFs, HAPFs and TCLC-HAPFs. However, the cost of PPF and SVC are lower than that of the active inverter part, thus the reduction of the active inverter part rating can lead to a decrease in the total cost of APFs, HAPFs and TCLC-HAPFs. Therefore, this chapter aims to propose comprehensive comparisons in terms of DC-link voltage, cost, reliability, power loss and tracking performance of the APFs, HAPFs and TCLC-HAPFs.

The layout of this chapter can be described as following. First of all, the structures of APF, HAPF and TCLC-HAPF are introduced in Sect. 2.2. Then, the comparisons among different power quality compensators are given in terms of V-I characteristic (Sect. 2.3), cost (Sect. 2.4), reliability (Sect. 2.5), power loss (Sect. 2.6) and tracking performance (Sect. 2.7). After that, the experimental verifications are provided in Sect. 2.8. Finally, the summary is drawn in Sect. 2.9.

2.2 Structures of APF, HAPF and TCLC-HAPF

Figures 2.1 and 2.2 show the structures of APF, HAPF and TCLC-HAPF, in which the subscript "x" stands for phase a, b, and c in the following analysis. v_{sx}, v_x and v_{invx} are the source voltage, load voltage and inverter output voltage, respectively; i_{sx}, i_{Lx}, and i_{cx} are the source, load, and compensating currents, respectively. L_s is the transmission line impedance.

For low voltage level applications, the two-level power quality compensators as shown in Fig. 2.1 can be used to solve power quality issues. In Fig. 2.1, the TCLC part of the TCLC-HAPF is composed of a coupling inductor L_c, a parallel capacitor C_{PF}, and a thyristor-controlled reactor with L_{PF}. The coupling component of the APF is an inductor L. And, the coupling components of the HAPF are an inductor L in series with a capacitor C. For all three power quality compensators, the active inverter parts are composed of a voltage source inverter with a DC-link capacitor C_{DC}.

For the high voltage applications, both coupling LC part of the HAPF and TCLC part of the TCLC-HAPF can provide large voltage drops, so that the DC-link voltages can be significantly reduced. However, the minimum required DC-link voltage for the APFs needs to be higher than the peak value of the line-to-line load voltage ($V_{\rm DC} > \sqrt{2} \cdot V_{LL}$) according to [17–19]. When the voltage level is higher than the maximum rating of power switches, the cascade H bridge technique can be

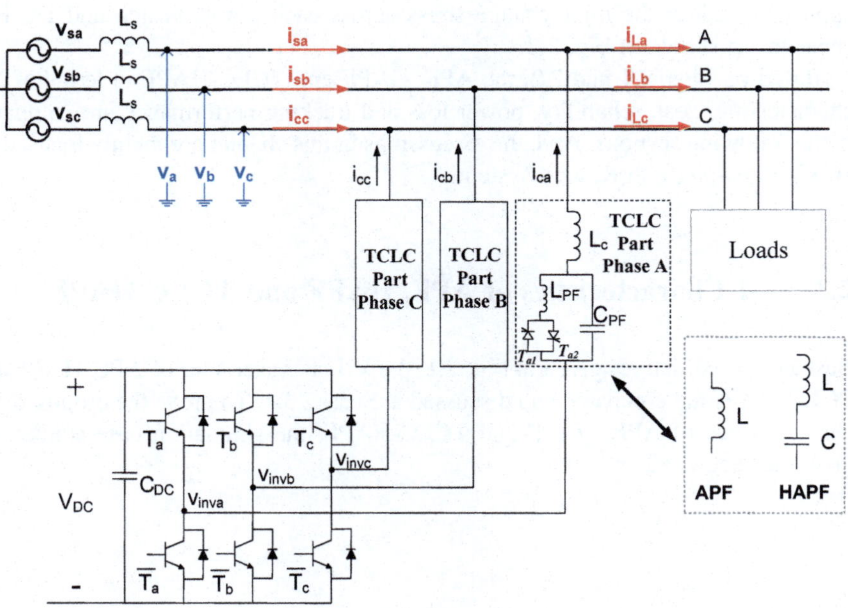

Fig. 2.1 Structures of two-level APF, HAPF and TCLC-HAPF

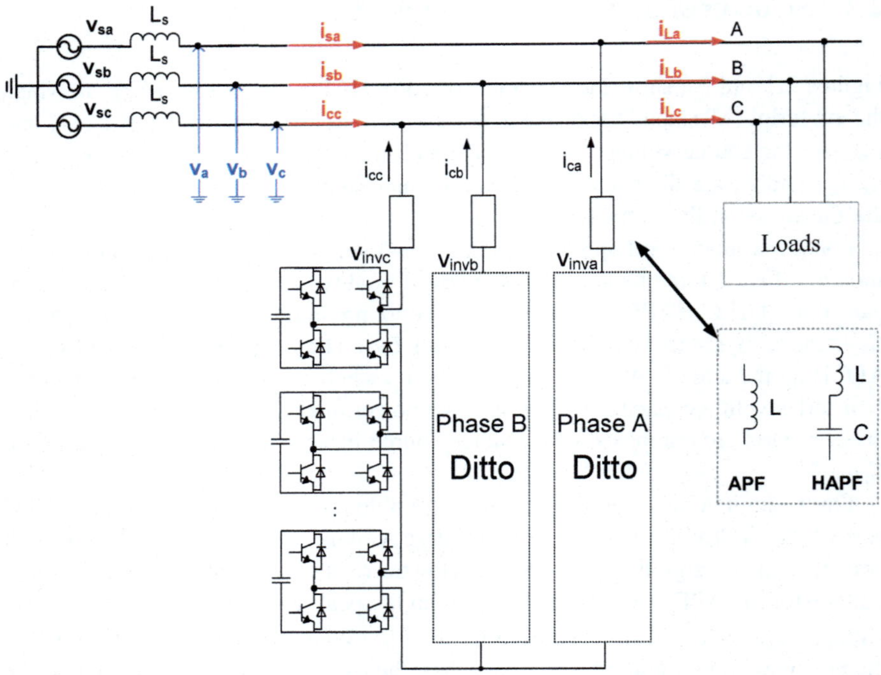

Fig. 2.2 Structures of cascade H bridge APF and HAPF

applied to reduce the high voltage stress across each power switch and DC-link capacitor as shown in Fig. 2.2.

Based on Figs. 2.1 and 2.2, the APF, HAPF and TCLC-HAPF in terms of V-I characteristic, cost, reliability, power loss and tracking performance are compared in the following sections. And, the discussions in this chapter are mainly focused on 10 kV three-phase three-wire systems.

2.3 V-I Characteristics of APF, HAPF and TCLC-HAPF

Based on the structures given in Fig. 2.1, the V-I characteristics of APF, HAPF and TCLC-HAPF are compared and discussed in Sect. 2.3.1. To prove the proposed V-I characteristics of APF, HAPF and TCLC-HAPF, the simulation case studies are given in Sect. 2.3.2.

2.3.1 V-I Characteristics of APF, HAPF and TCLC-HAPF

Under the fundamental frequency, the purpose of the TCLC-HAPF is to provide the same amount of reactive power as the loading (Q_{Lx}) consumed with the opposite polarity ($Q_{cx}= -Q_{Lx}$). The compensating reactive power of TCLC-HAPF Q_{cx} is the sum of the reactive power Q_{TCLC} provided by the TCLC part and the reactive power Q_{invx} provided by the active inverter part. Based on above discussions, the relationship among Q_{Lx}, Q_{TCLC}, and Q_{invx} can be expressed as:

$$Q_{Lx} = -Q_{cx} = -(Q_{TCLC} + Q_{invx}) \tag{2.1}$$

The reactive power can also be provided in terms of voltage and current as

$$Q_{Lx} = V_x I_{Lqx} = -\left(X_{TCLC}(\alpha_x) I_{cqx}^2 + V_{invx(TCLC-HAPF)} I_{cqx}\right) \tag{2.2}$$

where V_x and $V_{invx(TCLC\text{-}HAPF)}$ are the RMS values of the coupling point and the inverter voltages of TCLC-HAPF; α_x is the corresponding firing angle; $X_{TCLC}(\alpha_x)$ is the coupling impedance of the TCLC part; and I_{Lqx} and I_{cqx} are the RMS value of the load and compensating reactive currents, where $I_{Lqx}= I_{cqx}$. By further simplifying the (2.2), the $V_{invx(TCLC\text{-}HAPF)}$ can be simplified as:

$$V_{invx(TCLC-HAPF)} = V_x + X_{TCLC}(\alpha_x) I_{Lqx} \tag{2.3}$$

In (2.3), the TCLC part impedance $X_{TCLC}(\alpha_x)$ can be expressed as:

$$X_{TCLC}(\alpha_x) = \frac{X_{TCR}(\alpha_x) X_{C_{PF}}}{X_{C_{PF}} - X_{TCR}(\alpha_x)} + X_{L_c} = \frac{\pi X_{L_{PF}} X_{C_{PF}}}{X_{C_{PF}}(2\pi - 2\alpha_x + \sin 2\alpha_x) - \pi X_{L_{PF}}} + X_{L_c} \tag{2.4}$$

In (2.4), X_{Lc}, X_{LPF} and X_{CPF}, are the fundamental impedances of L_c, L_{PF}, and C_{PF}, respectively. Equation (2.4) shows that the TCLC part impedance is controlled by firing angle α_x. The minimum inductive and capacitive impedances (absolute value) of the TCLC part can be obtained by substituting the firing angles $\alpha_x = 90°$ and $\alpha_x = 180°$, respectively. In the following discussions, the minimum value for impedances stands for its absolute minimum value. The minimum inductive ($X_{Ind(min)}> 0$) and capacitive ($X_{Cap(min)}< 0$) TCLC part impedances can be expressed as:

$$X_{Ind(min)}(\alpha_x = 90°) = \frac{X_{L_{PF}} X_{C_{PF}}}{X_{C_{PF}} - X_{L_{PF}}} + X_{L_c} \tag{2.5}$$

$$X_{Cap(min)}(\alpha_x = 180°) = -X_{C_{PF}} + X_{L_c} \tag{2.6}$$

As shown in (2.4), if the $X_{TCLC}(\alpha_x)$ is controlled to be $V_x \approx X_{TCLC}(\alpha_x) I_{Lqx}$, the minimum inverter voltage can be obtained ($V_{invx(TCLC\text{-}HAPF)} \approx 0$). In this case, the power loss and switching noise can be significantly reduced. A small active inverter voltage $V_{invx(min)(TCLC\text{-}HAPF)}$ of the TCLC-HAPF is necessary to avoid mistuning the firing angles, to absorb the harmonic current generated by the TCLC part, and to prevent a resonance problem. Moreover, if the loading capacitive current or inductive current is outside the TCLC part compensating range, the active inverter voltage $V_{invx(TCLC\text{-}HAPF)}$ can be used to further enlarge the compensation range.

The coupling impedances for APFs and HAPFs, as shown in Fig. 2.1, are fixed as X_L and $X_L - X_C$. The relationships among the load reactive current I_{Lqx}, the inverter voltage V_{invx}, the load voltage V_x and the coupling impedance of APF and HAPF can be expressed as:

$$V_{invx(APF)} = V_x + X_L I_{Lqx} \tag{2.7}$$

$$V_{invx(HAPF)} = V_x + (X_L - X_C) I_{Lqx} \tag{2.8}$$

where $X_C \gg X_L$. Based on (2.3)–(2.8), the V-I characteristics of the APF, HAPF, and TCLC-HAPF can be plotted as shown in Fig. 2.3.

As shown in Fig. 2.3a, when the loading is inductive, the required active inverter voltage $V_{invx(APF)}$ of the APF is larger than V_x. In contrast, when the loading is

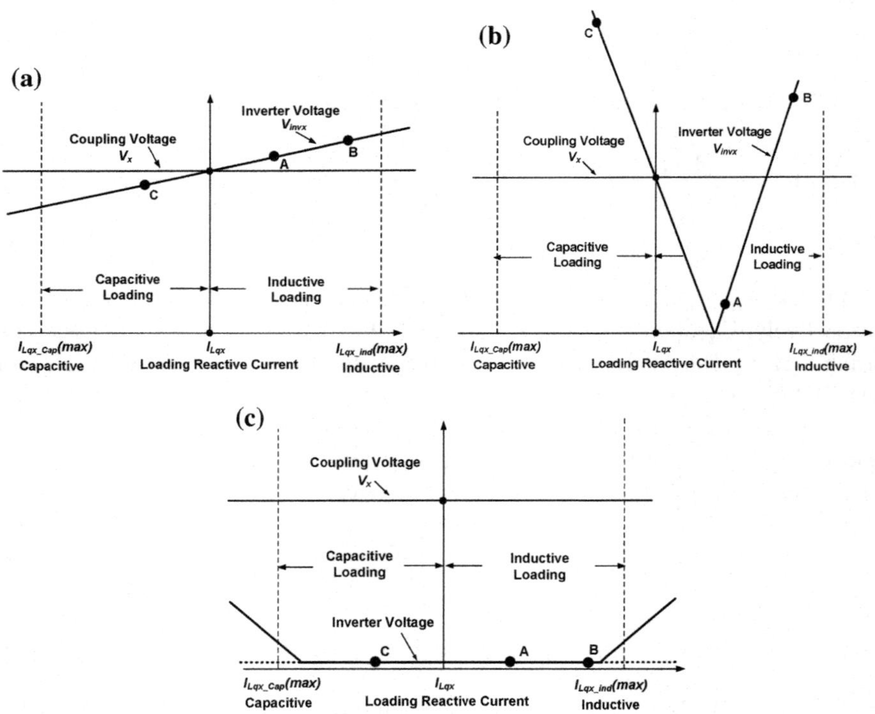

Fig. 2.3 V-I characteristics of **a** APF, **b** HAPF, and **c** TCLC-HAPF

capacitive, the required $V_{invx(APF)}$ is smaller than V_x. Due to the small value of coupling inductor L [20–24], the required inverter voltage $V_{invx(APF)}$ is close to the coupling load voltage V_x.

In Fig. 2.3b, it is shown that the required active inverter voltage $V_{invx(HAPF)}$ of the HAPF is lower than V_x under a small inductive loading range. And, when the coupling capacitive impedance can fully compensate the loading reactive current, the required $V_{invx(HAPF)}$ of the HAPF can be as low as zero. In contrast, when the loading is capacitive or outside its small inductive loading range, $V_{invx(HAPF)}$ is larger than V_x. Therefore, the HAPF only has the small inductive compensation range.

As shown in Fig. 2.3c, the required active inverter voltage $V_{invx(TCLC\text{-}HAPF)}$ of the TCLC-HAPF can be maintained at a minimum level ($V_{invx(min)(TCLC\text{-}HAPF)}$) for a large inductive and capacitive reactive current range. Moreover, when the loading reactive current is outside the compensation range of the TCLC part, the $V_{invx(TCLC\text{-}HAPF)}$ is slightly increasing to further enlarge the compensating range.

Compared with APF in Fig. 2.3a and HAPF in Fig. 2.3b, the proposed TCLC-HAPF in Fig. 2.3c has the superior V-I characteristics of a wider compensation range with a low inverter voltage. In addition, three cases represented by points A, B, and C in Fig. 2.3 are simulated in Sect. 2.3.2.

2.3.2 *Simulation Case Studies*

In this section, the simulation results are provided and compared among APF, HAPF, and the TCLC-HAPF. The previous discussions of the required inverter voltages (or DC-link voltage $V_{DC} = \sqrt{2} \cdot \sqrt{3} \cdot V_{invx}$) are also verified by simulations for different power quality compensators. The simulation studies are carried out with PSCAD/EMTDC. Table 2.2 shows the simulation system parameters for APF, HAPF, and TCLC-HAPF. In addition, three different cases of loading are built for testing: (A) inductive and light loading, (B) inductive and heavy loading, and (C) capacitive loading. These three testing cases are also represented by points *A*, *B*, and *C* in Fig. 2.3. The detailed simulation results are summarized in

Table 2.2 Simulation and experimental parameters for APF, HAPF and TCLC-HAPF

	Parameters	Physical values
System parameters	v_x, f, L_s	110 V, 50 Hz, 0.1 mH
Traditional APF	L	5 mH
HAPF	L, C	5 mH, 80 uF
TCLC-HAPF	L_c, L_{PF}, C_{PF}	5 mH, 30 mH, 160 uF
Case A: inductive and light loading	L_{L1}, R_{L1}	30 mH, 14 Ω
Case B: inductive and heavy loading	L_{L2}, R_{L2}	30 mH, 9 Ω
Case C: capacitive loading	C_{L3}, R_{L3}	200 uF, 20 Ω

Table 2.3 Simulation results for inductive and capacitive reactive power compensation of APF, HAPF and TCLC-HAPF

Loading Type	Without and With Comp.	i_{sx}(A)	*DPF*	$THDi_{sx}$ (%)	V_{DC}(V)
Case A: inductive and light loading	Before Comp.	6.50	0.83	0.01	--
	APF	5.55	1.00	7.22	300
	HAPF	5.48	1.00	2.01	80
	TCLC-HAPF	5.48	1.00	1.98	50
Case B: inductive and heavy loading	Before Comp.	8.40	0.69	0.01	--
	APF	5.95	1.00	6.55	300
	HAPF	6.30	0.85	17.5	50
	HAPF	5.90	0.98	7.02	300
	TCLC-HAPF	5.89	1.00	2.10	50
Case C: capacitive loading	Before Comp.	4.34	0.78	0.01	--
	APF	3.67	1.00	7.61	250
	HAPF	7.10	0.57	23.5	50
	HAPF	5.02	0.99	10.6	500
	TCLC-HAPF	3.41	1.00	3.01	50

Note The shaded areas indicate undesirable results

Table 2.3. Finally, the discussions in this section will be summarized in part *D*. With the consideration of IEEE standard 519-2014 [25], the below results evaluate the compensation performance by setting THD < 15%.

A. *Inductive and Light Loading*

When the loading is inductive and light, APF requires a high DC-link voltage ($V_{DC(APF)} > \sqrt{2} \cdot V_{L-L} = 269$ V, $V_{DC(APF)}$= 300 V) for compensation. After compensation, the source current i_{sx} is reduced to 5.55 A from 6.50 A. And, the source-side displacement power factor (DPF) is compensated from 0.83 to unity. In addition, the source current total harmonics distortion (THD_{isx}) is 7.22% after compensation, which can satisfy the international standard [25] (THD_{isx} < 15%).

For HAPF compensation, the coupling impedance contributes a large voltage drop between the load voltage and the inverter voltage, so that the required DC-link voltage can be small ($V_{DC(HAPF)}$= 80 V). The i_{sx}, DPF and THD_{isx} are compensated to 5.48 A, unity, and 2.01%, respectively.

For the proposed TCLC-HAPF compensation, the i_{sx}, DPF, and THD_{isx} are compensated to 5.48 A, unity, and 1.98%, respectively. As discussed in the previous section, the low DC-link voltage ($V_{DC(TCLC\text{-}HAPF)}$= 50 V) of TCLC-HAPF is used to avoid mistuning of firing angles, prevent resonance problems, and reduce the injected harmonic currents.

B. *Inductive and Heavy Loading*

To compensate the inductive and heavy loading, the APF still requires a high DC-link voltage of $V_{DC(APF)}$= 300 V for compensation. The APF can obtain acceptable results (DPF = 1.00 and THD_{isx} = 6.55%). The i_{sx} is reduced to 5.95 A from 8.40 A after compensation.

With a low dc-link voltage ($V_{DC(HAPF)}$= 50 V), the HAPF cannot provide satisfactory compensation results (*DPF* = 0.85 and THD_{isx} = 17.5%). However, when the DC-link voltage is increased to V_{DC}= 300 V, the compensation results (DPF = 1.00 and THD_{isx} = 7.02%) can satisfy the international standard [25] (THD_{isx} < 15%). The i_{sx} is reduced to 5.90 A from 8.40 A after compensation.

In contrast, the proposed TCLC-HAPF can still obtain acceptable compensation results (DPF = 1.00 and THD_{isx} = 3.01%) with a low DC-link voltage of $V_{DC(TCLC\text{-}HAPF)}$= 50 V. The i_{sx} is reduced to 5.89 A from 8.40 A after compensation.

C. *Capacitive Loading*

When the loading is capacitive, the compensation results of APF are acceptable with $V_{DC(APF)}$= 250 V ($V_{DC(APF)} < \sqrt{2}\,V_{L-L} = 269$ V). The DPF and THD_{isx} are compensated to unity and 7.61%. And, the i_{sx} is also reduced to 3.67 A from 4.34 A after compensation.

For HAPF with $V_{DC(HAPF)}$ = 50 V compensation, the i_{sx} increases to 7.10 A from the original 4.34 A. And, the compensation performances (DPF = 0.57 and THD_{isx} = 23.5%) cannot satisfy the international standard [25] (THD_{isx} < 15%). However, when V_{DC} is increased to 500 V, the DPF is improved to 0.99 and the THD_{isx} is reduced to 10.6%, which can be explained by its V-I characteristic. However, the compensated i_{sx} = 5.02 A is still larger than i_{sx}= 3.73 A before compensation.

With the lowest DC-link voltage ($V_{DC(TCLC\text{-}HAPF)}$= 50 V) among three power quality compensators, TCLC-HAPF can still obtain the superior compensation results with *DPF* = 1.00 and THD_{isx} = 3.01%. In addition, the i_{sx} is reduced to 3.41 A from 4.34 A after compensation.

D. *Summary of the V-I Characteristics and Simulation Results*

Based on the above discussions and simulation results, a summary can be drawn as follows:

1. The APF can compensate both inductive and capacitive reactive currents with a high DC-link voltage due to small coupling inductor impedance.
2. Compared with TCLC-HAPF, the APF obtains the poor source current THD_{isx} (caused by switching noise) due to the high dc-link voltage.
3. The HAPF has a low DC-link voltage characteristic only under a narrow inductive loading range. However, when the loading current is outside its designed range, the HAPF requires the high DC-link operating voltage due to the large coupling capacitor impedance.
4. The TCLC-HAPF obtains the superior performances among the three power quality compensators under both inductive and capacitive loadings.
5. The TCLC-HAPF has the desirable characteristics of wide compensation range and low DC-link voltage.

2.4 Cost Comparison Among APF, HAPF and TCLC-HAPF

In this part, the cost study of APF, HAPF and TCLC-HAPF will be provided and compared under different voltage levels namely: low voltage level applications (≤ 10 kV) and high voltage level applications (>10 kV).

2.4.1 Cost Comparison Among APF, HAPF and TCLC-HAPF Under Low Voltage Levels

For the low voltage level applications, the cost study is done under the following assumptions:

1. The cost comparisons are given under the commonly used low voltage levels: 0.38, 3, 6 and 10 kV.
2. The devices and components costs are dominated for the low voltage level applications. Therefore, the total cost is calculated as the sum of all components in each compensator.
3. When the voltage level is higher than the maximum rating of power switches, the cascade H-bridge voltage source inverter (VSI)/APF is applied to reduce the high voltage stress across each power switch and DC-link capacitor.

If the voltage rating of the IGBT is higher than the required DC-link voltage, the structure of APF is applied as Fig. 2.1. On the other hand, if the voltage rating of the IGBT is lower than the required DC-link voltage, the APF requires cascade H-bridge VSI structure to release the voltage stress on IGBT as shown in Fig. 2.2. Based on above discussion, the estimation cost equations of the APF and cascade H-bridge APF can be expressed as:

$$Cost_{\mathrm{APF}} = 6 \cdot Cost_{IGBT} + Cost_{DC\_Cap} + 3 \cdot Cost_{AC\_ind} + Cost_{Contr} \tag{2.9}$$

$$Cost_{Multi\,\mathrm{APF}} = 12n \cdot Cost_{IGBT} + 3n \cdot Cost_{DC\_Cap} + 3 \cdot Cost_{AC\_ind} + Cost_{Contr} \tag{2.10}$$

where $Cost_{IGBT}$, $Cost_{DC\_Cap}$, and $Cost_{AC\_Ind}$ are costs of each IGBT and its driver, DC capacitor and AC inductor, respectively. $Cost_{contr}$ is the cost of controller such as DSP and PFGA. n is the number of stage in each phase which depends on the ratio of minimum required DC-link voltage and voltage rating of power switches. For the single-stage APF as shown in Fig. 2.1, there are six IGBTs, one DC capacitor and three AC inductors. However, for cascade H-bridge APF as shown in Fig. 2.2, the numbers of required IGBTs and DC-link capacitors are related to n, which are $12n$ and $3n$, respectively.

The HAPF is made up of a coupling LC filter and a VSI. The coupling LC filter can provide the large voltage drop so that the DC-link voltage of active VSI is low and the cascade H-bridge structure can be avoided. Therefore, the total cost of the HAPF can be expressed as:

$$\begin{aligned} Cost_{HAPF} = & 6 \cdot Cost_{IGBT} + 3 \cdot Cost_{DC\_Cap} + 3 \cdot Cost_{AC\_ind} \\ & + 3 \cdot Cost_{AC\_Cap} + Cost_{Contr} \end{aligned} \quad (2.11)$$

where $Cost_{IGBT}$, $Cost_{DC\_Cap}$, $Cost_{AC\_in}$ and $Cost_{AC\_Cap}$ are costs of each IGBT and its driver, DC capacitor, AC inductor and AC capacitor, respectively. And $Cost_{Contr}$ is the cost of controller such as DSP and PFGA. In coupling LC part, there are three AC inductors and three AC capacitors. And there are one DC capacitor and six IGBTs in the active VSI part of HAPF.

As for the structure of TCLC-HAPF, the DC-link voltage of the active VSI is low due to the significant voltage drop of TCLC part. The total cost of the TCLC-HAPF can be expressed as:

$$\begin{aligned} Cost_{TCLC-HAPF} = & 6 \cdot Cost_{IGBT} + Cost_{DC\_Cap} + 6 \cdot Cost_{Thy} \\ & + 6 \cdot Cost_{AC\_ind} + 3 \cdot Cost_{AC\_Cap} + Cost_{Contr} \end{aligned} \quad (2.12)$$

where $Cost_{IGBT}$, $Cost_{DC\_Cap}$, $Cost_{Thy}$, $Cost_{AC\_Ind}$ and $Cost_{AC\_Cap}$ are costs of each IGBT, DC capacitor, thyristor, AC inductor and AC capacitor, respectively. And $Cost_{Contr}$ is the cost of controller like DSP and FPGA. TCLC-HAPF is made up of TCLC part and an active inverter part. In TCLC part, there are six AC inductors and three capacitors. And there are one DC capacitor and six IGBTs in the active inverter part of the TCLC-HAPF.

With the cost study of each component [26–29], the cost comparisons among APF, HAPF and TCLC-HAPF under 0.38, 3, 6 and 10 kV can be summarized in Table 2.4.

Table 2.4 Cost comparisons among APF, HAPF and TCLC-HAPF under 0.38, 3, 6 and 10 kV

Voltage (V_{LL}) (kV)	Power and current ratings	Cost comparisons
0.38	–	APF < HAPF < TCLC-HAPF
3	<250 kVA (50 A)	HAPF < TCLC-HAPF < APF (< 250 kVA)
	>250 kVA (50 A)	HAPF < APF < TCLC-HAPF (> 250 kVA)
6	<700 kVA (70 A)	HAPF < TCLC-HAPF < Cascade H-bridge APF (<700 kVA)
	>700 kVA (70 A)	HAPF < Cascade H-bridge APF < TCLC-HAPF (>700 kVA)
10	–	HAPF < TCLC-HAPF < Cascade H-bridge APF

2.4.2 Cost Comparisons Among APF, HAPF and TCLC-HAPF Under High Voltage Levels

For the high voltage level cost comparison, the components cost may not be dominated cost. Many aspects need to be considered as below [27–29]: (a) Field construction cost, (b) Auxiliary components cost, (c) Magnetic components cost, (d) Passive components cost and (e) Solid-state devices cost. Each aspect has its corresponding portion of total cost.

In the industrial market, the costs of PPF, SVC and APF in terms of power rating are available in [28, 29] (shown as Fig. 2.4a, b). Even the detailed costs of the HAPF and TCLC-HAPF cannot be found, the cost of the HAPF and TCLC-HAPF can be approximately calculated by using the costs of PPF, SVC and APF [28, 29].

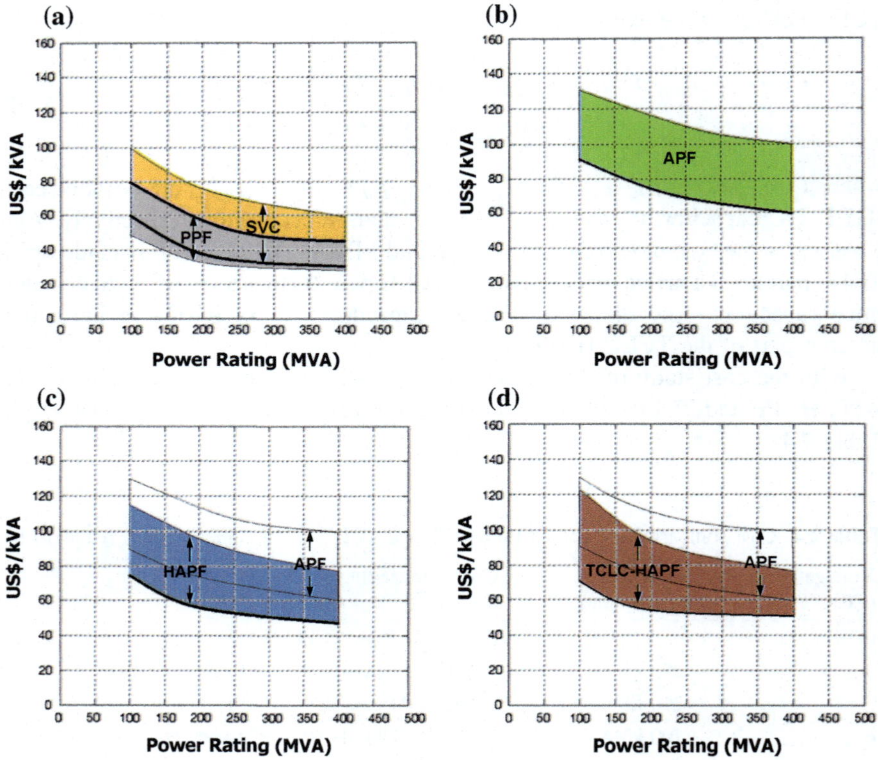

Fig. 2.4 Typical high voltage investment costs for **a** PPF and SVC **b** APF, **c** HAPF, **d** TCLC-HAPF

The HAPF can be considered as the PPF in series with APF (active inverter part). However, the required rating of the active inverter part of HAPF is much lower than APF. The cost of HAPF can be approximately calculated as:

$$Cost_{HAPF} = k_{HAPF} \cdot Cost_{APF} + Cost_{PPF} \tag{2.13}$$

where $Cost_{APF}$ and $Cost_{PPF}$ are the costs of APF and PPF respectively. The k_{HAPF} is the power ratio of active inverter part between HAPF and APF. If the k_{HAPF} is small enough, the HAPF is cheaper than APF. Taking an example of the simulation case studies as shown in Sect. 2.3.2 (Table 2.4), the capacity of the active inverter part (or DC-link voltage) of HAPF is only k_{HAPF}= 27% of the capacity of the APF.

The TCLC-HAPF can be considered as the SVC part (TCLC part) in series with active inverter part. The cost of TCLC-HAPF can be approximately calculated as:

$$Cost_{TCLC-HAPF} = k_{TCLC-HAPF} \cdot Cost_{APF} + Cost_{SVC} \tag{2.14}$$

where $Cost_{APF}$ and $Cost_{SVC}$ are the costs of APF and SVC respectively. The $k_{TCLC\text{-}HAPF}$ is the power ratio of active inverter part between TCLC-HAPF and APF. Taking an example of the simulation case studies as shown in Sect. 2.3.2 (Table 2.4), the capacity of the active inverter part (or DC-link voltage) of the TCLC-HAPF is only $k_{TCLC\text{-}HAPF}$= 17% of the capacity of the APF for wide range compensation (both inductive and capacitive).

Based on (2.13), (2.14) and the cost study in [28, 29], the costs of the PPF, SVC, APF, HAPF and TCLC-HAPF can be plotted as in Fig. 2.4.

In Fig. 2.4, there are two limit lines for each compensator. The lower limit cost line indicates the equipment cost, while the upper limit means the total investment cost. It is clearly shown that at low power rating, the cost per MVA is higher than at the high power rating. By comparing the cost in Fig. 2.4, it can be seen that the cost of both HAPF and TCLC-HAPF are around 7.7–23% cheaper than the APF for high voltage level applications.

Based on above cost study in Table 2.4 and Fig. 2.4, it can be said that the HAPF and TCLC-HAPF offer cost-effective ways for high voltage applications ($\geq$ 10 kV) compared with the conventional cascade H-bridge APF.

2.5 Reliability Comparison Among APF, HAPF and TCLC-HAPF

In order to have a high-quality performance, all of the components of power quality compensators are required to work safely and well-functioning. In the practical systems, all of the components have specific failure rate and repair/replacement time. Among them, the switching components like IGBTs and thyristors have the highest failure rates, so that only switching components are focused for the reliability comparison among different power quality compensators. In the following, a reliability comparison among APF, HAPF and TCLC-HAPF under a medium voltage application (10 kV) is discussed.

For a medium voltage application (10 kV), the required DC-link voltage of the HAPF and TCLC-HAPF are around 4.0 kV (27% of APF rating) and 2.6 kV (17% of APF rating), respectively (based on simulations in Sect. 2.3.2). In contrast, the minimum required DC-link voltage of APF is about 15 kV ($V_{DC} > \sqrt{2} \cdot V_{LL}$), which is higher than the voltage rating of any existed IGBT devices, so that the APF requires a cascade H-bridge structures as shown in Fig. 2.2 to reduce the high voltage stress across each power switch and DC-link capacitor. To reduce the DC-link voltage level at each level to be similar as TCLC-HAPF (about <3 kV), the HAPF and APF require two stages and five stages H-bridge structures in each phase, respectively. And, the total required switching components of the HAPF and APF are 24 and 60. As for TCLC-HAPF, the total required switching components is just 12. Considering the system can work if and only if all the components are functioning well, the system reliability can then be roughly calculated as [30–33]:

$$Re_{tot} = \mathrm{e}^{-\lambda_{tot} t} \tag{2.15}$$

where Re_{tot} and λ_{tot} are the total reliability and total failure rate of the system. The λ_{tot} is the sum of failure rate of different components as shown in Table 2.5. The detailed failure rate calculation steps of IGBT, thyristor, inductor, AC capacitor and DC-link capacitor in Table 2.5 are provided in Appendix A–E, respectively.

Based on (2.15) and Table 2.5, the system reliability of APF, HAPF and TCLC-HAPF can be plotted as Fig. 2.5.

From Fig. 2.5, it can be clearly observed that the TCLC-HAPF has much higher reliability than the cascade H-bridge HAPF and APF. Therefore, the potential maintenance time and costs of the TCLC-HAPF can be significantly reduced.

Table 2.5 The failure rate of each component in power quality compensators [31–33]

Component	λ_p (failure/10^6 h)	APF	HAPF	TCLC-HAPF
IGBT	$\lambda_{p(IGBT)} = 12.0$	720 (n = 60)	288 (n = 24)	72 (n = 6)
Thyristor	$\lambda_{p(Thyristor)} = 16.4$	0	0	98.4 (n = 6)
AC inductor	$\lambda_{p(L)} = 0.0718$	0.215 (n = 3)	0.215 (n = 3)	0.215 (n = 3)
AC capacitor	$\lambda_{p(C)} = 1.25$	0	3.75 (n = 3)	3.75 (n = 3)
DC-link capacitor	$\lambda_{p(C)} = 0.102$	1.52 (n = 15)	0.612 (n = 6)	0.102 (n = 1)
Whole system	λ_{tot}	721.7	292.6	174.46

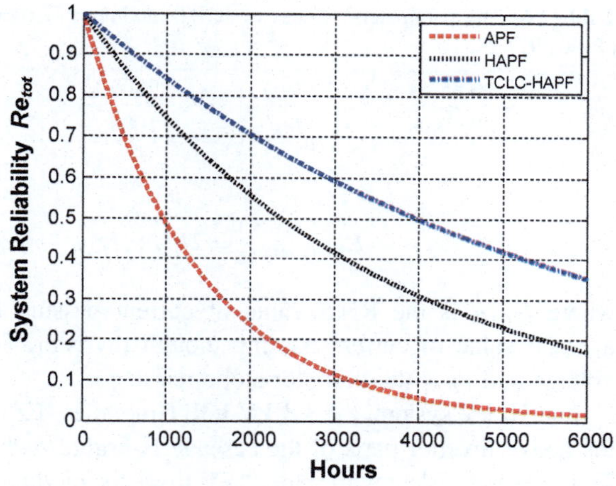

Fig. 2.5 The system reliability of APF, HAPF and TCLC-HAPF

2.6 Power Loss Comparison Among APF, HAPF and TCLC-HAPF

The power loss of the APF, HAPF and TCLC-HAPF are mainly from IGBTs and/or thyristors (switching loss), DC-link capacitor, coupling inductor and/or coupling capacitor.

A. *Switching Loss from IGBTs and/or Thyristors*

The switching loss of the APF and HAPF are mainly from the switching components (IGBTs) in the active inverter part. In contrast, the switching loss of the TCLC-HAPF consists of two parts: (1) switching loss from IGBTs in active inverter part and (2) switching loss from thyristors in TCLC part.

The total turn-on and turn-off switching loss from IGBTs in the active inverter part can be calculated as [34]:

$$P_{sw(IGBT)} = V_{DC} I_{CM} f_{sw} \left(\frac{1}{8} t_{rN} \frac{I_{CM}}{I_{CN}} + t_{fN} \left(\frac{1}{3\pi} + \frac{1}{24} \cdot \frac{I_{CM}}{I_{CN}} \right) \right) \tag{2.16}$$

where V_{DC}, I_{CM}, I_{CN}, t_{rN}, t_{fN} and f_{sw} are the DC-link voltage, maximum collector current, rated collector current, rated rise time, rated fall time, and switching frequency, respectively.

The thyristors in TCLC part only turn on once in a fundamental period which is different with IGBTs in the active inverter part. In TCLC-HAPF applications, the on-state switching loss is the dominant part for thyristors that can be easily calculated as [35]:

Table 2.6 The parameters' values of IGBT modules (FZ1000R33HL3) and thyristor modules (T1901N)

t_{rn}	0.55 μs	V_{DC}	2800 V	r_T	0.44 mΩ
t_{fn}	0.40 μs	I_{CN}	1000 A	U_{TO}	1.24 V

$$P_{sw(thyristor)} = U_{TO} \times I_{TAV} + r_T \times I_{TRMS}^2 \tag{2.17}$$

where I_{TRMS} is the RMS value of current passing though thyristor, I_{TAV} is the average value of current passing though thyristor, U_{TO} is the thyristor threshold voltage and r_T is the thyristor slope resistance.

In a 10 kV system, the 3.3 kV IGBT modules (FZ1000R33HL3) are selected for the active inverter parts of the cascade H-bridge APF and HAPF, and also for the TCLC-HAPF. Meanwhile, the 8 kV thyristor modules (T1901N [36]) are used for the TCLC part of the TCLC-HAPF. The parameters' values of IGBT modules (FZ1000R33HL [37]) and thyristor modules (T1901N) are provided in Table 2.6.

B. *Power Loss from Coupling Inductor, Coupling Capacitor and DC-link Capacitor*

The power loss of the coupling inductor, coupling capacitor and DC-link capacitor are mainly caused by the equivalent series resistance (ESR) in them. The power loss of the coupling inductor, coupling capacitor and DC-link capacitor can be expressed as:

$$P_{\mathrm{loss}(L)} = ESR_L \cdot I_{cx}^2 \tag{2.18}$$

$$P_{\mathrm{loss}(C)} = ESR_C \cdot I_{cx}^2 \tag{2.19}$$

$$P_{\mathrm{loss}(C_{DC})} = ESR_{C_{DC}} \cdot I_{cx}^2 \tag{2.20}$$

where ESR_L, ESR_C and ESR_{CDC} are the equivalent series resistances of the coupling inductor, coupling capacitor and DC-link capacitor, respectively. And I_{cx} is the compensating current. According to [38–40], reasonable assumptions have been given as: ESR_L= 2.5 mΩ, ESR_C = 20 mΩ and ESR_{CDC}= 30 mΩ.

Based on (2.16)–(2.20), Table 2.6 and the assumptions of $I_{CM} = I_{TRMS} = I_{TAV} = I_{cx}$ and f_{sw}= 5 kHz, the total power loss of the APF, HAPF and TCLC-HAPF can be plotted as:

From Fig. 2.6, it can be clearly observed that the TCLC-HAPF has much lower power loss than the cascade H-bridge APF and HAPF, especially under a higher current situation.

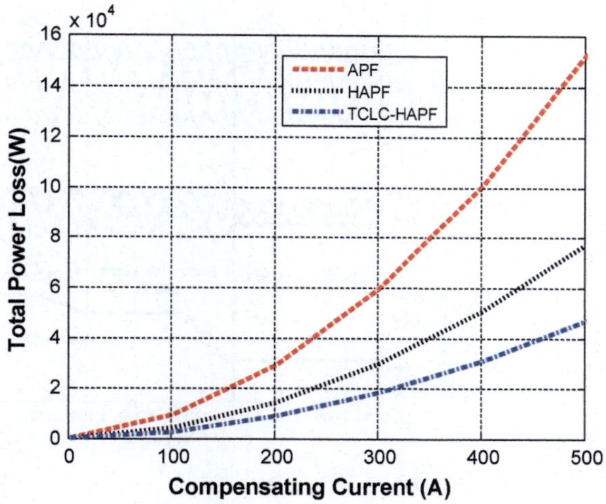

Fig. 2.6 The power loss of APF, HAPF and TCLC-HAPF

2.7 Tracking Performance Comparison Among APF, HAPF and TCLC-HAPF

With the similar active inverter part control method, the compensator's response time and tracking performance mainly depends on the structure characteristics and DC-link voltage. In this part, the response time and tracking performance among the APF, HAPF and TCLC-HAPF are compared. Figure 2.7 shows the tracking performances by applying APF, HAPF and TCLC-HAPF under different loadings cases.

Based on Fig. 2.7a, the transient response of APF is less than half cycle to settle down, when the load reactive power changes from capacitive to inductive. However, when the load reactive power is changing from light inductive loads to heavy inductive load, the transient time is very short and the source current waveforms are smooth. Meanwhile, the source reactive power is compensated to around zero even during the transient time.

From Fig. 2.7b, the HAPF can only compensate small fixed amount of reactive power (inductive light loads). And, the HAPF cannot provide satisfactory compensation results when the reactive power is capacitive or heavy inductive. The fundamental reactive power cannot be compensated to around zero for capacitive loads and heavy inductive loads. Therefore, the tracking performance of the HAPF is poor due to its inherent structure characteristic.

From Fig. 2.7c, the TCLC-HAPF takes less than one cycle to settle down, when the load reactive power changes from capacitive to inductive. However, when the load reactive power is changing within the inductive range, the transient time is significantly reduced and the source current waveforms are also very smooth.

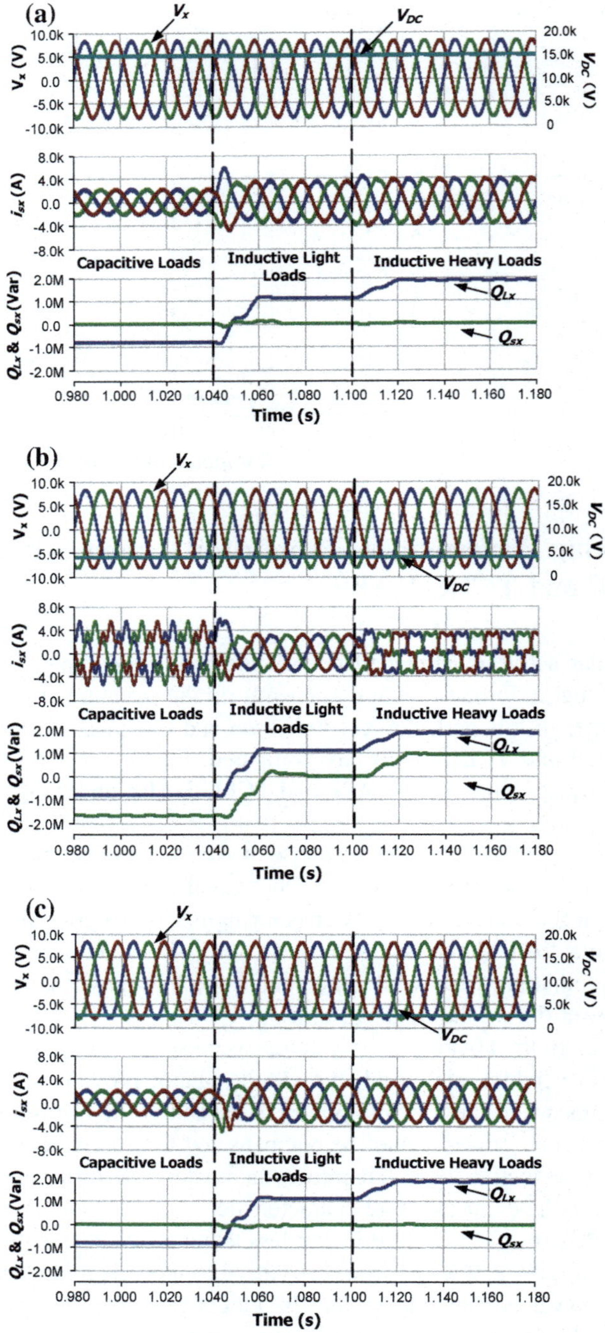

Fig. 2.7 Dynamic compensation waveforms of load voltage, dc-link voltage, source current, and load and source reactive power by applying APF, HAPF and TCLC-HAPF under different loadings cases

Based on simulation results and discussions, both APF and TCLC-HAPF obtain fast response and good tracking performance, while the HAPF has poor tracking performance.

2.8 Experimental Results

This part aims to prove the DC-link voltage characteristics, tracking performances and compensation performances of APF, HAPF and TCLC-HAPF. The three-phase three-wire hardware experimental prototypes of the APF, HAPF and TCLC-HAPF are constructed in the laboratory. The following experimental results are preformed under the same voltage level V_{LL}= 191 V with same hardware prototype rating of 5 kVA.

2.8.1 Experimental Results of APF

Figure 2.8 shows the three-phase three-wire experimental system voltage current source voltage, current and DC-link voltage before and after APF compensation. From Fig. 2.8 and Table 2.7, after APF compensation, it can be seen that the experimental PF and THD_{isx} can be improved to 0.98, 9.5% for the worst phase for the 1st loading compensation, and 0.99 and 8.7 for the 1st and 2nd loading compensation. The required DC-link voltage for APF is about 340 V.

For the transient performance as shown in Fig. 2.8b, when the load suddenly changes, the response time of the conventional APF is less than one fundamental period.

2.8.2 Experimental Results of HAPF

Figure 2.9 and Table 2.7 show the experimental compensation performance of the HAPF. From Fig. 2.9a and Table 2.7, it can be seen that the experimental *PF* and THD_{isx} are compensated to 0.85 and 12.5% for the worst phase after the HAPF compensation. Due to the loading falls outside its operational range, thus the PF even becomes worse after compensation. During the start-up process, the HAPF requires about five fundamental periods to charge the DC-link voltage to its reference value. And, the DC-link charging period mainly depends on the design (like P or PI gain) of the controller.

From Fig. 2.9b and Table 2.7, the experimental PF and THD_{isx} are improved to 0.99 and 7.5% for the worst phase during the 1st and 2nd loads compensation. And, the required DC-link voltage for the HAPF is about 105 V. For the transient performance as shown in Fig. 2.9b, when the load suddenly changes, the response time of the conventional HAPF is about 7 ms, which is less than one fundamental period.

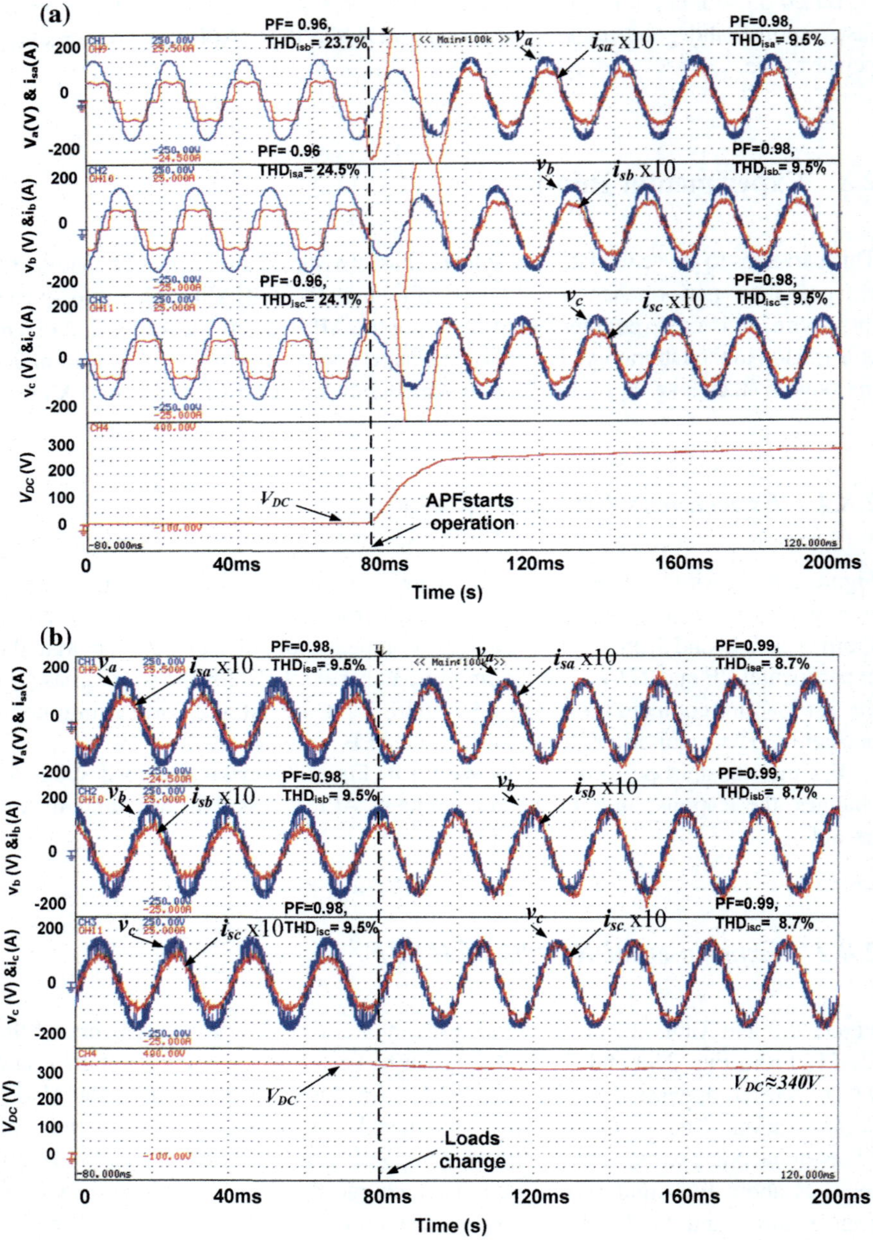

Fig. 2.8 Waveforms of source voltage, current, natural current and DC-link voltage: **a** before and after APF starts operation during 1st load and **b** when 2nd load is connected

Table 2.7 Experimental results by using HAPF and SVC-HAPF

Load Type	Before/after Comp. for worst phase	i_{sx}(A)	*PF*	$THDi_{sx}$(%)	V_{DC}(V)
1st nonlinear load	Before Comp.	6.6	0.96	24.5	--
	APF	6.8	0.98	9.5	340
	HAPF	7.1	0.85	12.5	105
	SVC-HAPF	6.5	0.99	6.3	60
1st and 2nd nonlinear loads	Before Comp.	8.5	0.94	21.1	--
	APF	9.4	0.99	8.7	340
	HAPF	9.4	0.99	7.5	105
	SVC-HAPF	8.4	0.99	5.5	60

Note The shaded areas indicate undesirable results

Based on Fig. 2.9 and Table 2.7, it can be verified that the HAPF has narrow compensation range, thus causing over-compensation for the 1st loading compensation. And, the PF and THD_{isx} compensation performance is relatively poor for the 1st inductive loading case.

2.8.3 Experimental Results of TCLC-HAPF

Figure 2.10 shows the experimental compensation performance of the TCLC-HAPF. From Fig. 2.10 and Table 2.7, it can be seen that the experimental *PF* and THD_{isx} can be improved to 0.99, 6.3% (the worst phase) for the 1st load compensation. And the experimental PF and THD_{isx} are improved to 0.90, 5.5% (the worst phase) for the 1st and 2nd loads compensation. To compensation both loads, the TCLC-HAPF required DC-link voltage is about 60 V. And, from Fig. 2.10b, the response time during the loads changing is short and the tracking performance is good.

2.9 Summary

In this chapter, the comparisons among the APF, HAPF and TCLC-HAPF have been provided in terms of V-I characteristic (compensation range and require DC-link voltage) cost, reliability, power loss and tracking performance. Compared TCLC-HAPF with APF and HAPF, it has higher reliability and lower power loss than both APF and HAPF. Besides, the TCLC-HAPF can be more cost effective than the APF for medium/high voltage level applications ($\geq$ 10 kV). In addition, the TCLC-HAPF has wider reactive current compensation range than the HAPF and lower DC-link voltage than the HAPF. Therefore, the TCLC-HAPF has a large potential to be further developed for medium/high voltage level applications.

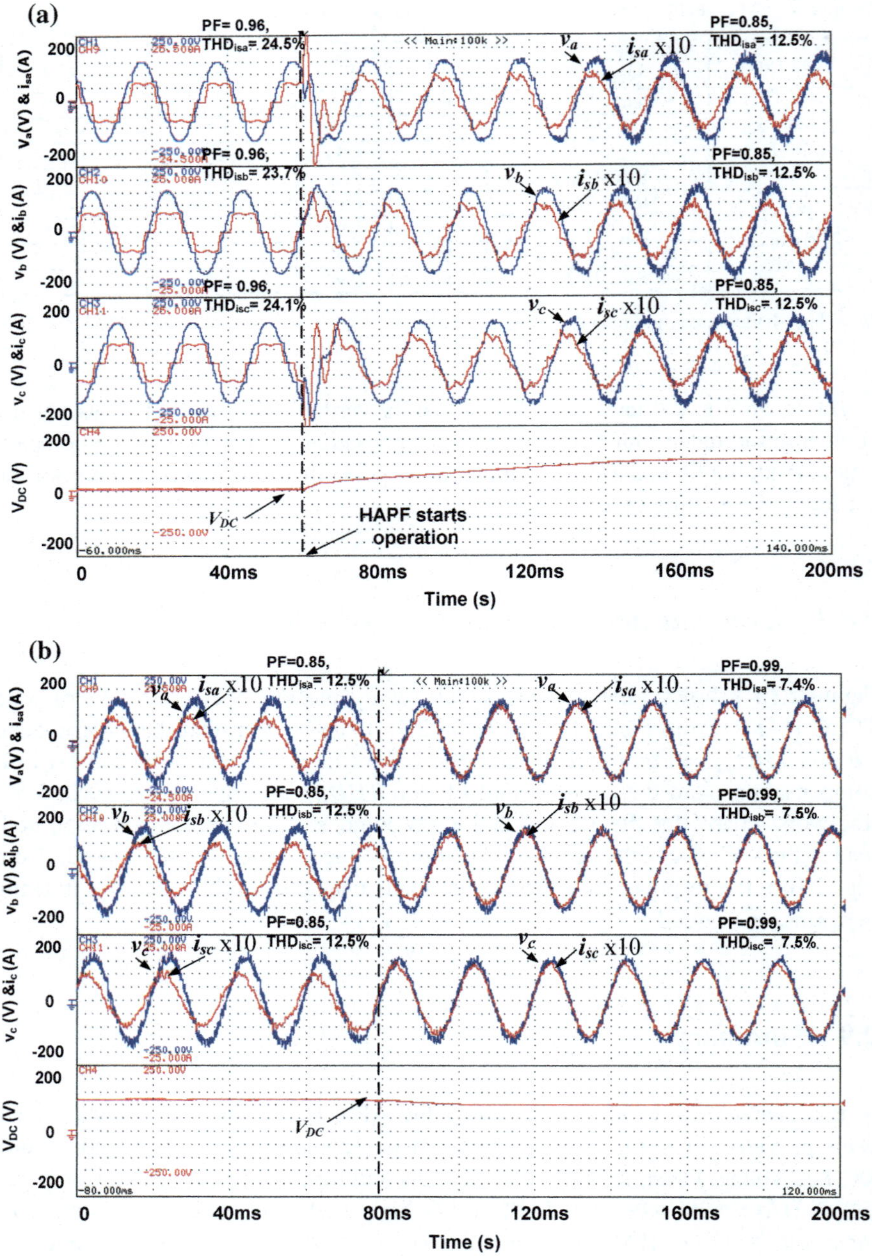

Fig. 2.9 Waveforms of source current, natural current and DC-link voltage: **a** before and after HAPF starts operation during 1st load and **b** when 2nd load is connected

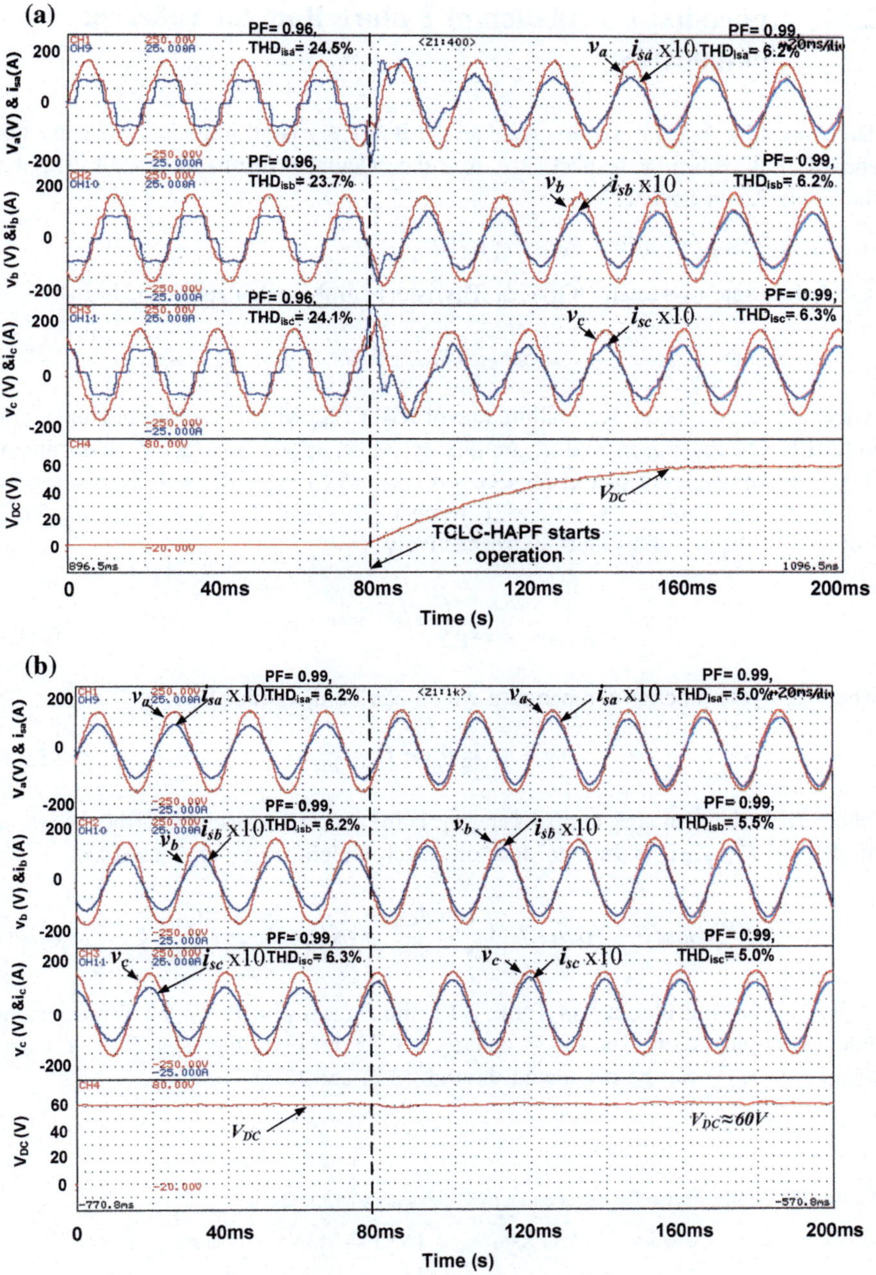

Fig. 2.10 Waveforms of source current, natural current and DC-link voltage: **a** before and after TCLC-HAPF starts operation during 1st load and **b** when 2nd load is connected

2.10 Appendix: Calculation of Failure Rate for Different Components

The Appendix A–E is used to calculate the IGBT,thyristor, inductor, AC capacitor and DC-link capacitor, respectively. And the calculated results are used to support the Table 2.5 in Sect. 2.5.

A. *Calculation of Failure Rate for IGBT*

The failure rate of switch (IGBT or Thyristor) can be calculated as [31–33]:

$$\lambda_{\mathrm{P(IGBT\ or\ Thyristor)}} = \lambda_{b(IGBT\ \mathrm{or\ Thyristor})} \times \pi_{\mathrm{Q}} \times \pi_{\mathrm{A}} \times \pi_{\mathrm{E}} \times \pi_{\mathrm{T}} \tag{2.21}$$

where the $\lambda_{b(IGBT)}$ is the base failure rate of IGBT (λ_{b}) which is a constant and equal to 0.012 [32]. π_{Q}, π_{A}, π_{E} and π_{T} are quality factor, application factor, environment factor and temperature factor, respectively. π_{Q} (=5.5), π_{A}(=10), and π_{E}(=6) can be obtained from Tables 2.8–2.10 [32], respectively. And the π_{T} can be calculated from (2.22) from Tables and 2.9, respectively.

$$\pi_{\mathrm{T}(IGBT)} = \exp^{\left(-1925\times\left(\frac{1}{T_j+273}-\frac{1}{298}\right)\right)} \tag{2.22}$$

where T_j is the junction temperature which can be determined by:

$$\mathrm{T}_j = T_C + \theta_{jc} \times P_{sw(IGBT)} \tag{2.23}$$

where T_{C} is the heat sink temperature, θ_{jc} is the thermal resistance of the switch or diode, and $P_{\mathrm{sw(IGBT)}}$ is the total loss of switch which can be expressed as:

$$P_{sw(IGBT)} = V_{\mathrm{DC}} I_{CM} f_{sw} \left(\frac{1}{8} t_{rN} \frac{I_{CM}}{I_{CN}} + t_{fN}\left(\frac{1}{3\pi} + \frac{1}{24} \cdot \frac{I_{CM}}{I_{CN}}\right)\right) \tag{2.24}$$

By substituting the data in Table 2.12, the $P_{\mathrm{sw(IGBT)}}$ = 166.7 W. With reasonable assumptions of T_C= 45 °C and θ_{jc} = 0.25 in [31], the T_j (=86.7 °C) and $\pi_{\mathrm{T(IGBT)}}$ (=3.03) can be calculated through (2.2) and (2.3).

Table 2.8 Quality factor for semiconductors, capacitors and coils

Semiconductor	Quality	JANTXV	JANTX	JAN	Plastic	Lower	–
	π_{Q}	0.7	1.0	2.4	8	5.5	–
Capacitor	Quality	MIL-C-15305	S	R	P	Lower	M
	π_{Q}	4.0	0.030	0.10	0.30	20	1.0
Coil	Quality	Non-Est. Rel.	S	R	P	Lower	M
	π_{Q}	3.0	0.030	0.1	0.30	10	1.0

Table 2.9 Application factor

Application (P_r, rated output power)	π_A
Linear application ($P_r < 2$ W)	1.5
Small signal switching	0.7
Non-linear ($P_r \geq 2$ W)	
$2 \leq P_r < 5$ W	2.0
$5 \leq P_r < 50$ W	4.0
$50 \leq P_r < 250$ W	8.0
$P_r \geq 250$ W	10

$$\lambda_{P(IGBT)} = \lambda_{b(IGBT)} \times \pi_Q \times \pi_A \times \pi_E \times \pi_T$$
$$= 0.012 \times 5.5 \times 10 \times 6 \times 3.03 = 12.0\,(failure/10^6\,h)$$

B. *Calculation of Failure Rate for Thyristor*

The basic data for calculating of failure rate of thyristor are similar to IGBT as shown in below:

λ_b	0.012	π_Q	5.5	π_A	10
π_E	6	T_C	45 °C	θ_{jc}	0.25

The $P_{sw(Thyristor)}$ is the total loss of switch which can be expressed as:

$$P_{sw(thyristor)} = U_{TO} \times I_{TAV} + r_T \times I_{TRMS}^2 \tag{2.25}$$

By substituting the data in Table 2.12 the $P_{sw(Thyristor)}$ = 256.6 W. By substituting the T_j (=109.15 °C) into (2.22), the $\pi_{T(IGBT)}$ can be calculated as (4.15). With above parameters, the $\lambda_{b(Thyristor)}$ can be calculated through (2.17) as:

$$\lambda_{P(Thyristor)} = \lambda_{b(Thyristor)} \times \pi_Q \times \pi_A \times \pi_E \times \pi_T$$
$$= 0.012 \times 5.5 \times 10 \times 6 \times 4.15 = 16.4\,(failure/10^6\,h)$$

C. *Calculation of Failure Rate for Inductor*

The failure rate of inductor can be calculated as [32]:

$$\lambda_{P(L)} = \lambda_{b(L)} \times \pi_C \times \pi_Q \times \pi_E \tag{2.26}$$

where $\lambda_{b(L)}$ is the base failure rate of inductor. π_Q, π_A and π_E are quality factor and contact construction factor, respectively. The π_Q (=20), π_E (=4), and π_C (=1) can be obtained from Tables 2.8, 2.10 and 2.11 [32], respectively. The $\lambda_{b(L)}$ is the base failure rate of inductor which can be expressed as:

Table 2.10 Environment factor for semiconductors, inductors, and capacitors

Environment	π_E			Environment	π_E		
	Semiconductor	Inductor	Capacitor		Semiconductor	Inductor	Capacitor
G_B	1.0	1.0	1.0	A_{UC}	20	6.0	28
G_M	9.0	12	12	A_{RW}	24	25	27
G_F	6.0	4.0	2.0	A_{UF}	43	8.0	35
N_S	9.0	5.0	6.0	S_F	0.5	0.5	0.5
N_U	19	16	17	M_F	14	13	14
A_{IF}	29	7.0	12	C_L	320	610	690
A_{IC}	13	5.0	10	M_L	32	34	38

Table 2.11 Contact construction factor

Contact construction	π_C
Metallurgically bonded	1.0
Non-metallurgically bonded and spring loaded contacts	2.0

$$\lambda_{b(L)} = 0.00035 \times \exp^{\left(\frac{T_{HS}+273}{329}\right)^{15.6}} \quad (2.27)$$

where T_{HS} is the hot spot temperature of the inductor (in °C) and it can be calculated, as follows:

$$T_{HS} = T_A + 1.1 \times \Delta T \quad (2.28)$$

where T_A is the device ambient operating temperature (in °C) and ΔT is the average temperature rise (in °C). Referenced to [32] (T_A= 22.7 and ΔT = 30), the T_{HS} can be calculated as 55.7 °C. Through (2.27), the $\lambda_{b(L)}$ is calculated to be 8.98×10^{-4}. Through (2.26), $\lambda_{P(L)}$ can be calculated as:

$$\begin{aligned}\lambda_{P(L)} &= \lambda_{b(L)} \times \pi_C \times \pi_Q \times \pi_E = 8.98 \times 10^{-4} \times 20 \times 1 \times 4 \\ &= 0.0718\,(failure/10^6\,\text{h})\end{aligned}$$

D. *Calculation of Failure Rate for AC Capacitor*

The failure rate of AC capacitor can be calculated as [32]:

$$\lambda_{P(C)} = \lambda_{b(C)} \times \pi_{CV} \times \pi_Q \times \pi_E \quad (2.29)$$

where $\lambda_{b(C)}$ is the base failure rate of capacitor. π_{CV}, π_Q and π_E are capacitance factor, quality factor and environment factor, respectively. π_Q (=10) and π_E (=2) can be obtained from Table 2.9, respectively. The $\lambda_{b(C)}$ is the base failure rate of AC capacitor which can be expressed as:

$$\lambda_{b(C)} = 0.00254 \times \left[\left(\frac{S}{0.5} \right)^3 + 1 \right] \exp^{5.09 \times \left(\frac{T_A + 273}{378} \right)^5} \tag{2.30}$$

where S is the ratio of operating voltage to the rated voltage, T_A is the device ambient operating temperature (in °C). For the ac capacitor in both HAPF and TCLC-HAPF, the AC capacitor is design to provide the major voltage drop. Therefore, the S is approximately equal to 1. Substituting S = 1 and T_A= 22.7 °C into (2.30), the $\lambda_{b(C)}$= 0.10. The capacitance factor π_{CV} can be expressed as:

$$\pi_{CV} = 0.34 \times C^{0.12} \tag{2.31}$$

where C is the capacitance in microfarad (μF). According to case study in section, the C is assumed to be 160 μF. By using (2.31), π_{CV} can be calculated as 0.625. Through (2.29), $\lambda_{P(C)}$ can be calculated as:

$$\lambda_{P(C)} = \lambda_{b(C)} \times \pi_{CV} \times \pi_Q \times \pi_E = 0.10 \times 0.625 \times 10 \times 2 = 1.25$$

E. *Calculation of Failure Rate for DC-link Capacitor*

The calculation of failure rate for dc-link capacitor can basically follow the steps in above section D. However, for the dc-link capacitor; the capacitance of 3000 μF is selected. The capacitance factor π_{CV} can be calculated as:

$$\pi_{CV} = 0.34 \times C^{0.12} = 0.779$$

Based on (2.30), the S in (2.30) for dc-link capacitor is equal to 2800/ 10,000 = 0.28. Substituting the S = 0.28 and T_A= 22.7 °C into (2.30), the $\lambda_{P(C)}$ can be expressed as:

$$\lambda_{b(C)} = 0.00254 \times \left[\left(\frac{0.28}{0.5} \right)^3 + 1 \right] \exp^{5.09 \times \left(\frac{22.7 + 273}{378} \right)^5} = 6.539 \times 10^{-3}$$

Through (2.29), $\lambda_{P(C)}$ can be calculated as:

$$\lambda_{P(C)} = \lambda_{b(C)} \times \pi_{CV} \times \pi_Q \times \pi_E = 6.539 \times 10^{-3} \times 0.779 \times 10 \times 2 = 0.102$$

In 10 kV system, the 3.3 kV IGBT modules (FZ1000R33HL3) are selected for active inverter parts of cascade H-bridge APF, HAPF and TCLC-HAPF. Meanwhile, the 8 kV thyristor modules (T1901N [36]) are used for TCLC part of the TCLC-HAPF. The parameters' values of IGBT modules (FZ1000R33HL3 [37]) and thyristor modules (T1901N) are provided in Table 2.12.

Table 2.12 The parameters' values of IGBT modules (FZ1000R33HL3) and thyristor modules (T1901N)

t_{rn}	0.55 μs	V_{DC}	2800 V	r_T	0.44 mΩ
t_{fn}	0.40 μs	I_{CN}	1000 A	U_{TO}	1.24 V
f_{sw}	5 kHz		$I_{CM} = I_{TRMS} = I_{TAV}$		200 A

References

1. J.C. Das, Passive filters—potentialities and limitations. IEEE Trans. Ind. Appl. **40**(1), 232–241 (2004)
2. R.N. Beres, X. Wang, M. Liserre, F. Blaabjerg, C.L. Bak, A review of passive power filters for three-phase grid connected voltage source converters. IEEE J. Emerg. Sel. Topics Power Electron. **4**(1), 54–69 (2016)
3. L. Wang, C.S. Lam, M.C. Wong, Design of a thyristor controlled *LC* compensator for dynamic reactive power compensation in smart grid. IEEE Trans. Smart Grid. **8**(1), 409–417 (2017)
4. S.D. Swain, P.K. Ray, K.B. Mohanty, Improvement of power quality using a robust hybrid series active power filter. IEEE Trans. Power Electron. **32**(5), 3490–3498 (2017)
5. J. Chen, X. Zhang, C. Wen, Harmonics attenuation and power factor correction of a more electric aircraft power grid using active power filter. Trans. Ind. Electron. **63**(12), 7310–7319 (2016)
6. J. Fang, G. Xiao, X. Yang, Y. Tang, Parameter design of a novel series-parallel-resonant LCL filter for single-phase half-bridge active power filters. IEEE Trans. Power Electron. **32**(1), 200–217 (2017)
7. L. Wang, C.S. Lam, M.C. Wong et al., Non-linear adaptive hysteresis band pulse-width modulation control for hybrid active power filters to reduce switching loss. IET Power Electron. **8**(11), 2156–2167 (2015)
8. İ. Yılmaz, E. Durna, M. Ermiş, Design and implementation of a hybrid system for the mitigation of pq problems of medium-frequency induction steel-melting furnaces. Trans. Ind. Appl. **52**(3), 2700–2713 (2016)
9. K.W. Lao, M.C. Wong et al., Analysis in the effect of co-phase traction railway HPQC coupled impedance on its compensation capability and impedance-mapping design technique based on required compensation capability for reduction in operation voltage. IEEE Trans. Power Electron. **32**(4), 2631–2646 (2017)
10. L. Wang, C.S. Lam, M.C. Wong, A SVC-HAPF with wide compensation range and low dc-link voltage. IEEE Trans. Ind. Electron. **63**(6), 3333–3343 (2016)
11. S. Rahmani, A. Hamadi, K. Al-Haddad, A combination of shunt hybrid power filter and thyristor-controlled reactor for power quality. IEEE Trans. Ind. Electron. **61**(5), 2152–2164 (2014)
12. L. Wang, C.S. Lam, M.C. Wong, An unbalanced control strategy for a thyristor controlled LC-coupling hybrid active power filter (SVC-HAPF) in three-phase three-wire systems. IEEE Trans. Power Electron. **32**(2), 1056–1069 (2017)
13. L. Wang, C.S. Lam, M.C. Wong, Hardware and software design of a low dc-link voltage and wide compensation range thyristor controlled LC-coupling hybrid active power filter, in *TENCON 2015 IEEE Region 10 Conference proceedings*, Nov 2015
14. L. Wang, C.S. Lam, M.C. Wong, Modeling and parameter design of thyristor controlled LC-coupled hybrid active power filter (SVC-HAPF) for unbalanced compensation. IEEE Trans. Ind. Electron. **64**(3), 1827–1840 (2017)

15. C.S. Lam, L. Wang, S.I. Ho, M.C. Wong, Adaptive thyristor controlled LC—hybrid active power filter for reactive power and current harmonics compensation with switching loss reduction. IEEE Trans. Power Electron. **32**(10), 7577–7590 (2017)
16. L. Wang, C.S. Lam, M.C. Wong, Selective compensation of distortion, unbalanced and reactive power of a thyristor controlled LC-coupling hybrid active power filter (SVC-HAPF). IEEE Trans. Power Electron. **32**(12), 9065–9077 (2017)
17. C.S. Lam, X.X. Cui, W.H. Choi, M.C. Wong, Y.D. Han, Minimum inverter capacity design for three-phase four-wire LC-hybrid active power filters. IET, Power Electron. **5**(7), 956–968 (2012)
18. C.-S. Lam, W.-H. Choi, M.-C. Wong, Y.-D. Han, Adaptive dc-link voltage controlled hybrid active power filters for reactive power compensation. IEEE Trans. Power Electron. **27**(4), 1758–1772 (2012)
19. C.S. Lam, M.C. Wong, W.-H. Choi, X.-X. Cui, H.-M. Mei, J.-Z. Liu, Design and performance of an adaptive low-dc-voltage-controlled LC-Hybrid active power filter with a neutral inductor in three-phase four-wire power systems. IEEE Trans. Power Electron. **61**(6), 2635–2647 (2014)
20. L.K. Haw, M.S. Dahidah, H.A.F. Almurib, A new reactive current reference algorithm for the STATCOM system based on cascaded multilevel inverters. IEEE Trans. Power Electron. **30** (7), 3577–3588 (2015)
21. J.A. Munoz, J.R. Espinoza, C.R. Baier, L.A. Moran, J.I. Guzman, V.M. Cardenas, Decoupled and modular harmonic compensation for multilevel STATCOMs. IEEE Trans. Ind. Electron. **61**(6), 2743–2753 (2014)
22. V. Soares, P. Verdelho, An instantaneous active and reactive current component method for active filters. IEEE Trans. Power Electron. **15**(4), 660–669 (2000)
23. T.L. Lee, S.H. Hu, Y.H. Chan, D-STATCOM with positive-sequence admittance and negative-sequence conductance to mitigate voltage fluctuations in high-level penetration of distributed-generation systems. IEEE Trans. Ind. Electron. **60**(4), 1417–1428 (2013)
24. M. Hagiwara, R. Maeda, H. Akagi, Negative-sequence reactive-power control by a PWM STATCOM based on a modular multilevel cascade converter (MMCC-SDBC). IEEE Trans. Ind. Appl. **48**(2), 720–729 (2012)
25. IEEE Recommended Practices and Requirements for Harmonic Control in Electrical Power Systems, 2014, IEEE Standard 519-2014
26. Galco Industrial Electronics, Component price list (Sept 2015) [Online]. Available: www.galco.com
27. C. Kawann, A.E. Emanuel, Passive shunt harmonic filters for low and medium voltage: a cost comparison study. IEEE Trans. Power Syst. **11**(4), 1825–1831 (1996)
28. K. Habur, D. O'Leary, FACTS—For cost effective and reliable transmission of electrical energy [Online]. Available: http://www.worldbank.org/html/fpd/em/transmission/facts_siemens.pdf
29. L.M. Tolbert, T.J. King et al., Power electronics for distributed energy systems and transmission and distribution applications: assessing the technical needs for utility applications. ORNL/TM-2005/230, Oak Ridge Nat. Lab., Dec 2005
30. F. Richardeau, T.T.L. Pham, Reliability calculation of multilevel converters: theory and applications. IEEE Trans. Ind. Electron. **60**(10), 4225–4233 (2013)
31. X. Yu, A.M. Khambadkone, Reliability analysis and cost optimization of parallel-inverter system. IEEE Trans. Ind. Electron. **59**(10), 3881–3889 (2012)
32. B. Abdi, A.H. Ranjbar, G.B. Gharehpetian, J. Milimonfared, Reliability considerations for parallel performance of semiconductor switches in high-power switching power supplies. IEEE Trans. Ind. Electron. **56**(6), 2133–2139 (2009)
33. Reliability Prediction of Electronic Equipments (MIL-HDBK-217), Relex Software Corporation, Greensburg, PA, 1990
34. M.-C. Wong, J. Tang, Y.-D. Han, Cylindrical coordinate control of three-dimensional PWM technique in three-phase four-wired trilevel inverter. IEEE Trans. Power Electron. **18**(1), 208–220 (2003)

35. E. Baussan, E. Bouquerel, M. Dracos et al., Study of the pulse power supply unit for the four-horn system of the CERN to Frejus neutrino super beam. J. Instrum. T07006 (Aug 2013)
36. Infineon Technologies AG: http://www.infineon.com/. Datasheet: T1901N (2011). [Online]. Available: http://www.infineon.com/eupec
37. Infineon Technologies AG: http://www.infineon.com/. Datasheet: FZ1000R33HL3 (2013). [Online]. Available: http://www.infineon.com/eupec
38. Z. Dang, J.A.A. Qahouq, Evaluation of high-current toroid power inductor with NdFeB magnet for DC–DC power converters. IEEE Trans. Ind. Electron. **62**(11), 6868–6876 (2015)
39. A. Braham, A. Lahyani, P. Venet, N. Rejeb, Recent developments in fault detection and power loss estimation of electrolytic capacitors. IEEE Trans. Power Electron. **25**(1), 33–43 (2010)
40. X.S. Pu, T.H. Nguyen, D.C. Lee, K.B. Lee, J.M. Kim, Fault diagnosis of DC-link capacitors in three-phase AC/DC PWM converters by online estimation of equivalent series resistance. IEEE Trans. Ind. Electron. **60**(9), 4118–4127 (2013)

Chapter 3
Mitigation of the Harmonic Injection in TCLC Part and Nonlinear Hysteresis PWM Control in Active Inverter Part of Thyristor Controlled LC-Coupling Hybrid Active Power Filter (TCLC-HAPF)

Abstract In this chapter, two techniques are proposed to improve the performance and reduce the inverter loss of the thyristor controlled LC-coupled hybrid active power filter (TCLC-HAPF). Specifically, for the TCLC part, a design method is proposed to mitigate its self-harmonic injection of the thyristor pairs, which can improve the TCLC-HAPF performance for linear loading compensation. Compared with the traditional static var compensators (SVCs) like a fixed capacitor-thyristor controlled reactor (FC-TCR), which generates low order harmonic current, the proposed TCLC design method can significantly mitigate the injection of harmonic current. The design of the TCLC parameters is investigated with the considerations of its reactive power compensation range and harmonic current rejection. And, representative simulation and experimental results of the proposed three-phase three-wire TCLC are presented to show its effectiveness in dynamic reactive power compensation in comparison with the traditional FC-TCR. For active inverter part, the nonlinear pulse width modulation (PWM) control method is proposed to reduce the inverter loss of the TCLC-HAPF and keep the total harmonic distortion at an acceptable level. As the coupling TCLC impendence of the TCLC-HAPF can yield nonlinear compensating current, the quasi-linear and nonlinear regions are focused with special compensation characteristics under relatively low switching frequency. An approximated harmonic distortion (*ATHD*) index is proposed to reflect the *THD* in a simplified way, which results in speed up the TCLC-HAPF system response. A non-linear adaptive hysteresis band PWM controller is proposed for TCLC-HAPF to reduce switching loss. Finally, the performance of the nonlinear adaptive hysteresis band controller is verified by simulation results in comparison with the conventional hysteresis PWM.

Keywords Harmonic current rejection · Reactive power compensation Smart grid · Thyristor controlled *LC* (TCLC) compensator · Pulse width modulations (PWMs) · Active power filters · Hysteresis PWM

L. Wang et al., *Adaptive Hybrid Active Power Filters*, Power Systems,
https://doi.org/10.1007/978-981-10-8827-8_3

3.1 Introduction

To improve the performance and reduce the inverter loss of thyristor controlled LC-coupled hybrid active power filter (TCLC-HAPF), two techniques are proposed in this chapter. Specifically, for the thyristor controlled LC (TCLC) part, the design technique is proposed to mitigate its harmonic injection by the thyristor pair, which can improve the TCLC-HAPF performance for linear loading compensation. For active inverter part, the nonlinear PWM control method is proposed to reduce the inverter loss of the TCLC-HAPF.

The phenomenon of harmonic current injection by the traditional static var compensators (SVCs) was firstly presented in 1985 [1]. However, there was no solution provided at that time. Years later, the authors in [2, 3] suggested using passive power filters (PPFs) to reduce the harmonic current injection by the SVCs. However, the resonance problem between SVCs and PPFs may affect the compensation performance. Afterwards, the authors in [4, 5] proposed the combined systems of the SVC and active power filter (APF), which can eliminate the harmonic current injection by the SVC and compensate both the reactive power and harmonic current of the nonlinear loads. By doing so, the initial cost of the combined systems can be very high. To reduce the initial cost, D. B. Kulkarni et al. [6] proposed an artificial neural network (ANN) approach to find the optimum trigger angles for the thyristor controlled reactor and thyristor switched capacitor (TCR-TSC) in order to reduce its harmonic current injection and compensate reactive power. However, the reactive power compensation ability is weakened as the firing angles are probably not match with the required compensating reactive power. Due to the limitations among the existing literatures, the Sect. 3.3 aims to analyze the TCLC part of TCLC-HAPF [5, 7–12] for reactive power compensation and the mitigation of its harmonic current injection in comparison with traditional SVCs [1, 13].

Pulse width modulations (PWMs) are widely used in active power filter (APF) applications [14]. However, there are seldom PWMs proposed in different hybrid active power filters (HAPFs) applications due to its non-linear current characteristic. Section 3.4 proposes a non-linear adaptive hysteresis band PWM controller for TCLC-HAPF to reduce the inverter loss and keep the total harmonic distortion at an acceptable level. As the coupling TCLC part impendence of TCLC-HAPF can yield nonlinear compensating current, the quasi-linear and non-linear regions are focused to obtain the low switching frequency. Besides, an approximated harmonic distortion (*ATHD*) index is proposed to reflect the THD in a simplified way, which results in speed up TCLC-HAPF system response.

In this chapter, the circuit configuration and modeling of a TCLC-HAPF is given in Sect. 3.2. In Sect. 3.3, the mitigation method of TCLC part harmonic current injection is proposed (Sect. 3.3.1). And, the corresponding simulated and experimental results to verify the proposed method are given in Sect. 3.3.3. In Sect. 3.4, a non-linear adaptive hysteresis band controller for TCLC-HAPF is proposed to reduce the inverter loss (Sect. 3.4.2). And the simulated results are reported to verify the proposed adaptive hysteresis band controller (Sect. 3.4.2.3) in comparisons with the conventional hysteresis PWM method. Finally, a summary is presented in the last section.

3.2 Circuit Configuration and Modeling of TCLC Part and TCLC-HAPF

The three-phase three-wire loading system with a shunt connected TCLC part and TCLC-HAPF is shown in Fig. 3.1. The subscript "*x*" denotes phase *a*, *b* and *c*. v_{sx}, v_x and v_{invx} are the system voltage, load voltage and inverter output voltage respectively; L_s is the system inductance; i_{sx}, i_{Lx} and i_{cx} are the source, load and compensated current respectively. C_{DC} and V_{DC} are dc-link capacitor and DC-link voltage; L_c is the coupling inductor; L_{PF} and C_{PF} are the TCLC part inductor and capacitor.

To analyze the TCLC part harmonic current injection, the circuit on figuration of TCLC part in Fig. 3.1a is applied. Based on Fig. 3.1a, the mitigation of harmonic current injection is proposed in next section.

3.3 Mitigation of the Harmonic Injection in TCLC Part

In this section, the analysis of harmonic current injection by the TCLC part is provided in Sect. 3.3.1. And the method to mitigate the injected harmonic current is proposed in Sect. 3.3.2. To verify the proposed method, the simulation and experimental results are given in Sect. 3.3.3.

3.3.1 Mitigation of the Harmonic Injection in TCLC Part

In Fig. 3.1a, the thyristors (T_{x1} and T_{x2}) for each phase of the proposed TCLC part can be considered as a pair of bidirectional switches, which would generate low order harmonic current when the switches change their states. Through the harmonic current rejection analysis, the harmonic current orders generated can then be deduced in term of L_c. Therefore, with appropriate design of L_c, the harmonic current injection by the controlled TCLC part can be reduced.

Figure 3.2 illustrates the equivalent single-phase TCLC part model for harmonic current rejection analysis. From Fig. 3.2, when the switch *S* is turned off, the differential equation of $i_{cx}(t)$ can be obtained as:

$$L_c \frac{di_{cx}(t)}{dt} + \frac{1}{C_{PF}} \int i_{cx}(t)dt = v_x(t) \tag{3.1}$$

On the other hand, when the switch *S* is turned on, the following relationships can be obtained:

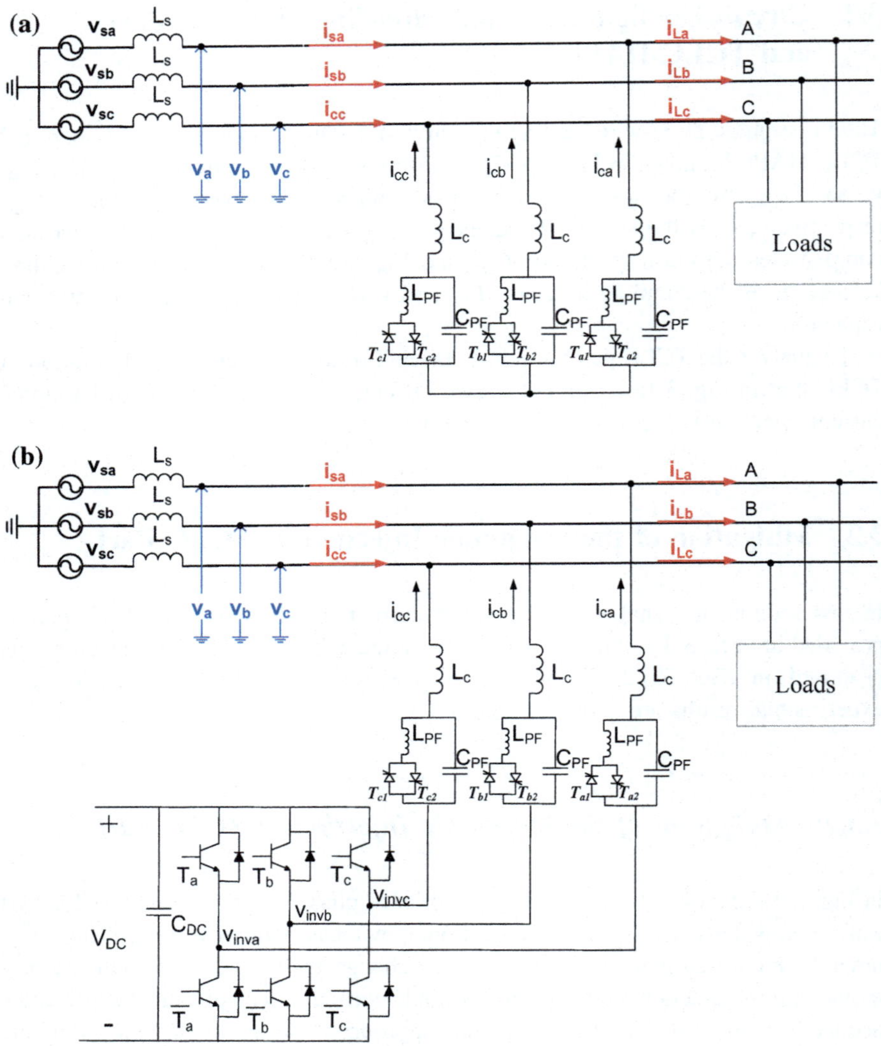

Fig. 3.1 Circuit configurations of the three-phase three-wire **a** TCLC part and **b** TCLC-HAPF

$$L_c \frac{di_{cx}(t)}{dt} + \frac{1}{C_{PF}} \int i_{C_{PF}}(t)dt = v_x(t) \tag{3.2}$$

$$L_{PF} \frac{di_{L_{PF}}(t)}{dt} = \frac{1}{C_{PF}} \int i_{C_{PF}}(t)dt \tag{3.3}$$

$$i_{L_{PF}} + i_{C_{PF}} = i_{cx} \tag{3.4}$$

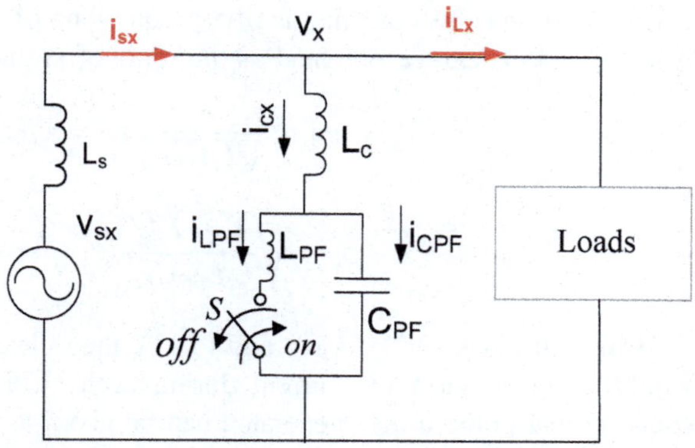

Fig. 3.2 Equivalent single-phase TCLC part model for harmonic current rejection analysis

Combining (3.2)–(3.4), the equation in term of $i_{cx}(t)$ can be obtained as:

$$L_c L_{PF} C_{PF} \frac{di_{cx}^3(t)}{dt^3} - L_{PF} C_{PF} v_x(t) \frac{di_{cx}^2(t)}{dt^2} + (L_c + L_{PF}) \frac{di_{cx}(t)}{dt} = v_x(t) \tag{3.5}$$

In (3.1) and (3.5), the compensating current i_{cx} (i_{cx_off}) is a second order equation when the switch S is turned off, while i_{cx} (i_{cx_on}) is a third order equation when the switch S is turned on. i_{cx_off} and i_{cx_on} can be deduced by solving their differential equations (or through Laplace Transform) in (3.1) and (3.5). Then i_{cx_off} and i_{cx_on} can be expressed in the forms as:

$$\begin{aligned} i_{cx\_off} = & \underbrace{A_1 \sin(\omega t - \alpha)}_{Fundamental} + \underbrace{K_1 \sin(\omega_1 t + \varphi_1)}_{Harmonic} \end{aligned} \tag{3.6}$$

$$\begin{aligned} i_{cx\_on} = & \underbrace{A_2 \sin(\omega t + \alpha)}_{Fundamental} + \underbrace{K_2 \cos(\omega_2 t + \varphi_2) + K_3}_{Harmonic} \end{aligned} \tag{3.7}$$

where A_1 and A_2 are the peak values of the fundamental compensating current during each turn off and turn on; K_1, K_2, K_3, ϕ_1 and ϕ_2 are constants during each switching cycle, and they depend on the initial conditions of the compensating current and load voltage values; ω is the system fundamental angular frequency with $\omega = 2\pi f$ and f is the system frequency; ω_1 and ω_2 are the harmonic angular frequencies which can be obtained as the roots (r_1 and r_2) of the homogeneous equations of (3.1) and (3.5):

$$r_1^2 + \frac{1}{L_c C_{PF}} = 0 \tag{3.8}$$

$$L_c L_{PF} C_{PF} \cdot r_2^3 + (L_c + L_{PF}) \cdot r_2 = 0 \tag{3.9}$$

where (3.8) and (3.9) are the auxiliary equations of (3.1) and (3.5), respectively. And ω_1 and ω_2 can be obtained as the roots of r_1 and r_2:

$$\omega_1 = \frac{1}{\sqrt{L_c C_{PF}}}\ (r_1 = \pm j\omega_1) \tag{3.10}$$

$$\omega_2 = \sqrt{\frac{L_c + L_{PF}}{L_c L_{PF} C_{PF}}}\ (r_2 = \pm j\omega_2) \tag{3.11}$$

Based on (3.6), (3.7), (3.10) and (3.11), they clearly show that the TCLC part would generate harmonic current during each switching cycle. Then the corresponding harmonic orders generated can be given as:

$$n_1 = \frac{\omega_1}{2\pi f} = \frac{1}{2\pi f\sqrt{L_c C_{PF}}} \tag{3.12}$$

$$n_2 = \frac{\omega_2}{2\pi f} = \frac{1}{2\pi f}\sqrt{\frac{L_c + L_{PF}}{L_c L_{PF} C_{PF}}} \tag{3.13}$$

For a three-phase three-wire system, the characteristic harmonic orders are 6n $\pm$ 1 (≥ 5) with n =1, 2, 3…. To avoid the harmonic current injection by the TCLC part, it is suggested to tune the harmonic orders n_1 and n_2 generated to be away from the 6n $\pm$ 1th order. Based on the above discussions, the design criteria of L_c can be given as:

$$L_c = \frac{1}{(2\pi f n_1)^2 C_{PF}} \text{ and } L_c = \frac{1}{(2\pi f n_2)^2 C_{PF} - 1/L_{PF}} \tag{3.14}$$

In (3.14), they can easily be satisfied simultaneously as long as n_1 and n_2 are away from the harmonic orders k = 6n $\pm$ 1th (n = 1, 2, 3…). Therefore, by carefully designing the L_c, the harmonic current injection can be mitigated. In the following part, the selection of the n_1 and n_2 through the design of L_c will be discussed.

3.3.2 The Selection of n_1 and n_2 Through the Design of L_c

To select the optimal n_1 and n_2 values, the simulation case studies are provided. Figure 3.3 shows the total injected harmonic current by the TCLC part with varying of n_1 and n_2. The simulation is done at 10 kV medium voltage level application, with the L_{PF}= 30 mH and C_{PF}= 160 μF.

Based on Fig. 3.3, it can be seen that the total harmonic current can be significantly reduced if the n_1 and n_2 tuned to be smaller than 5th (<5) order harmonic. However, if n_1 and n_2 designed to be closed to 5th order, the injected harmonic current is the largest. Based on discussions in Sects. 3.3.1 and 3.3.2, by satisfying (3.14), L_c is designed to tune the n_1 and n_2 smaller than 5th order, and at the same time L_c is suggested to be as small as possible considering the increasing cost when

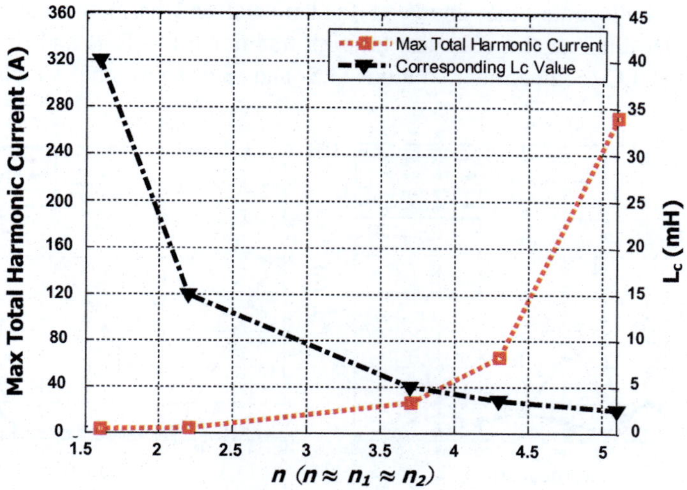

Fig. 3.3 The total injected harmonic current by TCLC part and corresponding L_c value with varying of $n \approx n_1 \approx n_2$

L_c gets larger. Therefore, based on above discussions, the TCLC part with L_c= 5 mH and $n \approx n_1 \approx n_2 \approx 3.7$ are selected.

In the following section, the TCLC part will be verified by both simulations and experiments in comparison with the traditional FC-TCR [1, 13].

3.3.3 *Simulation and Experimental Verifications of the Mitigation of the Harmonic Injection in TCLC Part*

In this section, the compensating performances of the proposed TCLC compensator are verified by simulations and experiments in comparison with the traditional FC-TCR [1, 13]. The simulations are carried out by using PSCAD/EMTDC. In experiments, 110 V–5 kVA three-phase three-wire TCLC compensator and FC-TCR experimental prototypes were designed and constructed in the laboratory. Table 3.1 shows the system parameters of the power grid, traditional FC-TCR and the proposed TCLC compensator (n_1 = 3.56 and n_2 = 3.84, which are designed according to (3.14)) for both simulations and experiments.

Table 3.1 System parameters of power system and compensators

	Parameters	Physical values
Power grid	v_x, f, L_s	110 V, 50 Hz, 1 mH
FC-TCR	L_{PF}, C_{PF}	30 mH, 160 μF
TCLC part	L_c, L_{PF}, C_{PF}	5 mH, 30 mH, 160 μF

Figure 3.4 illustrates the simulated fundamental and harmonic compensating current under different firing angles: (a) with traditional FC-TCR and (b) with the proposed TCLC compensator. Figures 3.5 and 3.6 show the simulated and

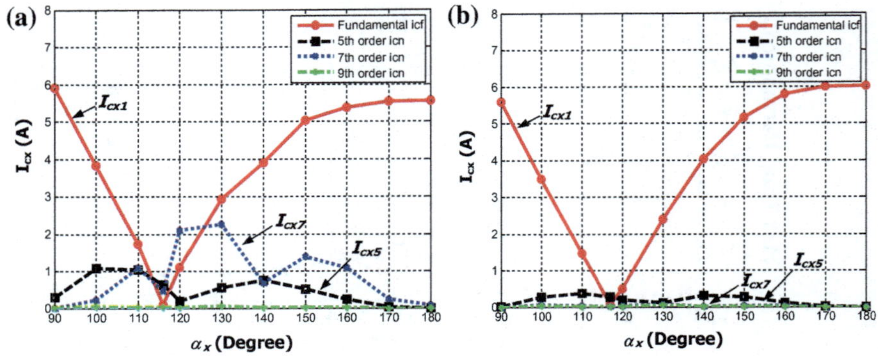

Fig. 3.4 Simulated fundamental and harmonic compensating current under different firing angles with. **a** Traditional FC-TCR. **b** Proposed TCLC part

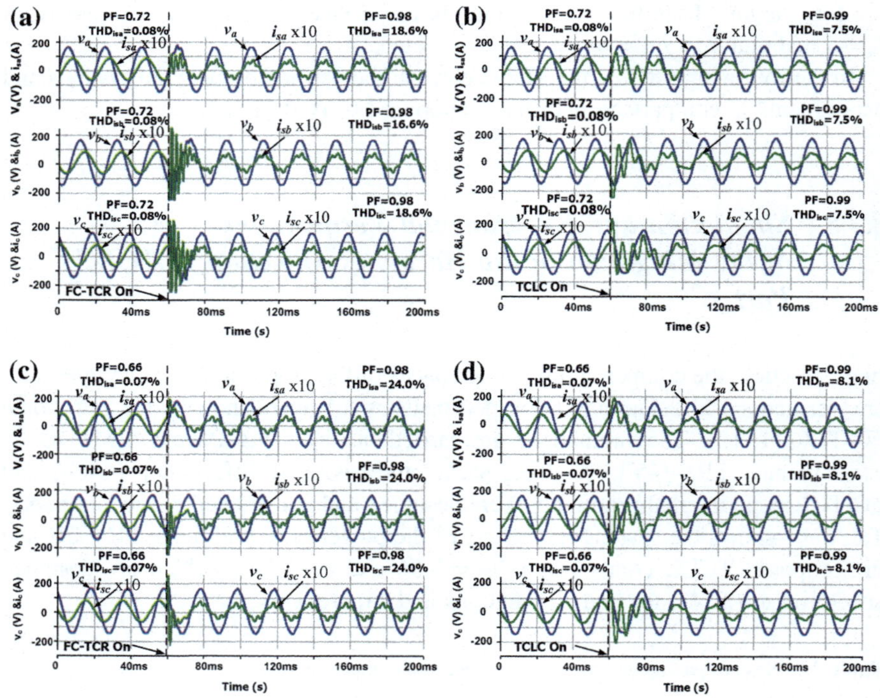

Fig. 3.5 Simulation results of dynamic compensation process of load voltage and source current. **a** By applying FC-TCR for inductive reactive power compensation. **b** By applying TCLC for inductive reactive power compensation. **c** By applying FC-TCR for capacitive reactive power compensation. **d** By applying TCLC for capacitive reactive power compensation

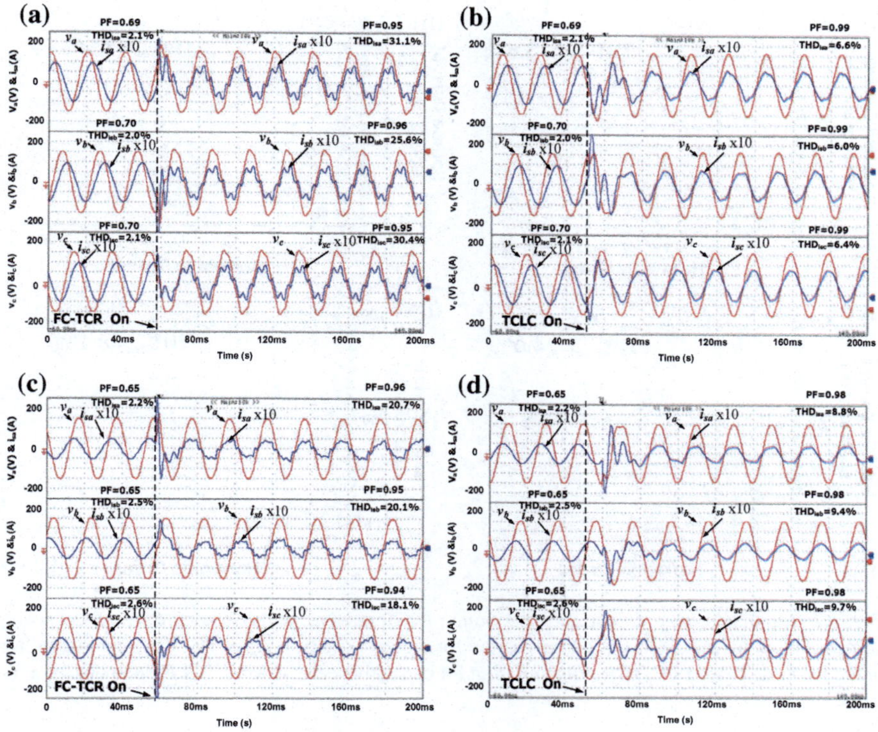

Fig. 3.6 Experimental results of dynamic compensation process of load voltage and source current. **a** By applying FC-TCR for inductive reactive power compensation. **b** By applying TCLC for inductive reactive power compensation. **c** By applying FC-TCR for capacitive reactive power compensation. **d** By applying TCLC for capacitive reactive power compensation

experimental grid voltage and system current waveforms before and after the traditional FC-TCR and the proposed TCLC compensations under both inductive and capacitive reactive power cases. Figures 3.7 and 3.8 illustrate the simulated and experimental grid system current spectrums after the FC-TCR and TCLC compensations in both inductive and capacitive reactive power cases. Figures 3.9 and 3.10 give the simulated and experimental dynamic reactive power compensation waveforms by using the proposed TCLC compensator. Tables 3.2 and 3.3 summarize the simulation and experimental results.

From Fig. 3.4a, when the traditional FC-TCR compensator is applied, the large 5th and 7th order harmonic current are injected into the power grid. On the other hand, when the proposed TCLC part is applied, the injected harmonic current are significantly reduced as shown in Fig. 3.4b. Therefore, Fig. 3.4 verifies the effectiveness of the proposed TCLC part in reducing the current harmonics injection during operation.

When FC-TCR is applied, Figs. 3.5a, c and 3.6a, c and Tables 3.2 and 3.3, show that the simulated power factors (PFs) of the worst phase have been compensated from the original 0.72 (inductive PF) and 0.66 (capacitive PF) to 0.98 for both

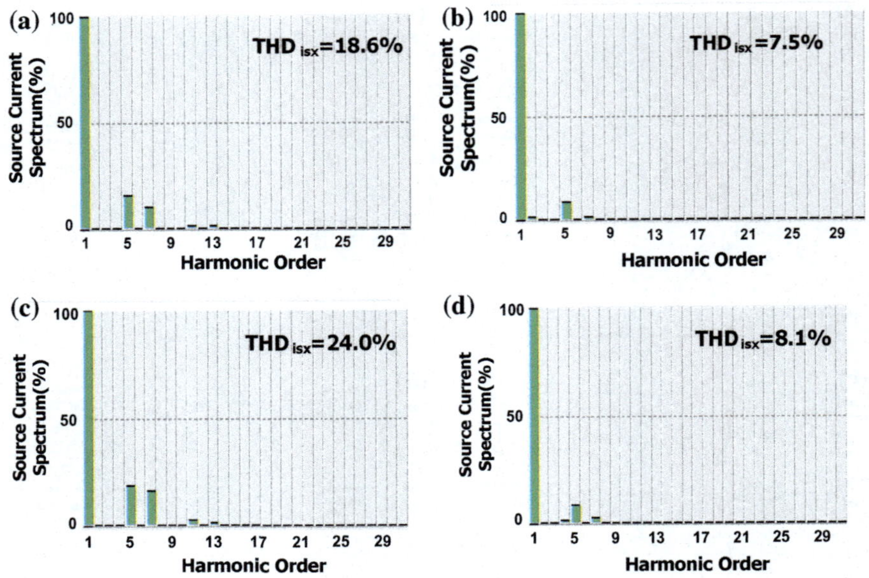

Fig. 3.7 Simulated source current spectrums for inductive and capacitive reactive power compensation. **a** After FC-TCR compensation for inductive case. **b** After TCLC compensation for inductive case. **c** After FC-TCR compensation for capacitive case. **d** After TCLC compensation for capacitive case

cases, while the experimental PFs have been compensated to 0.95 and 0.94 from the original 0.69 (inductive PF) and 0.65 (capacitive PF), respectively. As shown in Figs. 3.7a, c and 3.8a, c, source current total harmonic distortions (THD_i) are increased after FC-TCR compensation. According to IEEE 519-2014 standard [15] with the ratio of short circuit and demand current in the range of 100–1000, the source current total harmonic distortions are required to satisfy the condition of $THD_{isx} < 15\%$. However, the worst phase simulated source current THD_{isx} ($THD_{isx} = 18.6\%$ for inductive case and $THD_{isx} = 24.0\%$ for capacitive case), experimental source current THD_{isx} ($THD_{isx} = 31.1\%$ for inductive case and $THD_{isx} = 20.7\%$ for capacitive case) cannot satisfy the IEEE 519-2014 standard [15]. Moreover, Figs. 3.7a, c and 3.8a, c clearly show that large 5th and 7th order harmonic current are injected into the power grid system after compensation, which agrees with the simulation results as shown in Fig. 3.4a.

When the TCLC part is applied, Figs. 3.5b, d and 3.6b, d and Tables 3.2 and 3.3, show that the simulated and experimental PFs of worst phase have been compensated to closed to unity (0.99 and 0.98) for inductive and capacitive reactive power compensations. As shown in Figs. 3.5b, d and 3.6b, d, the source current THD_{isx} of the worst phase have been compensated to 8.1% (for simulations)

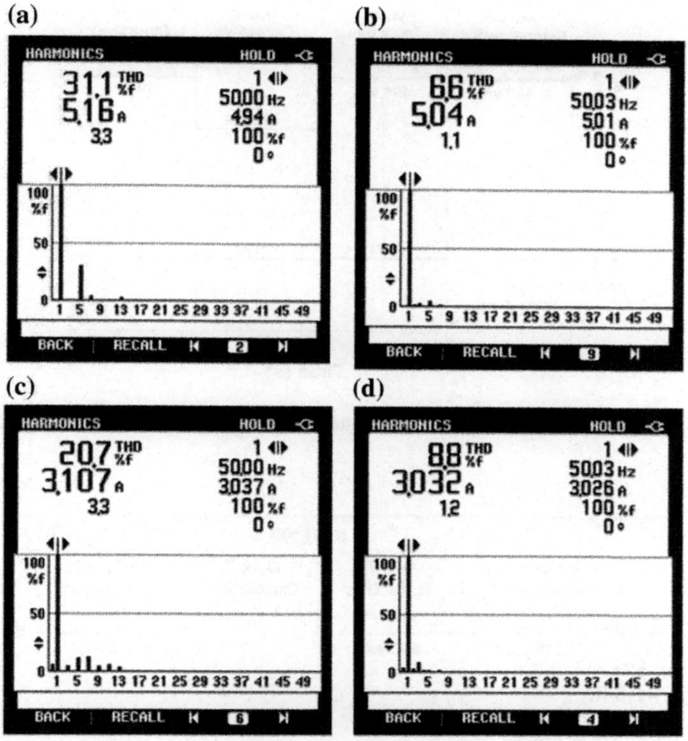

Fig. 3.8 Experimental source current spectrums of *phase a* for inductive and capacitive reactive power compensation. **a** After FC-TCR compensation for inductive case. **b** After TCLC compensation for inductive case. **c** After FC-TCR compensation for capacitive case. **d** After TCLC compensation for capacitive case

and 9.7% (for experiments), which can satisfy the IEEE 519-2014 standards [15]. Moreover, Figs. 3.7b, d and 3.8b, d clearly show that much smaller 5th and 7th order harmonic current are injected into the power grid system after TCLC compensation, which agrees with the simulation results as shown in Fig. 3.4b and verifies the effectiveness of the proposed TCLC part in reducing the current harmonics injection during operation.

Based on the simulation and experimental results, it is proved that the TCLC part can provide better dynamic reactive power compensation with much less harmonic current injection, while the traditional FC-TCR injects significant low order current harmonics into the power grid system during operation.

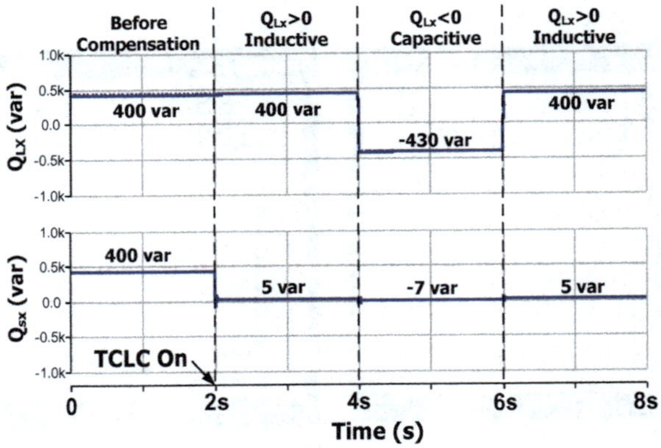

Fig. 3.9 Simulation results of TCLC compensator dynamic reactive power compensation

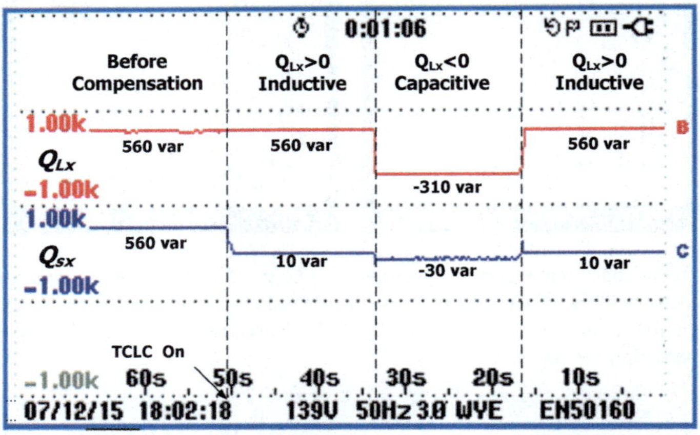

Fig. 3.10 Experimental results of TCLC dynamic reactive power compensation

Table 3.2 Simulation results for inductive and capacitive reactive power compensations by using traditional FC-TCR and proposed TCLC

	Inductive reactive power			Capacitive reactive power		
	Before Comp.	FC-TCR	TCLC	Before Comp.	FC-TCR	TCLC
$THDv_x$(%)	0.1	1.2	0.5	0.1	1.5	0.3
i_{sx}(A)	6.1	3.9	3.9	5.0	3.6	3.5
Q_{sx}(var)	400	9	5	-430	-9	-7
PF	0.72	0.98	0.99	0.66	0.98	0.99
$THDi_{sx}$(%)	0.1	18.6	7.5	0.1	24.0	8.1

Note The shaded area means unsatisfactory results

Table 3.3 Experimental results for inductive and capacitive reactive power compensations by using traditional FC-TCR and proposed TCLC

		Inductive reactive power			Capacitive reactive power		
		Before Comp.	FC-TCR	TCLC	Before Comp.	FC-TCR	TCLC
$THDv_x$ (%)	A	1.0	4.5	1.1	1.0	1.7	1.0
	B	1.1	4.4	1.2	1.1	1.8	1.3
	C	1.3	4.5	1.3	1.1	1.7	1.2
i_{sx} (A)	A	6.9	5.0	5.2	3.6	3.1	3.0
	B	6.9	5.1	5.2	3.6	2.9	2.8
	C	6.9	5.3	5.3	3.6	2.9	2.9
Q_{sx} (var)	A	560	20	10	-310	-20	-30
	B	550	-10	-30	-320	-30	-20
	C	550	20	-30	-320	-30	-40
PF	A	0.69	0.95	0.99	0.65	0.96	0.98
	B	0.70	0.96	0.99	0.65	0.95	0.98
	C	0.70	0.95	0.99	0.65	0.94	0.98
$THDi_{sx}$ (%)	A	2.1	31.1	6.6	2.2	20.7	8.8
	B	2.0	25.6	6.0	2.5	20.1	9.0
	C	2.1	30.4	6.4	2.6	18.1	9.7

3.4 Nonlinear Hysteresis PWM Control in Active Inverter Part of TCLC-HAPF

In this section, considering the users' requirements of THD_{isx} (*THD* oriented), a non-linear adaptive hysteresis band controller for the TCLC-HAPFs is proposed. The key points of the proposed method are:

- Compared with the state-of-the-art hysteresis PWM control methods [16–22], the proposed method allows the compensating current of the TCLC-HAPF to operate in nonlinear or quasi-linear regions in some circumstances to reduce the inverter loss, instead of always staying in a linear region.
- A fast performance index: an approximated *THD* (*ATHD*) is proposed for HAPFs to instantaneously assess the *THD* value. The *ATHD* can be controlled to track the reference *ATHD** instantaneously according to users' requirements.
- Based on the deduced *ATHD* and nonlinear characteristics of the TCLC-HAPF, a non-linear adaptive hysteresis band PWM controller for HAPFs is proposed to reduce the inverter loss and keep the *THD* at an acceptable level.

In this section, the mathematical model and non-linear current characteristics are introduced in Sect. 3.4.1. In Sect. 3.4.2, the suggested *ATHD* are discussed and the non-linear adaptive hysteresis band control for TCLC-HAPF is addressed. In Sect. 3.4.2.3, the simulated and experimental results are reported to verify the proposed approach. And comparisons with other recently developed methods are also provided.

3.4.1 Compensating Current Characteristics of TCLC-HAPF

The TCLC part has two back-to-back connected thyristors T_{x1} and T_{x2} for each phase, and they are triggered alternately in every half cycle according to deduced firing angle. When the thyristors is connected, the coupling TCLC part is an L_c in series with C_{PF} (*LC* mode). When the thyristors is disconnected, the coupling TCLC part is an L_c in series with the parallel combination of L_{PF} and C_{PF} (*LCL* mode). Compared with TCLC part control frequency (50 Hz), the PWM control frequency in active inverter is much faster (few kHz). Therefore, the modeling for PWM control study can be plotted as shown in Fig. 3.11.

From Fig. 3.11a, a mathematical model for a *LC* mode can be deduced. The relationships among the compensating current i_{cx}, coupling inductor L_c, TCLC part capacitor C_{PF} and inverter voltage $v_{invx}(t)$ of *LC* mode are calculated as:

$$L_c\frac{di_{cx}(t)}{dt}+\frac{1}{C_{PF}}\int i_{cx}(t)dt=v_x(t)-v_{invx}(t) \tag{3.15}$$

Taking the derivative of (3.15), it yields:

$$L_c\frac{d^2i_{cx}(t)}{dt^2}+\frac{1}{C_{PF}}i_{cx}(t)=\frac{d[v_x(t)-v_{invx}(t)]}{dt} \tag{3.16}$$

It can be clearly observed that (3.16) is a second order equation in terms of $i_{cx}(t)$. The time domain solution of $i_{cx}(t)$ is deduced below to show its characteristics.

Take the Laplace transform of (3.16) and simplify it. Next, substitute the initial compensating current and capacitor voltage conditions into the characteristic Laplace equation. Finally, the time domain solution $i_{cx}(t)$ can be obtained by taking the inverse Laplace transformation:

$$i_{cx}(t)=A_1\cdot\sin(\omega_1 t+\phi_1) \tag{3.17}$$

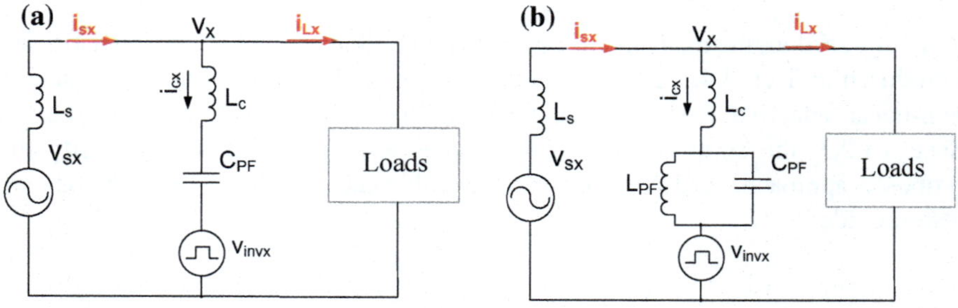

Fig. 3.11 Modeling of TCLC-HAPF PWM control study. **a** *LC* mode and **b** *LCL* model

where $A_1 = \sqrt{i_{cx}^2(0) + \frac{C_{PF}}{L_c} \cdot [v_x(t) - v_{invx}(t) - v_{C_{PF}}(0)]^2}$, $\omega_1 = \frac{1}{\sqrt{L_c \cdot C_{PF}}}$ and $\phi_1 = \tan^{-1}\left(\frac{\sqrt{L_c}}{\sqrt{C_c}} \cdot \frac{i_{cx}(0)}{v_x(t) - v_{invx}(t) - v_{C_C}(0)}\right)$. The compensating current $i_{cx}(t)$ is expressed in terms of the coupling inductor L_c, TCLC part capacitor C_{PF}, initial compensating current $i_{cx}(0)$, initial capacitor voltage $v_{C_{PF}}(0)$, load voltage $v_x(t)$ and inverter voltage $v_{invx}(t)$. The value of $v_{invx}(t)$ can be obtained from the DC-link voltage V_{DC} and switching states.

From Fig. 3.11b, a mathematical model for a *LCL* mode can be deduced. The relationships among the compensating current i_{cx}, coupling inductor L_c, TCLC part capacitor C_{PF} and inverter voltage $v_{invx}(t)$ of *LCL* are calculated as:

$$L_c \frac{di_{cx}(t)}{dt} + \frac{1}{C_{PF}} \int i_{cx}(t)dt = v_x(t) - v_{invx}(t) \tag{3.18}$$

$$L_{PF} \frac{di_{L_{PF}}(t)}{dt} = \frac{1}{C_{PF}} \int i_{C_{PF}}(t)dt \tag{3.19}$$

$$i_{L_{PF}} + i_{C_{PF}} = i_{cx} \tag{3.20}$$

Combining the (3.18)–(3.20) and taking the derivative, it yields:

$$L_c C_{PF} \cdot \frac{dv_{C_{PF}}^2(t)}{dt^2} + (\frac{L_c + L_{PF}}{L_{PF}}) \cdot v_{C_{PF}}(t) = v_{invx}(t) \tag{3.21}$$

Applying Laplace transform to (3.21), the simplified equations for *LCL* mode can be deduced as:

$$i_{cx}(t) = A_2 \sin(\omega_2 t + \phi_2) + A_3 \cdot t \tag{3.22}$$

where $A_2 = (C_{PF}\omega_2 - \frac{1}{L_{PF}\omega_2}) \cdot \sqrt{(v_{C_{PF}}(0) - \frac{L_{PF}(v_x(t) - v_{invx}(t))}{L_c + L_{PF}})^2 + (\frac{i_{cx}(0)}{C_{PF}\omega_2 - \frac{1}{L_{PF}\omega_2}})^2}$, $A_3 = \frac{v_x(t) - v_{invx}(t)}{L_c + L_{PF}}$ and $\omega_2 = \sqrt{\frac{L_c + L_{PF}}{L_c L_{PF} C_{PF}}}$. The compensating current $i_{cx}(t)$ is expressed in terms of the coupling inductor L_c, TCLC part capacitor C_{PF}, TCLC part inductor L_{PF}, initial compensating current $i_{cx}(0)$, initial capacitor voltage $v_{CPF}(0)$, load voltage $v_x(t)$ and inverter voltage $v_{invx}(t)$. The value of $v_{invx}(t)$ can be obtained from the dc-link voltage V_{DC} and switching states.

Based on (3.17) and (3.22), the compensating current $i_{cx}(t)$ solution in the time domain contains a sine function. And, the (3.17) and (3.22) aim to show the nonlinear characteristics and transient response of the compensating current in a TCLC-HAPF during each switching (on or off) interval. The value of ω_1 and ω_2 depends on the coupling components only, rather than the system frequency ($\omega_f = 2\pi f$). Moreover, the A_1, A_2 and A_3 are different with the different initial conditions at each moment. Based on the relationship between the A_1, A_2 and A_3 and hysteresis band H, three operating regions for the TCLC-HAPF compensating

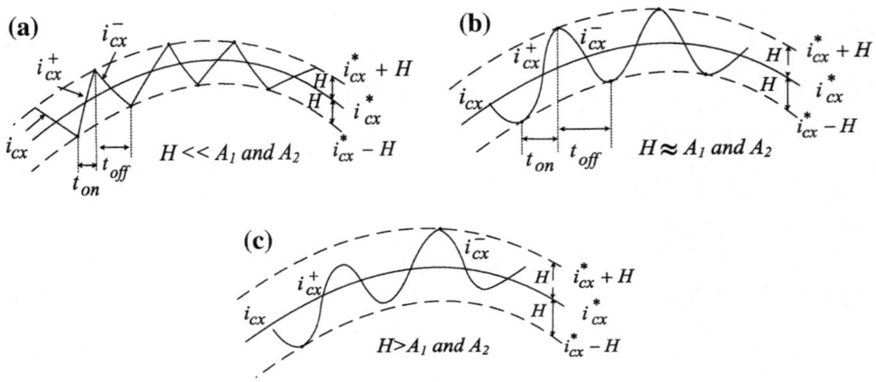

Fig. 3.12 Different operating regions for TCLC-HAPF compensating current based on the relationship between the hysteresis band H and amplitude A: **a** linear region; **b** quasi-linear region; **c** nonlinear region

current can be defined: the linear, quasi-linear and non-linear regions, as shown in Fig. 3.12.

In the linear region shown in Fig. 3.12a, the slope of the compensating current can be treated as linear if the hysteresis band is set small enough ($H \ll A_1$ and A_2). In the linear region, the switching time t_{sw} is relatively small, which is the key to linearising the current slope, as discussed in [23].

In the quasi-linear region (Fig. 3.12b), the slope polarity of the compensating current interchanges once i_{cx} touches the top or bottom hysteresis band with $H \approx A_1$ and A_2. Under this region, the switching time t_{sw} increases compared with the linear region case. Therefore, the switching frequency f_{sw} and inverter loss P_{loss} decrease.

As shown in Fig. 3.12c, the nonlinear region case can occur when the hysteresis band is set to be larger than the amplitude ($H \gg A_1$ and A_2). In the nonlinear region, the compensating slope polarity interchanges before i_{cx} touches the top or bottom hysteresis band. A large H value would enlarge the output current ripple and degrade the system performance. However, this region does not generate trigger signals to the switching devices; thus, f_{sw} and P_{loss} can be minimized.

It is interesting to further investigate the quasi-linear and nonlinear regions. In these two operation modes, less f_{sw} is required but the current tracking error is still within the set hysteresis band. Therefore, the reduction of the switching frequency can be achieved by enlarging the hysteresis band. However, when the hysteresis band is too large, the *THD* is poor and the system performance deteriorates. Thus, it is essential to investigate the conditions such that the HAPF compensation performance is within the requirements, with a reduction in the switching frequency to reduce unnecessary loss.

As the amplitude of the compensating current A_1 and A_2 is varying due to different initial conditions at different switching intervals, the compensating current may operate in three different regions. A simulation for the probability of the

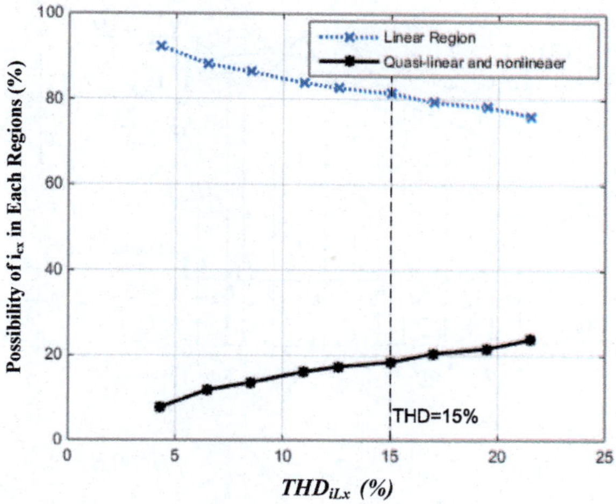

Fig. 3.13 Relationship between the operation probability of i_{cx} in each region and *THD*

compensating current falling in different regions is performed based on the system parameters defined in next section and is summarized in Fig. 3.13.

In Fig. 3.13, the probabilities of a TCLC-HAPF operating in the linear, quasi-linear and nonlinear regions are calculated by dividing the time duration in these regions (t_{linear}, t_{quasi}, $t_{nonlinear}$) by the fundamental period t_f. The curves of t_{linear}/t_f and $(t_{nonlinear}+ t_{quasi})/t_f$ are obtained by taking their average values over the entire simulated periods. The time duration in the linear region t_{linear} is the sum of all the switching time periods t_{sw} within the linear region over one fundamental period. If the TCLC-HAPF is not operating in a linear region, then it is operating in a nonlinear or quasi-linear region.

3.4.2 Non-linear Adaptive Hysteresis Band PWM Controller for TCLC-HAPF

In this section, the proposed non-linear adaptive hysteresis band PWM controller is discussed. First, the relationship between hysteresis bands and *ATHD* is addressed. Second, the control block diagram of the proposed non-linear adaptive hysteresis band PWM controller for TCLC-HAPF is illustrated.

3.4.2.1 Relationship Between Hysteresis Band *H* and *ATHD*

In this section, the relationship between the hysteresis band *H* and the harmonic distortion in a source current is deduced. The TCLC-HAPF injects compensating current to the point of common coupling (PCC). If the compensating current is exactly the same as its reference, the source current i_{sx} will be a sinusoidal wave

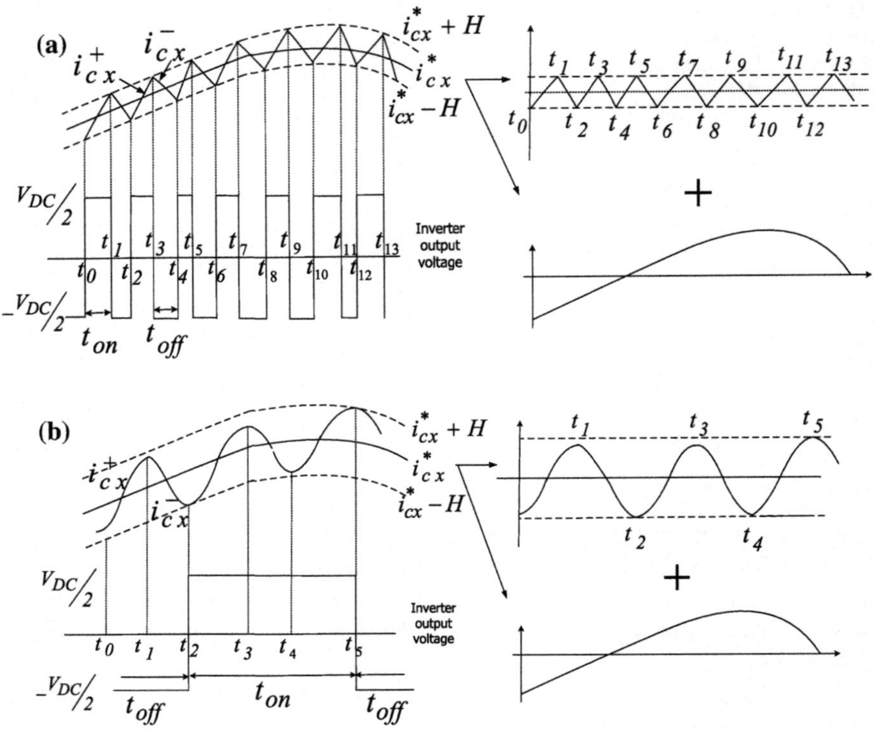

Fig. 3.14 Proposed current *THD* analysis method: **a** linear region and **b** nonlinear region

without distortion. However, the difference between the reference current and the compensating current varies in terms of the hysteresis band H, as discussed in Sect. 3.4.1. As a result, the compensating current waveform can be approximately decomposed into a reference compensating current $i_{cx}{}^{*}$ and current tracking error. The waveform of current tracking error is either an irregular triangle waveform (linear region) or a sinusoidal-like (nonlinear region) waveform, as shown in Fig. 3.14. The RMS value of the current error I_e can be estimated by:

$$I_e = \sqrt{\frac{1}{t_f}\left(\int_0^{t_1} i_e^2 dt + \int_{t_1}^{t_2} i_e^2 dt + \cdots + \int_{t_{n-1}}^{T_f} i_e^2 dt\right)} \approx \frac{H}{k} \quad (3.23)$$

where t_f is the fundamental period and k is a factor with a range of $k \in [\sqrt{2}, \sqrt{3}]$. The value of k is $\sqrt{3}$ when the compensating current is fully in the linear region. Its value is $\sqrt{2}$ when the compensating current is fully in the nonlinear region. When the compensating current falls in both linear region and nonlinear region during TCLC-HAPF operation, the value of k is calculated as the sum of the product of the probabilities and corresponding factors:

$$k = \sqrt{3} \cdot t_{linear}/t_f + \sqrt{2} \cdot (t_{nonliner} + t_{quasi})/t_f \tag{3.24}$$

where the probabilities can be obtained from Fig. 3.13. Hence, the value of k is a pre-defined constant and can be used to calculate RMS value of the current tracking errors when the target *THD** of the source current is set.

Generally, the harmonic distortion of the source current after compensation is evaluated by *THD,* which is defined as the ratio between the root mean square (RMS) values of the harmonic and fundamental components [24].

$$THD = \frac{\sqrt{\sum_{n=2}^{\infty} I_n^2}}{I_1} \times 100\% = \frac{I_h}{I_1} \times 100\% \tag{3.25}$$

where I_1 is the fundament current and I_h is the RMS value of harmonic current. During TCLC-HAPF operation, the fundamental source current I_{s1} consists of only active power component while the reactive power component is compensated by the TCLC-HAPF. Therefore, the fundamental source current value I_{s1} can be calculated as:

$$I_{s1} = \frac{\bar{p}_{\alpha\beta}}{\sqrt{3} \cdot \|v_L\|} \tag{3.26}$$

where

$$\|v_L\| = \sqrt{v_a^2 + v_b^2 + v_c^2} \tag{3.27}$$

in which $\|v_L\|$ is the norm or magnitude of the load voltage. $\bar{p}_{\alpha\beta}$ can be obtained by passing the loading instantaneous active power $p_{\alpha\beta}$ though a low pass filter (LPF).

To obtain I_h via fast Fourier transform (FFT) is complex and time-consuming. By using the RMS value of current tracking error I_e in (3.23) to replace I_h in (3.26), an approximated *THD* (*ATHD*) value is proposed in this chapter, which is expressed as:

$$ATHD = \frac{I_e}{I_{s1}} \times 100\% = \frac{H}{k \cdot I_{s1}} \times 100\% \tag{3.28}$$

In contrast to (3.25), a constant coefficient k and the adaptive hysteresis band H are used to evaluate the current distortion. The calculation is greatly simplified. The *ATHD* is proposed to simplify the calculation steps and to speed up the determination of the system performance in real-time. By using H in *ATHD*, the computation can avoid the time consumption required for actual current error calculations. However, the *ATHD* is an estimated *THD* index rather than the real *THD*. Based on the system parameters given in Table 3.1, a comparison between the *THD* and *ATHD* values is provided in Fig. 3.15.

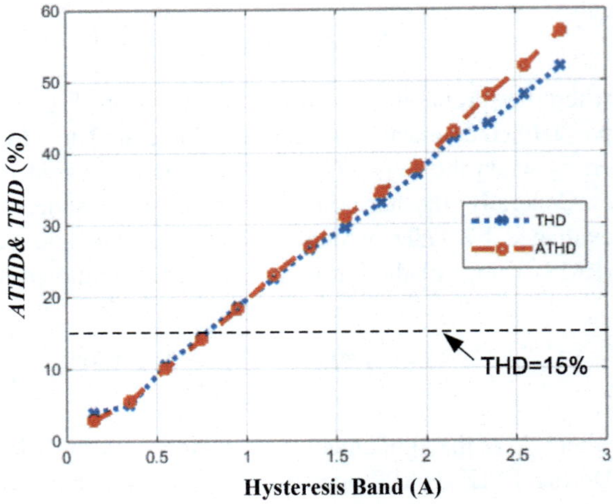

Fig. 3.15 Comparison of *THD* and *ATHD*

Both *THD* and *ATHD* increase as *H* increases. It can be concluded from Fig. 3.15 that the difference between *THD* and *ATHD* is small at *THD* < 15%. Thus, it is acceptable to use *ATHD* in place of *THD* if the source current distortion is kept below 15%.

Figures 3.13 and 3.15 are used to explain the relationship among different TCLC-HAPF operation regions, *H* and *ATHD* controller.

- When the *ATHD* reference is set to be small, the *H* is controlled to be small. Moreover, the possibility of the HAPF operating in the linear region is higher and that in the nonlinear region is lower, thus leading to a higher switching frequency (power loss).
- When *ATHD* reference is set to be large, the *H* is controlled to be large and the *THD* value becomes worse. The possibility of the HAPF operating in the nonlinear region is increasing and that in the linear region is decreasing, thus leading to a lower switching frequency (power loss).
- To achieve a balance between the power loss and system performance, a higher reference *ATHD* can be set according to the users' design requirement (satisfying IEEE Standards 519-2014). Therefore, even the *THD* compensation performance by the TCLC-HAPF may not be as good as it is designed to be when operating within the linear region [23]. When the system is operated in the nonlinear region, the switching loss can be significantly reduced.

In the following section, a control block diagram of the proposed non-linear adaptive hysteresis band PWM controller is given based on the above discussions.

3.4.2.2 Control Block Diagram of Proposed Non-linear Adaptive Hysteresis Band Controller for TCLC-HAPF

Based on the above discussions, a control block diagram of the proposed non-linear adaptive hysteresis PWM band controller for TCLC-HAPF is shown in Fig. 3.16. The purpose of adaptive hysteresis band control is to adaptively change the *H* value to maintain the *ATHD* (and thus the *THD*) at its reference value. Referring to the control block in Fig. 3.16, there are three input variables for the nonlinear adaptive hysteresis band controller: the fundamental source current I_{s1}, the target *ATHD**, a pre-defined *k* value. The output variable is the updated hysteresis band *H*. An initial value is assigned to *H* when the control system is powered on. The *ATHD* calculated by (3.28) is compared with the reference *ATHD**. This *ATHD* error is passed through a transform function (3.29) to obtain ΔH. After that, ΔH is added to the *H* value in previous cycle to adaptively change the *H* value.

$$\Delta H = k_H \cdot (ATHD^* - ATHD) \tag{3.29}$$

The *P* controller is chosen because it is simpler and faster than a *PI* controller. A limiter is applied to avoid the controller's overflow problem. A delay buffer is used to delay *H* for one sampling period.

In the steady-state, the calculated *ATHD* is approximately equal to *ATHD**, so the value of ΔH is close to zero based on (3.29). However, in the transient state (when there is a load change), *k* does not change, whereas I_{s1} (deduced from (3.26)) has a stepwise change. Therefore, the calculated *ATHD* in terms of I_{s1} is different from the reference *ATHD**. The value of ΔH is used to adaptively change the value of *H*. Then, the obtained *H* is feedback to calculate the *ATHD* in next cycle. The hysteresis band *H* in PWM control unit is updated accordingly. This control loop is continuously executed until *ATHD* gets close enough to its reference. Under this

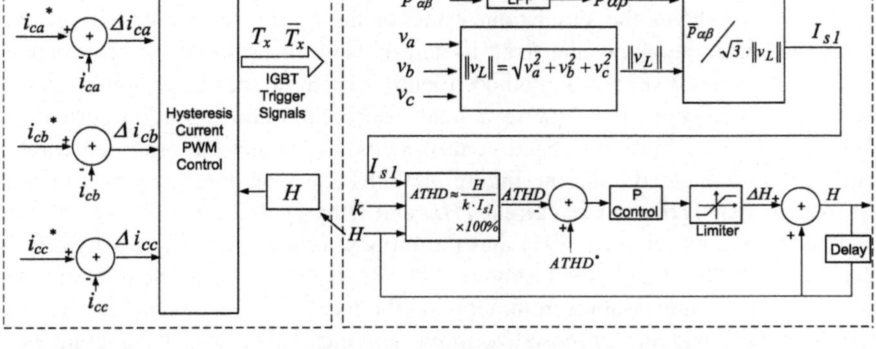

Fig. 3.16 Control block diagram of proposed non-linear adaptive hysteresis band PWM control for TCLC-HAPF

control, the switching frequency and inverter loss can be decreased. As a result, its thermal stress is reduced and the efficiency is increased.

3.4.2.3 Simulation Verifications of the Non-linear Adaptive Hysteresis Band PWM in Active Inverter Part of TCLC-HAPF

In this section, the proposed nonlinear adaptive hysteresis band PWM controller for TCLC-HAPFs is investigated by simulations. The simulated results for the proposed nonlinear adaptive hysteresis band are presented in comparison with the results for the fixed hysteresis band linearization proposed in [23]. In the simulations, the inverter loss of the active inverter part was obtained by measuring the collector-emitter voltage and current of each IGBT. The total inverter loss of the TCLC-HAPF can be found with the following equation [25]:

$$P_{Loss} = \sum_{n=0}^{n=6} P_{Loss,n} = \sum_{n=0}^{n=6} \frac{1}{t_f} \int_{0}^{T_f} v_{ce,n}(t) \cdot i_{ce,n}(t)dt \tag{3.30}$$

where $v_{ce,n}(t)$ and $i_{ce,n}(t)$ are the collector-emitter voltage and current of the IGBT, respectively; t_f is the fundamental period and n is the number of IGBTs. In the following simulations, the target $ATHD^*$ range is set as $ATHD^* = 10\%$ for the proposed nonlinear adaptive hysteresis band control. To keep the TCLC-HAPF operating in the linear region, the chosen H value should satisfy its linear limit H_{linear} [23]. Based on [23] and the system parameters in Table 3.1, H_{linear} can be calculated as $H_{linear} = 0.69$ A. Therefore, by choosing the hysteresis band $H < 0.69$ A, the TCLC-HAPF will mostly operate in the linear region. As a result, a comparison between the fixed hysteresis band linearization controller [23] and the proposed controller for the TCLC-HAPF can be performed. The simulation and experimental results of the fixed hysteresis band linearization control method [23] and the proposed non-linear adaptive hysteresis band control method for TCLC-HAPF are illustrated in Figs. 3.17, 3.18, 3.19 and 3.20, respectively.

Figure 3.17 shows the simulation results of the i_{cx} for the compensating nonlinear loading by applying the fixed hysteresis band linearization control method [23] and the proposed control method. Figure 3.18 show the simulation results of the dynamic process of the phase a load, source and compensating current by applying the fixed hysteresis band control method [23] and the proposed control method during the load transient. Figure 3.19 illustrates the dynamic waveforms of the load current THD, source current THD and H by applying the fixed hysteresis band linearization controller [23] and the proposed adaptive nonlinear hysteresis band controller to TCLC-HAPF. Figure 3.20 shows the curves of the total inverter loss during the compensation from applying the fixed hysteresis band linearization controller [23] and the proposed adaptive nonlinear hysteresis band controller. Table 3.4 summarizes all the simulation results.

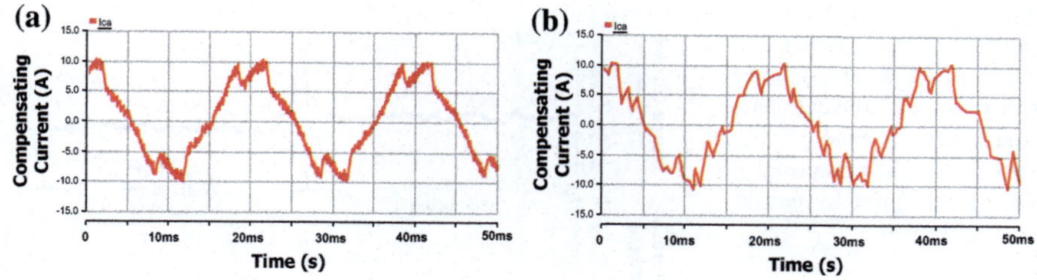

Fig. 3.17 Simulated i_{cx} for compensating nonlinear loading by applying: **a** fixed hysteresis band method [23], **b** proposed method

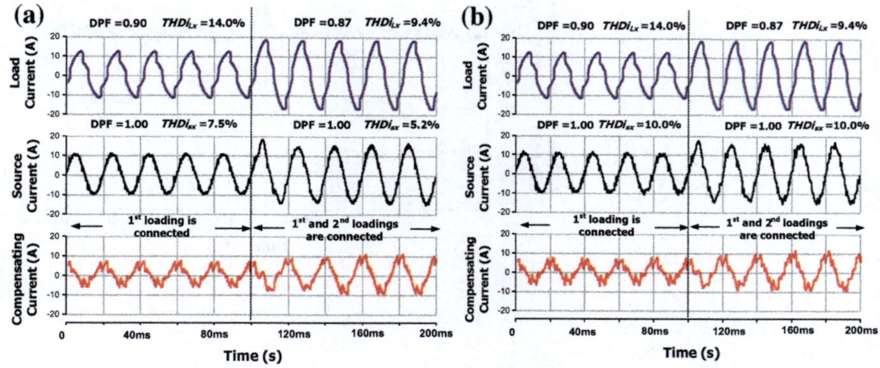

Fig. 3.18 Simulation waveforms of load, source and compensating current before and after 2nd loading connected by **a** applying fixed hysteresis band linearization controller [23]; **b** applying proposed hysteresis band controller

Under the fixed hysteresis band control as shown in Fig. 3.17a and Table 3.4, the compensating current i_{cx} mostly operates in the linear region and the simulated $(t_{quasi}+ t_{nonlinear})/t_f < 7\%$. From Fig. 3.18a, the simulated system *PF*s are unity during 1st loading and both the 1st and 2nd loading connected cases from original 0.90 and 0.87. And, the simulated *THD* are 7.5 and 5.2% after compensation from 14.0 to 9.4% before compensation.

From Fig. 3.20a and Table 3.4, after the fixed hysteresis band controlled HAPF [23] operates, the simulated average switching frequency (f_{sw}) and inverter loss (P_{VSI_loss}) of the TCLC-HAPF are f_{sw}= 0.80 kHz and P_{VSI_loss}= 13.5 W for the 1st loading case and f_{sw}= 1.6 kHz and P_{VSI_loss}= 55.8 W for the 1st and 2nd loadings case (Fig. 3.20a).

Compared with conventional fixed hysteresis band control, Fig. 3.17 and Table 3.4 show that using the proposed non-linear adaptive hysteresis band control method, the possibility of the compensating current i_{cx} falling in the quasi-linear and nonlinear regions ($(t_{quasi} + t_{nonlinear})/t_f > 14.0\%$) is higher than that resulting from the fixed hysteresis band method (<7%). From Fig. 3.18b the simulated *PF* is

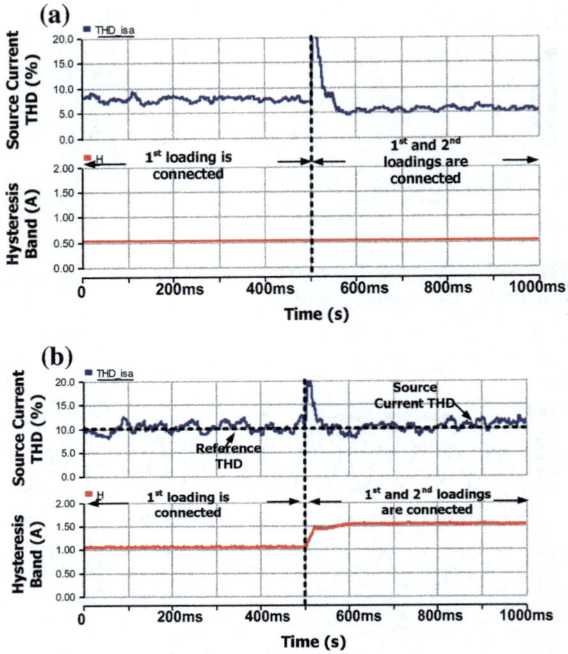

Fig. 3.19 Simulated THD_{isb} and H of phase b during TCLC-HAPF compensation by applying: **a** conventional fixed hysteresis band controller [23], **b** proposed nonlinear adaptive hysteresis band controller

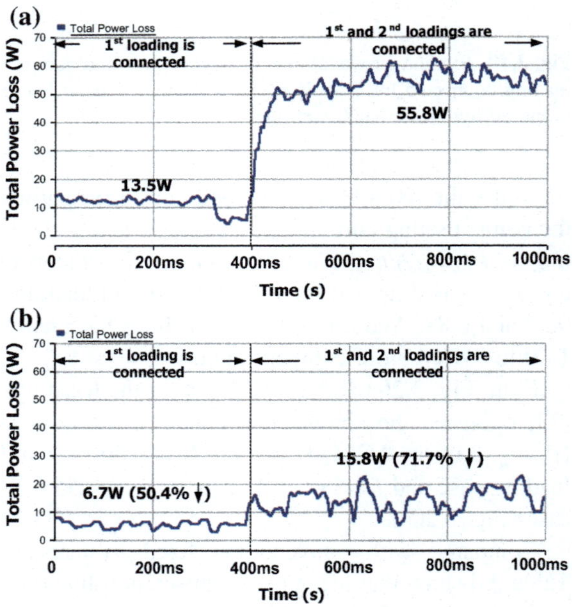

Fig. 3.20 Simulated curves of total inverter loss during the compensation by applying: **a** conventional fixed hysteresis band linearization controller [23], **b** proposed nonlinear adaptive hysteresis band controller

Table 3.4 Simulated results for fixed hysteresis band linearization controller [23] and proposed non-linear adaptive hysteresis band controller before and after compensation

	1st loading			1st and 2nd loading		
	Before comp.	Linearization method [23]	Proposed method	Before comp.	Linearization method [23]	Proposed method
$(t_{nonlinear}+t_{quasi})/t_f$ (%)	–	<7	>14	–	<7	>14
PF	0.90	1.00	1.00	0.87	1.00	1.00
THD_{isx} (%)	14.0	7.5	10.0	9.4	5.2	10.0
Average f_{sw} (kHz)	–	0.8	1.6	–	0.38	0.39
P_{VSI_loss} (W)	–	13.5	6.7 (↓50.4%)	–	30.2	15.8 (↓71.7%)

compensated to unity during both loading cases. Figure 3.19b shows that the value of H adaptively changes to maintain the THD value around its reference value (satisfying the IEEE standard [15]) when the loading varies.

Based on Fig. 3.20b and Table 3.4, the simulated average switching frequency (f_{sw}) and inverter loss (P_{VSI_loss}) of the TCLC-HAPF are f_{sw} = 0.38 kHz and P_{VSI_loss}= 6.7 W (50.4%↓) for the 1st loading case and f_{sw}= 0.39 kHz and P_{VSI_loss}= 15.8 W (71.7%↓) for the 1st and 2nd loadings cases.

3.5 Summary

In this chapter, two techniques have been proposed for TCLC part and active inverter part, for the TCLC-HAPF, respectively. The technique for mitigating the TCLC part harmonic current injection during switching has been proposed based on transient analysis and the coupling inductor design. From the simulation and experimental results, it is proved that the proposed TCLC part could compensate the dynamic reactive power with acceptable source current THD levels, while the traditional FC-TCR may not satisfy the THD international standard due to the self-harmonics injection characteristic. On the other hand, a non-linear adaptive hysteresis band PWM controller for active inverter part is proposed to reduce the inverter loss and maintain the source current total harmonic distortion within user requirements (satisfying IEEE standards). Due to nonlinear characteristics of the TCLC-HAPF, its compensating current is discussed in relation to the linear, quasi-linear and nonlinear operating regions. With the proposed nonlinear controller, the compensating current can be controlled to operate in the nonlinear or quasi-linear regions, rather than always staying in the linear region. To simplify the calculation, an $ATHD$ index is proposed and deduced to speed up the TCLC-HAPF system response time. Finally, simulation results show the proposed adaptive controller can reduce the inverter loss (more than 50% than fixed hysteresis band controller) and maintain an acceptable source current *THD*.

References

1. S.E. Haque, N.H. Malik, W. Shepherd, Operation of a fixed capacitor thyristor controlled reactor (FC-TCR) power factor compensator, in IEEE Trans. Power App. Syst. **PAS-104**(6) (1985)
2. J.S. Benton, Virtual instrument measures harmonic filter duty. IEEE Comput. Appl. Power **8** (4), 43–46 (1995)
3. L. Zanotto, R. Piovan, V. Toigo, E. Gaio, P. Bordignon, T. Consani, M. Fracchia, Filter design for harmonic reduction in high-voltage booster for railway applications, IEEE Trans. Power Deliv. **20**(1), 258–263 (2005)
4. A. Luo, Z. Shuai, W. Zhu, Z. Shen, Combined system for harmonic suppression and reactive power compensation. IEEE Trans. Ind. Electron. **56**(2), 418–428 (2009)
5. S. Rahmani, A. Hamadi, K. Al-Haddad, A combination of shunt hybrid power filter and thyristor-controlled reactor for power quality. IEEE Trans. Ind. Electron. **61**(5), 2152–2164 (2014)
6. D.B. Kulkarni, G.R. Udupi, ANN-based SVC switching at distribution level for minimal-injected harmonics. IEEE Trans. Power Deliv. **25**(3), 1978–1985 (2010)
7. L. Wang, C.S. Lam, M.C. Wong, A SVC-HAPF with wide compensation range and low dc-link voltage. IEEE Trans. Ind. Electron. **63**(6), 3333–3343 (2016)
8. L. Wang, C.S. Lam, M.C. Wong, An unbalanced control strategy for a thyristor controlled LC-coupling hybrid active power filter (SVC-HAPF) in three-phase three-wire systems. IEEE Trans. Power Electron. **32**(2), 1056–1069 (2017)
9. L. Wang, C.S. Lam, M.C. Wong, Hardware and software design of a low dc-link voltage and wide compensation range thyristor controlled LC-coupling hybrid active power filter, in *TENCON 2015 IEEE Region 10 Conference Proceedings*, Nov 2015
10. L. Wang, C.S. Lam, M.C. Wong, Modeling and parameter design of thyristor controlled LC-coupled hybrid active power filter (SVC-HAPF) for unbalanced compensation. IEEE Trans. Ind. Electron. **64**(3), 1827–1840 (2017)
11. C.S. Lam, L. Wang, S.I. Ho, M.C. Wong, Adaptive thyristor controlled LC – hybrid active power filter for reactive power and current harmonics compensation with switching loss reduction. IEEE Trans. Power Electron. **32**(10), 7577–7590 (2017)
12. L. Wang, C.S. Lam, M.C. Wong, Selective compensation of distortion, unbalanced and reactive power of a thyristor controlled LC-coupling hybrid active power filter (SVC-HAPF). IEEE Trans. Power Electron. **32**(12), 9065–9077 (2017)
13. C. Chai, C. Chompoo-Inwai, M. Kittipong, W.J. Lee, Reactive compensation techniques to improve the ride-through capability of wind turbine during disturbance. IEEE Trans. Ind. Appl. **41**(3), 666–672 (2007)
14. B. Singh, R. Saha, A. Chandra, K. Al-Haddad, Static synchronous compensators (STATCOM): a review. IET Power Electron. **2**(4), 297–324 (2009)
15. IEEE Recommended Practices and Requirements for Harmonic Control in Electrical Power Systems, 2014, IEEE Standard 519-2014
16. S. Srianthumrong, H. Akagi, A medium-voltage transformerless ac/dc Power conversion system consisting of a diode rectifier and a shunt hybrid filter. IEEE Trans. Ind. Appl. **39**, 874–882 (2003)
17. P. Salmeron, S.P. Litrán, A control strategy for hybrid power filter to compensate four-wires three-phase systems. IEEE Trans. Power Electron. **25**(7), 1923–1931 (2010)

18. L. Wei, R.A. Lukaszewski, Pulse width modulation (PWM) rectifier with variable switching frequency. U.S. patent 7 190 143 132, Mar 2007
19. B. Angélico, L. Campanhol, S. da Silva, Proportional integral/proportional integral derivative tuning procedure of a single-phase shunt active power filter using Bode diagram. IET Power Electron. **7**(10), 2647–2659 (2014)
20. M.S. Hamad, M.I. Masoud, B.W. Williams, S. Finney, Medium voltage 12-pulse converter: ac side compensation using a shunt active power filter in a novel front end transformer configuration. IET Power Electron. **5**(8), 1315–1323 (2012)
21. A.F. Zobaa, Optimal multiobjective design of hybrid active power filters considering a distorted environment. IEEE Trans. Ind. Electron. **61**(5), 107–114 (2014)
22. J.C. Wu, H.L. Jou, H.H. Hsaio, S.T. Xiao, A new hybrid power conditioner for suppressing harmonics and neutral-line current in three-phase four-wire distribution power systems. IEEE Trans. Power Deliv. **29**(4), 1525–1532 (2014)
23. C.S. Lam, M.C. Wong, Y.D. Han, Hysteresis current control of hybrid active power filters. IET Power Electron. **5**(7), 1175–1187 (2012)
24. D. Shmilovitz, On the definition of total harmonic distortion and its effect on measurement interpretation. IEEE Trans. Power Deliv. **20**(1), 526–528 (2005)
25. C. Xiao, G. Chen, W.G. Odendaal, Overview of power loss measurement techniques in power electronics systems. IEEE Trans. Ind. Appl. **43**(3), 657–664 (2007)

Chapter 4
Modeling and Parameter Design Method of Thyristor Controlled LC-Coupling Hybrid Active Power Filter (TCLC-HAPF) for Balanced/Unbalanced Loading Compensation

Abstract In this chapter, a new modeling and parameter design method are proposed for the thyristor controlled LC-coupled hybrid active power filter (TCLC-HAPF), which includes the analysis of the inner flowed three-phase unbalanced power during unbalanced condition. Different from the conventional method which ignores the inner flowed three-phase unbalanced power, the proposed design method can achieve better performance during unbalanced compensation. To verify the validity and advantages of the proposed modeling and parameter design method for the TCLC-HAPF, the mathematical analysis, representative simulation and experimental results for compensating different unbalanced loading situation are presented in comparison with the conventional design method.

Keywords Active power · Current harmonics · Reactive power
Thyristor controlled LC (TCLC) compensator · Thyristor controlled LC-coupling hybrid active power filter (TCLC-HAPF) · Unbalanced compensation

4.1 Introduction

This chapter proposes the parameter design methods of three-phase three-wire thyristor controlled LC-coupling hybrid active power filter (TCLC-HAPF) for balanced and unbalance loading compensation. The balanced parameter design method of TCLC-HAPF is proposed based on maximum reactive power compensation among the three phases. The TCLC-HAPF designed by balanced parameter design method can work with wide compensation range and low DC-link voltage for balanced load compensation. However, if the balanced design method is directly applied to the TCLC-HAPF for unbalanced compensation, it either requires a higher DC-link voltage or obtains relatively poor compensating performance. Therefore, the unbalanced parameter design method of TCLC-HAPF is proposed as the general parameter design solution for power quality compensation.

L. Wang et al., *Adaptive Hybrid Active Power Filters*, Power Systems,
https://doi.org/10.1007/978-981-10-8827-8_4

Unbalanced problem is mainly caused by connecting of the unsymmetrical or single-phase loads to power system. Researchers have well noticed the importance of addressing unbalanced problem by continuously improving the structures and control methods of different compensators [1–18]. At the early stage, the static var compensators (SVCs) were controlled to match their impedance to phase loading impedance individually. However, SVCs do not have harmonic compensation ability and also inject harmonic to the power system during operation [1–4]. In the past decades, active power filters (APFs) have gradually replaced the SVCs for unbalanced compensation [5–9]. Moreover, their control methods have been continuously improved by considering different operation voltage and current conditions. In recent two years, by applying APFs for unbalanced compensation, many control method shave been proposed to define the current or voltage references such as direct power control (DPC) [5, 6], repetitive control [7], artificial neural networks (ANNs) based control [8] id-iq control [9], etc. However, due to structure limitations, the active inverter part of APF needs to keep at a large rating to perform unbalanced compensation, which drives up their cost. To reduce ratings of APF, the combined systems of SVC in parallel with APF (SVC//APF) [10], SVC in series with hybrid active power filter (SVC + HAPF) [11, 12], TCLC-HAPF [13–18] have been proposed. For medium–voltage level applications, the cost and control complexity of TCLC-HAPF is the cheapest and simplest, since SVC//APF requires multi-level structure and SVC + HAPF require matching transformers. Moreover, the TCLC-HAPF has the distinct characteristics of wide compensation range and low active inverter capacity. Therefore, the TCLC-HAPF has large potential to be further developed for unbalanced compensation.

Due to the limitations among the existing literatures [1–18], this chapter proposes the unbalanced parameter design method which has taken into consideration of the inner flowed three-phase unbalanced power with the help of an active impedance concept. The contributions of this chapter are to:

- Propose a three-phase modeling for unbalanced compensation analysis instead of using the traditional equivalent single-phase modeling [19, 20] without considering unbalanced power;
- Propose a parameter design method with considering three-phase unbalanced power as well, besides reactive power and current harmonics;
- With the proposed parameter design method, the TCLC-HAPF can operate at a low DC-link voltage even under unbalanced loading compensation, thus lowering the system cost and switching loss.

In the following, the circuit configuration of the three-phase three-wire TCLC-HAPF is presented. Then, the three-phase modeling for compensation analysis is proposed in Sect. 4.2. Based on models, the parameter design method for the balanced loading compensation is proposed in Sect. 4.3. In Sect. 4.4, the relationship between the required TCLC impedance and load power is deduced. Based on this relationship, the unbalanced parameter design method is proposed and discussed under three different types of unbalanced loading: unbalanced inductive loads, unbalanced capacitive loads, and mixed inductive and capacitive loads. In Sect. 4.5, the representative simulation case studies are provided to compare the above proposed balanced parameter design method (Sect. 4.3) with the unbalanced parameter method (Sect. 4.4). And the corresponding experimental results are given in Sect. 4.6. Finally, the summary is drawn in Sect. 4.7.

4.2 Circuit Configuration and Modeling of TCLC-HAPF

The circuit topology of a three-phase three-wire TCLC-HAPF is provided in Fig. 4.1. $v_{sx}v_x$ and v_{invx} are the source voltage, load voltage and inverter output voltage, respectively, where the subscript 'x' denotes phase $x = a, b, c$; i_{sx}, i_{Lx} and i_{cx} are source current, load current and compensating current, respectively. The TCLC part of the TCLC-HAPF consists of a coupling inductor L_c, a parallel capacitive C_{PF} and a thyristor controlled reactor (TCR) with an inductor L_{PF}. The active inverter part is a two-level voltage source inverter (VSI) with a dc-link capacitor C_{DC}.

The active inverter part can be considered as the adjustable active impedance to improve the TCLC part fundamental and harmonic current compensation ability [19, 20]. Therefore, the three-phase modeling is proposed in Fig. 4.2 for TCLC-HAPF unbalanced compensation analysis.

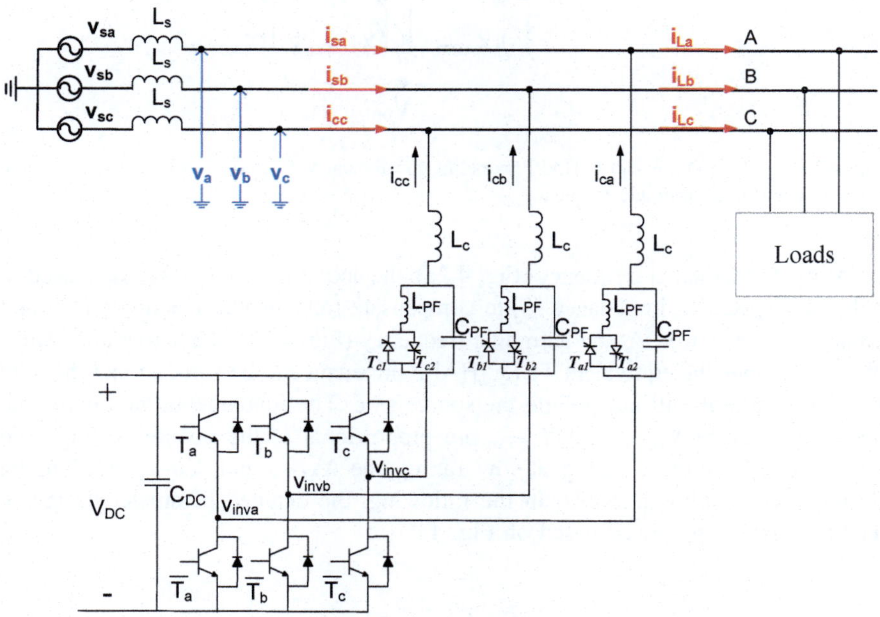

Fig. 4.1 Circuit configurations of the TCLC-HAPF

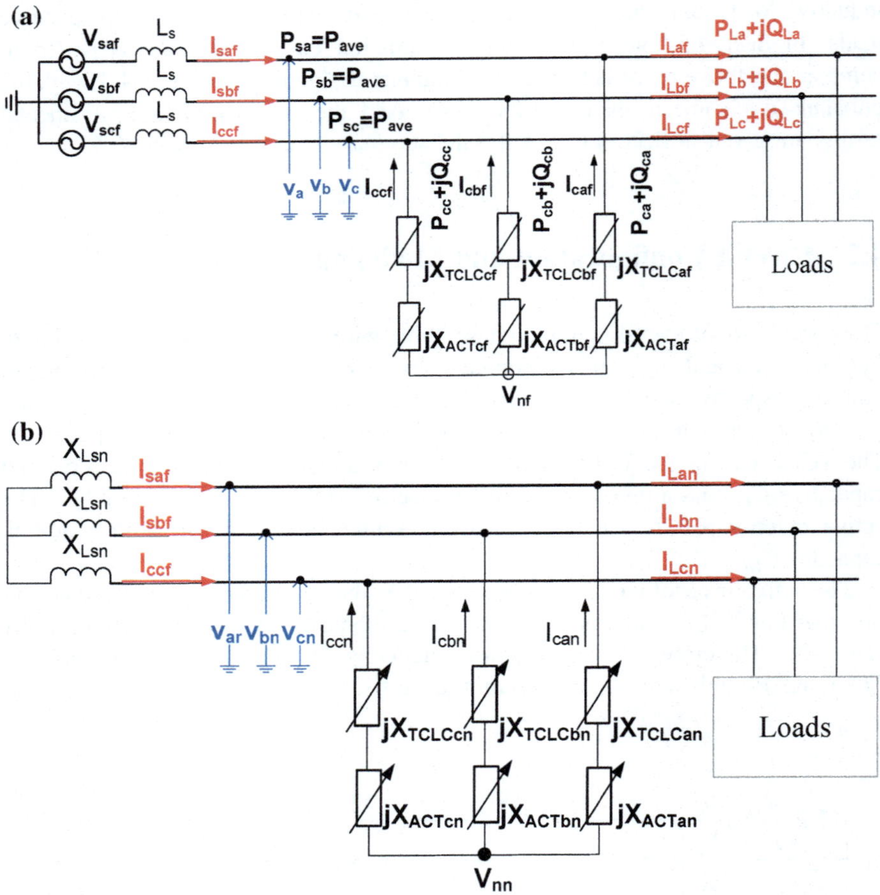

Fig. 4.2 The three-phase TCLC-HAPF modeling for unbalanced compensation **a** at fundamental frequency and **b** at harmonic frequency

At the fundamental frequency (Fig. 4.2a), the active impedance X_{ACTxf} is used to help the TCLC part impedance X_{af} to compensate fundamental reactive power and balance active power. At the harmonic frequency (Fig. 4.2b), the active impedance X_{ACTxn} changes the equivalent TCLC-HAPF impedance to be zero, so that the load harmonic current will not pollute the source side. The fundamental and harmonic active impedance X_{ACTxf} and X_{ACTxn} are proportional to the inverter voltage. To keep active inverter working at low rating, the X_{ACTxf} and X_{ACTxn} need to be designed as small as possible. In the following, the detailed parameter design of TCLC-HAPF is proposed based on Fig. 4.2.

4.3 Proposed TCLC-HAPF Parameter Design Method Design for Balanced Loads

The impedance of TCLC part can be calculated as an $L_c(X_{Lc})$ in series with the paralleled combination of a L_{PF} $(X_{TCR(x)})$ and a C_{PF} (X_{CPF}). And the TCLC impedances can be deduced as:

$$\begin{bmatrix} X_{TCLCaf}(\alpha_a) \\ X_{TCLCbf}(\alpha_b) \\ X_{TCLCcf}(\alpha_c) \end{bmatrix} = \begin{bmatrix} X_{TCR(\alpha_a)}//X_{C_{PF}} + X_{L_c} \\ X_{TCR(\alpha_b)}//X_{C_{PF}} + X_{L_c} \\ X_{TCR(\alpha_c)}//X_{C_{PF}} + X_{L_c} \end{bmatrix} = \begin{bmatrix} \frac{\pi X_{L_{PF}} X_{C_{PF}}}{X_{C_{PF}}(2\pi - 2\alpha_a + \sin 2\alpha_a) - \pi X_{L_{PF}}} + X_{L_c} \\ \frac{\pi X_{L_{PF}} X_{C_{PF}}}{X_{C_{PF}}(2\pi - 2\alpha_b + \sin 2\alpha_b) - \pi X_{L_{PF}}} + X_{L_c} \\ \frac{\pi X_{L_{PF}} X_{C_{PF}}}{X_{C_{PF}}(2\pi - 2\alpha_c + \sin 2\alpha_c) - \pi X_{L_{PF}}} + X_{L_c} \end{bmatrix} \tag{4.1}$$

where

$$X_{TCR}(\alpha_x) = \frac{\pi}{2\pi - 2\alpha_x + \sin 2\alpha_x} X_{L_{PF}} \tag{4.2}$$

In (4.1), X_{Lc}, X_{CPF}, X_{LPF} are the impedances of coupling inductor L_c, the paralleled capacitor C_{PF} and inductor L_{PF}. α_x is the firing angle of the thyristor. The TCLC part has two back-to-back connected thyristors T_{x1}, T_{x2}, and they are triggered alternately in every half cycle. When α_x = 180° (thyristors are opened for the whole cycle), the TCLC part has the maximum capacitive TCLC impedance $X_{Cap(Max)}$. On the other hand, when the firing angle α_x = 90° (one of thyristors is closed for whole cycle), the TCLC part has the minimum inductive TCLC impedance $X_{Ind(Min)}$. Therefore, $X_{Ind(Min)}$ and $X_{Cap(Max)}$ can be expressed as:

$$X_{Cap(Max)} = X_{L_c} - X_{C_{PF}} = \omega L_c - 1/\omega C_{PF} \tag{4.3}$$

$$X_{Ind(Min)} = \frac{X_{L_{PF}} X_{C_{PF}}}{X_{C_{PF}} - X_{L_{PF}}} + X_{L_c} = \frac{\omega L_{PF}}{1 - \omega^2 L_{PF} C_{PF}} + \omega L_c \tag{4.4}$$

where ω (=2π*f*) is the angular frequency.

Under balanced loading, the purpose of the TCLC part is to provide the same amount of the compensating reactive power Q_{cx} as the loading generated reactive power Q_{Lx}. The compensating reactive power Q_{cx} is related to the controllable fundamental impedance of the TCLC part $X_{TCLCxf}(\alpha_x)$ in (4.1), which can be expressed as:

$$Q_{cx} = \frac{V_x^2}{X_{TCLCxf}(\alpha_x)} = \frac{V_x^2}{\frac{\pi X_{L_{PF}} X_{C_{PF}}}{X_{C_{PF}}(2\pi - 2\alpha_x + \sin 2\alpha_x) - \pi X_{L_{PF}}} + X_{L_c}} \tag{4.5}$$

where V_x is the root mean square (RMS) phase load voltage. In (4.5), the phase compensating reactive power Q_{cx} depends on X_{Lc}, X_{LPF}, X_{CPF} and α_x. Hence, the required reactive power compensating range can be used to design the TCLC part parameters.

From Fig. 4.1, the TCLC part has two back-to-back connected thyristors T_{x1} and T_{x2} for each phase, and they are triggered alternately in every half cycle. When both thyristors $T_{x1}$and T_{x2} for each phase are turned off for the whole fundamental period (firing angle α_x = 180°), the TCLC part can be considered as a *LC* filter (L_c in series with C_{PF}). In this case, the TCLC part provides the maximum capacitive compensating reactive power $Q_{cx(MaxCap)}$. The expression of $Q_{cx(MaxCap)}$ can be expressed as:

$$Q_{cx(MaxCap)} = -\frac{V_x^2}{X_{C_{PF}} - X_{L_c}} \tag{4.6}$$

On the other hand, when one of the thyristors (T_{x1} or T_{x2}) is turned on for half of the fundamental period (firing angle α_x = 90°), the TCLC part can be considered as a coupling L_c in series with the parallel combination of L_{PF} and C_{PF}. In this case, the TCLC part provides the maximum inductive reactive power $Q_{cx(MaxInd)}$. The expression of $Q_{cx(MaxInd)}$ can be expressed as:

$$Q_{cx(MaxInd)} = \frac{V_x^2}{\frac{X_{L_{PF}} X_{C_{PF}}}{X_{C_{PF}} - X_{L_{PF}}} + X_{L_c}} \tag{4.7}$$

Based on (4.6), (4.7) and the system parameters as shown in Table 4.5, the relationships among the compensating reactive power Q_{cx}, firing angle α_x and the TCLC parameters (L_c, L_{PF} and C_{PF}) are shown in Fig. 4.3.

From Fig. 4.3a, it is clearly shown that Q_{cx} is not sensitive to the variation of L_c when α_x varies from 90° to 180°. Figure 4.3b illustrates that the inductive reactive power compensating range ($Q_{cx} > 0$) highly depends on L_{PF}, especially for $Q_{cx(MaxInd)}$ at α_x = 90°. However, the $Q_{cx(MaxCap)}$ at α_x = 180° is independent on L_{PF}. Moreover, as shown in Fig. 4.3c, the capacitive reactive power compensating range ($Q_{cx} < 0$) highly depends on C_{PF} when α_x changes. In addition, it is noted that when the inductive reactive power compensating range may disappear with C_{PF} increasing ($Q_{cx} < 0$ with $\alpha_x \in$ [90°, 180°]). This phenomenon can be explained by (4.7) that $Q_{cx(MaxInd)} < 0$ with the $X_{CPF} < X_{LPF}$. Therefore, it is necessary to design to $X_{CPF} > X_{LPF}$ if the users' requirement is to compensate both inductive reactive power and capacitive power.

From Fig. 4.3, the reactive power compensating range of the TCLC part depends mainly on the values of parallel C_{PF} and L_{PF}, and much less on the coupling L_c.

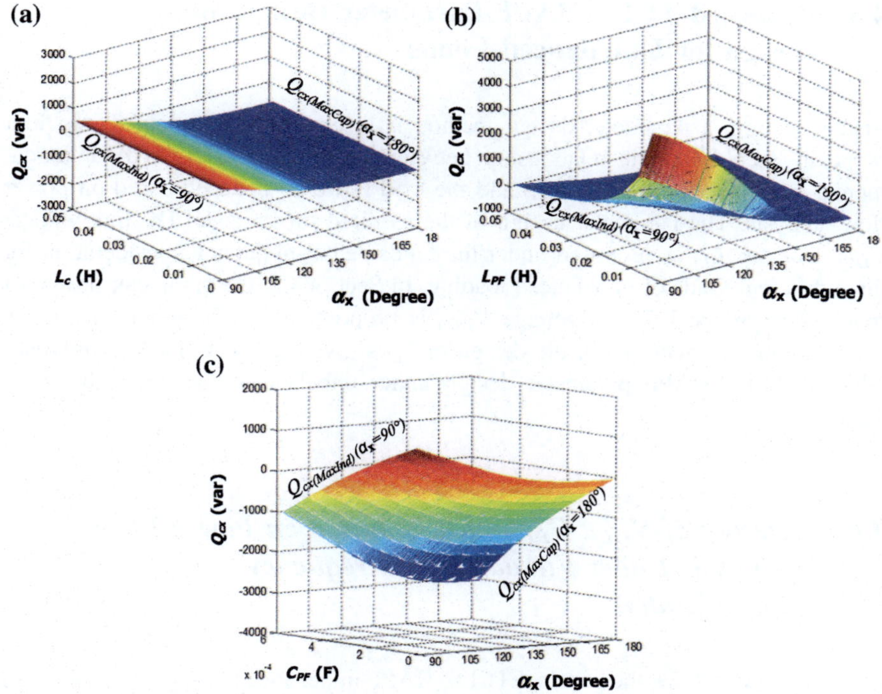

Fig. 4.3 Relationships among Q_{cx}, α_x and TCLC parameters (L_c, L_{PF} and C_{PF}). **a** L_c is varying while L_{PF} and C_{PF} are fixed. **b** L_{PF} is varying while L_c and C_{PF} are fixed. **c** C_{PF} is varying while L_c and L_{PF} are fixed

To compensate the reactive power, that is $Q_{cx} = Q_{Lx}$, C_{PF} and L_{PF} can be designed based on the maximum inductive reactive power $Q_{Lx(MaxInd)}$ ($= -Q_{cx(MaxCap)}$) and maximum capacitive reactive power $Q_{Lx(MaxCap)}$ ($= -Q_{cx(MaxInd)}$) generated by the loading.

From (4.6) and (4.7), the parallel capacitor C_{PF} and inductor L_{PF} can be designed as:

$$C_{PF} = \frac{Q_{Lx(MaxInd)}}{\omega^2 \cdot Q_{Lx(MaxInd)} \cdot L_c + \omega \cdot V_x^2} \tag{4.8}$$

$$L_{PF} = \frac{V_x^2 + \omega L_c Q_{Lx(MaxCap)}}{\omega^2 \cdot V_x^2 \cdot C_{PF} - \omega \cdot Q_{Lx(MaxCap)} + \omega^3 \cdot L_c \cdot C_{PF} \cdot Q_{Lx(MaxCap)}} \tag{4.9}$$

where ω is the fundamental angular frequency.

The purposed of L_c is to filter out the current ripple caused by the power switches of the active inverter part. And the detailed design of L_c will be included in Sect. 4.4.3. In the following section, the parameter design of TCLC-HAPF under unbalanced loading will be proposed.

4.4 Proposed TCLC-HAPF Parameter Design Method Design for Unbalanced Loads

In this section, a parameter design method is discussed and explained into four parts. In Sect. 4.4.1, the relationship between the required TCLC-HAPF fundamental impedances (X_{af} + X_{ACTxf}) and the load power is deduced based on power flow analysis. The parameter design of the required fundamental DC-link voltage V_{DCf}, C_{PF} and L_{PF} is proposed under the fundamental frequency consideration for three different kinds of unbalanced loading. In Sect. 4.4.2, the parameter design of required harmonic DC-link voltage V_{DCh} is proposed under harmonic frequency consideration. In Sect. 4.4.3, the design of L_c is given. In Sect. 4.4.4, a summary table of TCLC-HAPF parameter design under unbalanced loading is drawn in Table 4.2.

4.4.1 *Design of V_{DCf}, C_{PF} and L_{PF} Based on Power Flow Analysis Under Fundamental Frequency Consideration*

Referring to Fig. 4.2a, the required TCLC-HAPF impedance (X_{TCLCxf} + X_{ACTxf}) can be calculated by applying the Ohm's Law as:

$$X_{TCLCxf} + \mathrm{X}_{ACTxf} = \left(\vec{V}_{xf} - \vec{V}_{nf}\right) \Big/ \vec{I}_{cxf} \tag{4.10}$$

where $\vec{V}_{xf}$ and $\vec{I}_{cxf}$ are the fundamental load voltage and compensating current phasor respectively, where x stands for phase a, b and c. $\vec{V}_{nf}$ is the fundamental common point voltage. By using the Kirchhoff's circuit laws (KCL), the compensating current relationship can be expressed as:

$$\begin{aligned} \vec{I}_{caf} + \vec{I}_{cbf} + \vec{I}_{ccf} = & \frac{\vec{V}_a - \vec{V}_{nf}}{jX_{TCLCaf} + \mathrm{j}X_{ACTaf}} + \frac{\vec{V}_b - \vec{V}_{nf}}{jX_{TCLCbf} + \mathrm{j}X_{ACTbf}} \\ & + \frac{\vec{V}_c - \vec{V}_{nf}}{jX_{TCLCcf} + \mathrm{j}X_{ACTcf}} = 0 \end{aligned} \tag{4.11}$$

Simplifying (4.11), the expression of $\vec{V}_{nf}$ can be obtained as:

$$\vec{V}_{nf} = \frac{(X_{TCLCbf} + X_{ACTbf})(X_{TCLCcf} + X_{ACTcf})}{X_{Eqf}} \cdot \vec{V}_{af} + \frac{(X_{TCLCcf} + X_{ACTcf})(X_{af} + X_{ACTaf})}{X_{Eqf}} \cdot \vec{V}_{bf} + \frac{(X_{TCLCaf} + X_{ACTaf})(X_{TCLCbf} + X_{ACTbf})}{X_{Eqf}} \cdot \vec{V}_{cf} \tag{4.12}$$

where

$$\mathrm{X}_{Eqf} = (X_{TCLCaf} + X_{ACTaf})(X_{TCLCbf} + X_{ACTbf}) + (X_{TCLCbf} + X_{ACTbf})(X_{TCLCcf} + X_{ACTcf}) + (X_{TCLCcf} + X_{ACTcf})(X_{TCLCaf} + X_{ACTaf}) \tag{4.13}$$

The $\vec{I}_{cxf}$ can be expressed in terms of $\vec{V}_{xf}$ ($\vec{V}_x = \vec{V}_{xf}$) for v_x is assumed to be pure sinusoidal without harmonic components [18]) and the compensating active and reactive power P_{cx} and Q_{cx} as:

$$\vec{I}_{cxf} = \left[(P_{cx} + jQ_{cx}) \big/ \vec{V}_x \right]^* \tag{4.14}$$

where the note "*" denotes the conjugation. For unbalanced compensation, the TCLC-HAPF can provide the same amount of reactive power as the loading required but with opposite directions and balance the three-phase source active power to their average values $(P_{La} + P_{Lb} + P_{Lc})/3$ simultaneously. Thus, the compensating P_{cx} and Q_{cx} can be expressed as:

$$Q_{cx} = -Q_{Lx} \text{ and } P_{cx} = -\left(P_{Lx} - \frac{P_{La} + P_{Lb} + P_{Lc}}{3} \right) \tag{4.15}$$

where P_{Lx} and Q_{Lx} is the load active and reactive power of each phase.

Based on (4.10)–(4.15), the required TCLC-HAPF impedance $(X_{TCLCxf} + X_{ACTxf})$ can be deduced as:

$$\begin{bmatrix} X_{TCLCaf} + X_{ACTaf} \\ X_{TCLCbf} + X_{ACTbf} \\ X_{TCLCcf} + X_{ACTcf} \end{bmatrix} = \begin{bmatrix} \frac{3V_x^2 (Q_{Lc} - Q_{Lb} - Q_{La})^{-1} (Q_{Lb} - Q_{La} - Q_{Lc})^{-1}}{(Q_{Lc} - Q_{Lb} - Q_{La})^{-1} + (Q_{La} - Q_{Lb} - Q_{Lc})^{-1} + (Q_{Lb} - Q_{La} - Q_{Lc})^{-1}} \\ \frac{3V_x^2 (Q_{Lc} - Q_{Lb} - Q_{La})^{-1} (Q_{La} - Q_{Lb} - Q_{Lc})^{-1}}{(Q_{Lc} - Q_{Lb} - Q_{La})^{-1} + (Q_{La} - Q_{Lb} - Q_{Lc})^{-1} + (Q_{Lb} - Q_{La} - Q_{Lc})^{-1}} \\ \frac{3V_x^2 (Q_{La} - Q_{Lb} - Q_{Lc})^{-1} (Q_{Lb} - Q_{La} - Q_{Lc})^{-1}}{(Q_{Lc} - Q_{Lb} - Q_{La})^{-1} + (Q_{La} - Q_{Lb} - Q_{Lc})^{-1} + (Q_{Lb} - Q_{La} - Q_{Lc})^{-1}} \end{bmatrix} \tag{4.16}$$

where V_x is the root-mean-square (RMS) value of load voltage. From (4.16), the TCLC-HAPF impedances can be calculated based on Q_{Lx} and V_x. With the pre-measured variation ranges of the loading reactive power, the required TCLC-HAPF impedance $(X_{TCLCxf} + X_{ACTxf})$ can be obtained.

Referring to Fig. 4.2a, the fundamental inverter voltage V_{invxf} can be obtained as:

$$V_{invxf} = X_{ACTxf} \cdot I_{cxf} \tag{4.17}$$

where X_{ACTxf} and I_{cxf} are the fundamental active impedance and compensating current. The X_{ACTxf} is directly proportional to the required fundamental DC-link voltage (V_{DCf}), and V_{DCf} is proportional to the power loss and system cost. Therefore, the X_{ACTxf} can be designed to be zero ($X_{ACTxf} \approx 0$) in order to minimize them.

$$\begin{aligned} V_{DCf} &= \max\left(\sqrt{6} \cdot V_{invaf} \quad \sqrt{6} \cdot V_{invbf} \quad \sqrt{6} \cdot V_{invcf} \right) \\ &= \max\left(\sqrt{6} \cdot X_{ACTaf} \cdot I_{caf} \quad \sqrt{6} \cdot X_{ACTbf} \cdot I_{cbf} \quad \sqrt{6} \cdot X_{ACTcf} \cdot I_{ccf} \right) \approx 0 \end{aligned} \tag{4.18}$$

In (4.7), the scale of $\sqrt{6}(=\sqrt{2} \cdot \sqrt{3})$ can be explained by the following two reasons: *A.* to guarantee the sufficient V_{DCf}, the peak value of fundamental inverter voltage need to be considered $V_{invxf(\mathrm{p})} = \sqrt{2} \cdot V_{invxf}$. *B.* to transfer the phase voltage V_{invxf} to line-to-line dc-link voltage, the scale of $\sqrt{3}$ is required.

With such minimum V_{DCf} design, the TCLC part is mainly used to compensate reactive power and balance the active power, while the active inverter part is mainly used to improve the harmonic compensation ability of TCLC part (discussed in next part). Therefore, (4.16) can be simplified as:

$$\begin{bmatrix} X_{TCLCaf} \\ X_{TCLCbf} \\ X_{TCLCcf} \end{bmatrix} = \begin{bmatrix} \dfrac{3V_x^2 (Q_{Lc}-Q_{Lb}-Q_{La})^{-1} (Q_{Lb}-Q_{La}-Q_{Lc})^{-1}}{(Q_{Lc}-Q_{Lb}-Q_{La})^{-1} + (Q_{La}-Q_{Lb}-Q_{Lc})^{-1} + (Q_{Lb}-Q_{La}-Q_{Lc})^{-1}} \\ \dfrac{3V_x^2 (Q_{Lc}-Q_{Lb}-Q_{La})^{-1} (Q_{La}-Q_{Lb}-Q_{Lc})^{-1}}{(Q_{Lc}-Q_{Lb}-Q_{La})^{-1} + (Q_{La}-Q_{Lb}-Q_{Lc})^{-1} + (Q_{Lb}-Q_{La}-Q_{Lc})^{-1}} \\ \dfrac{3V_x^2 (Q_{La}-Q_{Lb}-Q_{Lc})^{-1} (Q_{Lb}-Q_{La}-Q_{Lc})^{-1}}{(Q_{Lc}-Q_{Lb}-Q_{La})^{-1} + (Q_{La}-Q_{Lb}-Q_{Lc})^{-1} + (Q_{Lb}-Q_{La}-Q_{Lc})^{-1}} \end{bmatrix} \tag{4.19}$$

The TCLC part is an L_c (X_{Lc}) in series with a paralleled combination of a L_{PF} (X_{LPF}) and a C_{PF} (X_{CPF}), in which the TCLC impedance can be deduced as:

$$\begin{bmatrix} X_{TCLCaf}(\alpha_a) \\ X_{TCLCbf}(\alpha_b) \\ X_{TCLCcf}(\alpha_c) \end{bmatrix} = \begin{bmatrix} \dfrac{\pi X_{L_{PF}} X_{C_{PF}}}{X_{C_{PF}}(2\pi - 2\alpha_a + \sin 2\alpha_a) - \pi X_{L_{PF}}} + X_{L_c} \\ \dfrac{\pi X_{L_{PF}} X_{C_{PF}}}{X_{C_{PF}}(2\pi - 2\alpha_b + \sin 2\alpha_b) - \pi X_{L_{PF}}} + X_{L_c} \\ \dfrac{\pi X_{L_{PF}} X_{C_{PF}}}{X_{C_{PF}}(2\pi - 2\alpha_c + \sin 2\alpha_c) - \pi X_{L_{PF}}} + X_{L_c} \end{bmatrix} \tag{4.20}$$

In (4.20), X_{Lc}, X_{CPF}, X_{LPF} are the fundamental impedances of coupling inductor L_c, the paralleled capacitor C_{PF} and inductor L_{PF}. α_x is the firing angle of the thyristor. The TCLC part has two back-to-back connected thyristors T_{1x}, T_{2x}, and they are triggered alternately in every half cycle. When $\alpha_x = 180°$ (thyristors are opened for the whole cycle), the TCLC part has the maximum capacitive TCLC impedance $X_{Cap(Max)}$ (<0). On the other hand, when the firing angle $\alpha_x = 90°$ (one of thyristors is closed for whole cycle), the TCLC part has the minimum inductive TCLC impedance $X_{Ind(Min)}$ (>0). Therefore, $X_{Cap(Max)}$ and $X_{Ind(Min)}$ can be expressed as:

$$X_{Cap(Max)} = X_{L_c} - X_{C_{PF}} = \omega L_c - 1/\omega C_{PF} \tag{4.21}$$

$$X_{Ind(Min)} = \frac{X_{L_{PF}} X_{C_{PF}}}{X_{C_{PF}} - X_{L_{PF}}} + X_{L_c} = \frac{\omega L_{PF}}{1 - \omega^2 L_{PF} C_{PF}} + \omega L_c \tag{4.22}$$

where ω (= $2\pi f$) is the angular frequency. To guarantee the TCLC part has inductive compensation range and capacitive compensation range, the basic conditions of $X_{Cap(Max)} < 0$ and $X_{Ind(Min)} > 0$ need to be satisfied. Thus, from (4.21) and (4.22), the following relationships can be obtained:

$$C_{PF} < \frac{1}{\omega^2 L_c} \tag{4.23}$$

$$L_{PF} < \frac{1}{\omega^2 C_{PF}} \tag{4.24}$$

In the following, the design of C_{PF} and L_{PF} will be separated into three different kinds of loading for discussion: (1) unbalanced inductive loads, (2) unbalanced capacitive loads and (3) mixed inductive and capacitive loads.

A. Unbalanced inductive loading compensation

For the unbalanced inductive loading compensation, the required TCLC impedance X_{TCLCxf} obtained in (4.19) are capacitive ($X_{TCLCxf} < 0$). To compensate inductive loads, the design maximum capacitive TCLC impedances $X_{Cap(\mathrm{Max})}$ in (4.21) is required to be equal to or larger than the required X_{TCLCxf} in (4.19). Therefore, the TCLC impedance design criteria for each phase can be given as:

$$X_{TCLCxf} \le X_{Cap(Max)} < 0 \tag{4.25}$$

Substituting (4.21) into (4.25), the design criteria of C_{PF} can be obtained as:

$$\frac{1}{\omega^2 L_c - \omega X_{TCLCxf}} \le C_{PF} < \frac{1}{\omega^2 L_c} \tag{4.26}$$

where the final C_{PF} can be obtained as:

$$C_{PF} = \max\left(C_{PF}|_{X_{TCLCaf}}, C_{PF}|_{X_{TCLCbf}}, C_{PF}|_{X_{TCLCcf}} \right) \tag{4.27}$$

To ensure the TCLC part can cover the entire inductive loading range, the compensating range needs to cover the capacitive boundary [left hand side term of (4.26)]. Then, the TCLC inductance L_{PF} can be designed by (4.24) accordingly.

B. Unbalanced capacitive loading compensation

For unbalanced capacitive loads compensation, the required TCLC impedances X_{TCLCxf} in (4.19) are inductive ($X_{TCLCxf} > 0$). To compensate capacitive loads, the design minimum inductive TCLC impedances $X_{Ind(min)}$ in (4.22) is required to be equal or smaller than X_{TCLCxf}. Therefore, the TCLC impendence design criterion for each phase can be given as:

$$X_{TCLCxf} \geq X_{Ind(Min)} > 0 \tag{4.28}$$

Substituting (4.22) into (4.28), the design criteria of L_{PF} can be obtained as:

$$\frac{L_c}{\omega^2 L_c C_{PF} - 1} < L_{PF} \leq \frac{X_{TCLCxf} - \omega L_c}{\omega^2 X_{TCLCxf} C_{PF} - \omega^3 L_c C_{PF} + \omega} \tag{4.29}$$

Similarly, to ensure the TCLC part can cover the entire capacitive loading range, the compensating range needs to reach the inductive boundary. Then, the TCLC capacitance C_{PF} can be designed by (4.23) accordingly. However, with satisfying (4.23), the left side boundary of (4.29) is a negative value, so that (4.29) can be further simplified as:

$$L_{PF} \leq \frac{X_{TCLCxf} - \omega L_c}{\omega^2 X_{TCLCxf} C_{PF} - \omega^3 L_c C_{PF} + \omega} \tag{4.30}$$

where the final L_{PF} can be obtained as:

$$L_{PF} = \min\left(L_{PF}\big|_{X_{TCLCaf}}, \; L_{PF}\big|_{X_{TCLCbf}}, \; L_{PF}\big|_{X_{TCLCcf}} \right) \tag{4.31}$$

C. Mixed inductive and capacitive loads

When the unbalanced loading is mixed inductive and capacitive loads, the phase load reactive power can be positive ($Q_{Lx} > 0$) or negative ($Q_{Lx} < 0$). Therefore, both (4.26) and (4.30) are required to be satisfied. If the $X_{TCLCxf} < 0$, the design criterion of C_{PF} can be obtained by substituting the calculated X_{TCLCxf} into (4.26). On the other hand, the design criterion of L_{PF} can be obtained by substituting the calculated $X_{TCLCxf} > 0$ into (4.30).

Based on the above analysis, the design of C_{PF} and L_{PF} under the different types of loads can be summarized in Table 4.2.

4.4.2 Design of V_{DCh} Based on Harmonic Frequency Analysis

Referred to the harmonic models given in Fig. 4.2b, the harmonic current I_{sxn} circulating in the source is provided as:

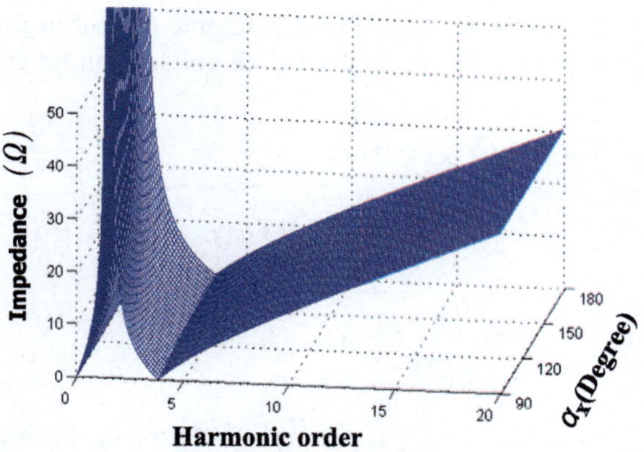

Fig. 4.4 Absolute value of X_{TCLCxn} and active impedance X_{ACTxn} under different harmonic order

$$I_{sxn} = \frac{X_{TCLCxn} + X_{ACTxn}}{X_{Lsn} + X_{TCLCxn} + X_{ACTxn}} \cdot I_{Lxn} \tag{4.32}$$

where I_{sxn} and I_{Lxn} are the source and load harmonic current, respectively. X_{Lsn} is the harmonic source side inductance. The purpose of the X_{ACTxn} is to reduce the harmonic current flowing into the source side. Ideally, the I_{sxn} can be reduced to be zero (I_{sxn}= 0). Therefore, the X_{ACTxn} can be controlled to be:

$$X_{ACTxn} = -X_{TCLCxn} \tag{4.33}$$

In (4.33), the X_{ACTxn} and X_{TCLCxn} are the harmonic active impedance and harmonic impedance of TCLC part. The absolute value of X_{ACTxn} and X_{TCLCxn} in each harmonic order can be expressed as:

$$|X_{ACTxn}(\alpha)| = |X_{TCLCxn}(\alpha)| = \left| \frac{\pi(n\omega L_{PF})}{(2\pi - 2\alpha + \sin 2\alpha) - \pi(n\omega)^2 \cdot L_{PF} C_{PF}} + n\omega L_c \right| \tag{4.34}$$

where n is the harmonic order. Based on (4.34), the absolute value of the X_{TCLCxn}(n) and active impedance X_{ACTxn}(n) under different harmonic order can be plotted as Fig. 4.4.

The required DC-link voltage V_{DCh} for harmonic current compensation can be expressed as:

$$\begin{aligned} V_{DCh} &= \max\left(\sqrt{6} \cdot V_{invah} \quad \sqrt{6} \cdot V_{invbh} \quad \sqrt{6} \cdot V_{invch} \right) \\ &= \max\left(\sqrt{6}\sqrt{\sum_{n=2}^{\infty} (X_{TCLCan} \cdot I_{Lan})^2} \; \sqrt{6}\sqrt{\sum_{n=2}^{\infty} (X_{TCLCbn} \cdot I_{Lbn})^2} \; \sqrt{6}\sqrt{\sum_{n=2}^{\infty} (X_{TCLCcn} \cdot I_{Lcn})^2} \right) \end{aligned} \tag{4.35}$$

In (4.35), I_{Lxn} is the load harmonic current in each order. Based on (4.18) and (4.35), the total required DC-link voltage can be given as:

$$V_{DC} = \sqrt{\mathrm{V}_{\mathrm{DCf}}^2 + \mathrm{V}_{\mathrm{DCh}}^2}$$
$$= \max\left(\sqrt{6}\sqrt{\sum_{n=2}^{\infty}(X_{\mathrm{TCLC}an}\cdot \mathrm{I}_{Lan})^2},\ \sqrt{6}\sqrt{\sum_{n=2}^{\infty}(X_{\mathrm{TCLCb}n}\cdot \mathrm{I}_{Lbn})^2},\right. \tag{4.36}$$
$$\left.\sqrt{6}\sqrt{\sum_{n=2}^{\infty}(X_{\mathrm{TCLC}cn}\cdot \mathrm{I}_{Lcn})^2}\right)$$

In (4.36), the I_{Lxn} is pre-measured from the loading and X_{TCLCxn} is obtained from Fig. 4.4. To guarantee a sufficient V_{DC} for TCLC-HAPF compensation within its compensation range, V_{DC} is designed based on worst case I_{Lxn} during the pre-measured period.

4.4.3 Design of L_c for Current Ripple Filtering

The purposed of L_c is to filter out the current ripple caused by the power switches of the active inverter part. The value of the L_c can be designed as:

$$L_c \geq \frac{V_{\mathrm{DC}}}{8\cdot f_s\cdot \Delta i_{L_c\max}} \tag{4.37}$$

where V_{DC} is the DC-link voltage, f_s is the switching frequency, $\Delta i_{L_c\max}$ is the maximum allowed output current ripple value.

4.4.4 Summary of TCLC-HAPF Parameter Design

Based on above discussions, the proposed parameter design method of TCLC-HAPF under three different types of loadings can be summarized in Table 4.2. Once the ranges of I_{Lxn} and Q_{Lx} are pre-measured, the TCLC-HAPF parameters can be designed accordingly by those equations in Table 4.2. In the following section, simulation case studies will be provided to show the validity of proposed parameter design method in comparison with the conventional design method.

4.5 Simulation Case Studies

In this section, simulation case studies are provided to verify the proposed unbalanced parameter design of TCLC-HAPF in Sect. 4.4 (Table 4.2) in comparison with the balanced parameter design of TCLC-HAPF in Sect. 4.3 (Table 4.1). The simulation studies are carried out by using PSCAD/EMTDC. The parameters of TCLC-HAPF will be designed according to Sects. 4.3 and 4.4 (Table 4.2) to compensate two sets of loads: (a) unbalanced inductive loads and (b) mixed inductive and capacitive loads. With RMS value of load voltage V_x= 110 V, the active and reactive power of two sets of testing loads are given as:

Table 4.1 Proposed balanced parameter design of TCLC-HAPF

Parameter	Expression	Equation no.
Design C_{PF}	$C_{PF}=\frac{Q_{Lx(MaxInd)}}{\omega^2\cdot Q_{Lx(MaxInd)}\cdot L_c+\omega\cdot V_x^2}$	(4.8)
Design L_{PF}	$L_{PF}=\frac{V_x^2+\omega L_c Q_{Lx(MaxCap)}}{\omega^2\cdot V_x^2\cdot C_{PF}-\omega\cdot Q_{Lx(MaxCap)}+\omega^3\cdot L_c\cdot C_{PF}\cdot Q_{Lx(MaxCap)}}$	(4.9)

where $Q_{Lx(MaxInd)}$ $(=-Q_{cx(MaxCap)})$ and $Q_{Lx(MaxCap)}$ $(=-Q_{cx(MaxInd)})$ can be found from (4.6) and (4.7)

Table 4.2 Proposed parameter design of TCLC-HAPF under three different types of loading

Parameter	Load types	C_{PF} design		L_{PF} design	
Design L_{PF} and C_{PF}	Unbalanced inductive loads	$\frac{1}{\omega^2 L_c-\omega X_{TCLCxf}}\le C_{PF}<\frac{1}{\omega^2 L_c}$	(4.26)	$L_{PF}<\frac{1}{\omega^2 C_{PF}}$	(4.24)
	Unbalanced capacitive loads	$C_{PF}<\frac{1}{\omega^2 L_c}$	(4.23)	$L_{PF}\le\frac{X_{TCLCxf}-\omega L_c}{\omega^2 X_{TCLCxf}C_{PF}-\omega^3 L_c C_{PF}+\omega}$	(4.30)
	Mixed inductive and capacitive loads	$\frac{1}{\omega^2 L_c-\omega X_{TCLCxf}}\le C_{PF}<\frac{1}{\omega^2 L_c}$	(4.26)	$L_{PF}\le\frac{X_{TCLCxf}-\omega L_c}{\omega^2 X_{TCLCxf}C_{PF}-\omega^3 L_c C_{PF}+\omega}$	(4.30)
	The final C_{PF} and L_{PF} can be expressed as: $C_{PF}=\max\left(C_{PF}\|_{X_{TCLCaf}},\ C_{PF}\|_{X_{TCLCbf}},\ C_{PF}\|_{X_{TCLCcf}}\right)$				(4.27)
	$L_{PF}=\min\left(L_{PF}\|_{X_{TCLCaf}},\ L_{PF}\|_{X_{TCLCbf}},\ L_{PF}\|_{X_{TCLCcf}}\right)$				(4.31)
Design L_c	$L_c\ge\frac{V_{DC}}{8\cdot f_s\cdot\Delta i_{L_c\max}}$				(4.37)
Design V_{DC}	$V_{DC}=\sqrt{V_{DCf}^2+V_{DCh}^2}$ $=\max\left(\sqrt{6}\sqrt{\sum_{n=2}^{\infty}(X_{TCLCan}\cdot I_{Lan})^2},\ \sqrt{6}\sqrt{\sum_{n=2}^{\infty}(X_{TCLCbn}\cdot I_{Lbn})^2},\ \sqrt{6}\sqrt{\sum_{n=2}^{\infty}(X_{TCLCcn}\cdot I_{Lcn})^2}\right)$				(4.36)
X_{TCLCxf} is obtained from (4.19) and x stands for phase a, b and c. $\begin{bmatrix}X_{TCLCaf}\\X_{TCLCbf}\\X_{TCLCcf}\end{bmatrix}=\begin{bmatrix}\frac{3V_x^2(Q_{Lc}-Q_{Lb}-Q_{La})^{-1}(Q_{Lb}-Q_{La}-Q_{Lc})^{-1}}{(Q_{Lc}-Q_{Lb}-Q_{La})^{-1}+(Q_{La}-Q_{Lb}-Q_{Lc})^{-1}+(Q_{Lb}-Q_{La}-Q_{Lc})^{-1}}\\\frac{3V_x^2(Q_{Lc}-Q_{Lb}-Q_{La})^{-1}(Q_{La}-Q_{Lb}-Q_{Lc})^{-1}}{(Q_{Lc}-Q_{Lb}-Q_{La})^{-1}+(Q_{La}-Q_{Lb}-Q_{Lc})^{-1}+(Q_{Lb}-Q_{La}-Q_{Lc})^{-1}}\\\frac{3V_x^2(Q_{La}-Q_{Lb}-Q_{Lc})^{-1}(Q_{Lb}-Q_{La}-Q_{Lc})^{-1}}{(Q_{Lc}-Q_{Lb}-Q_{La})^{-1}+(Q_{La}-Q_{Lb}-Q_{Lc})^{-1}+(Q_{Lb}-Q_{La}-Q_{Lc})^{-1}}\end{bmatrix}$					(4.19)

Table 4.3 Parameter design for (a) unbalanced inductive loads compensation

	Balanced design method (Sect. 4.3)	Unbalanced design method (Sect. 4.4)
Required X_{TCLCxf}	$\begin{bmatrix} X_{TCLCaf} \\ X_{TCLCbf} \\ X_{TCLCcf} \end{bmatrix} = \begin{bmatrix} \frac{V_x^2}{Q_{La}} \\ \frac{V_x^2}{Q_{Lb}} \\ \frac{V_x^2}{Q_{Lc}} \end{bmatrix} = \begin{bmatrix} -28.0 \\ -41.6 \\ -25.7 \end{bmatrix}$	$\begin{bmatrix} X_{TCLCaf} \\ X_{TCLCbf} \\ X_{TCLCcf} \end{bmatrix} = \begin{bmatrix} \frac{3V_x^2(Q_{Lc}-Q_{Lb}-Q_{La})^{-1}(Q_{Lb}-Q_{La}-Q_{Lc})^{-1}}{(Q_{Lc}-Q_{Lb}-Q_{La})^{-1}+(Q_{La}-Q_{Lb}-Q_{Lc})^{-1}+(Q_{Lb}-Q_{La}-Q_{Lc})^{-1}} \\ \frac{3V_x^2(Q_{Lc}-Q_{Lb}-Q_{La})^{-1}(Q_{La}-Q_{Lb}-Q_{Lc})^{-1}}{(Q_{Lc}-Q_{Lb}-Q_{La})^{-1}+(Q_{La}-Q_{Lb}-Q_{Lc})^{-1}+(Q_{Lb}-Q_{La}-Q_{Lc})^{-1}} \\ \frac{3V_x^2(Q_{La}-Q_{Lb}-Q_{Lc})^{-1}(Q_{Lb}-Q_{La}-Q_{Lc})^{-1}}{(Q_{Lc}-Q_{Lb}-Q_{La})^{-1}+(Q_{La}-Q_{Lb}-Q_{Lc})^{-1}+(Q_{Lb}-Q_{La}-Q_{Lc})^{-1}} \end{bmatrix} = \begin{bmatrix} -27.2 \\ -50.5 \\ -20.9 \end{bmatrix}$
C_{PF} Design	[107 μF, 74 μF, 116 μF]$_{max}$ $\leq C_{PF} \leq$ 2026 μF	[111 μF, 61 μF, 141 μF]$_{max}$ $\leq C_{PF} \leq$ 2026 μF
L_{PF} Design	$L_{PF} \leq$ [94 mH, 137 mH, 86 mH]$_{min}$	$L_{PF} \leq$ [92 mH, 165 mH, 71 mH]$_{min}$

Table 4.4 Parameter design for (b) mixed inductive and capacitive loads compensation

	Balanced design method (Sect. 4.3)	Unbalanced design method (Sect. 4.4)
Required X_{TCLCxf}	$\begin{bmatrix} X_{TCLCaf} \\ X_{TCLCbf} \\ X_{TCLCcf} \end{bmatrix} = \begin{bmatrix} \frac{V_x^2}{Q_{La}} \\ \frac{V_x^2}{Q_{Lb}} \\ \frac{V_x^2}{Q_{Lc}} \end{bmatrix} = \begin{bmatrix} -45.2 \\ 49.4 \\ -56.5 \end{bmatrix}$	$\begin{bmatrix} X_{TCLCaf} \\ X_{TCLCbf} \\ X_{TCLCcf} \end{bmatrix} = \begin{bmatrix} \frac{3V_x^2(Q_{Lc}-Q_{Lb}-Q_{La})^{-1}(Q_{Lb}-Q_{La}-Q_{Lc})^{-1}}{(Q_{Lc}-Q_{Lb}-Q_{La})^{-1}+(Q_{La}-Q_{Lb}-Q_{Lc})^{-1}+(Q_{Lb}-Q_{La}-Q_{Lc})^{-1}} \\ \frac{3V_x^2(Q_{Lc}-Q_{Lb}-Q_{La})^{-1}(Q_{La}-Q_{Lb}-Q_{Lc})^{-1}}{(Q_{Lc}-Q_{Lb}-Q_{La})^{-1}+(Q_{La}-Q_{Lb}-Q_{Lc})^{-1}+(Q_{Lb}-Q_{La}-Q_{Lc})^{-1}} \\ \frac{3V_x^2(Q_{La}-Q_{Lb}-Q_{Lc})^{-1}(Q_{Lb}-Q_{La}-Q_{Lc})^{-1}}{(Q_{Lc}-Q_{Lb}-Q_{La})^{-1}+(Q_{La}-Q_{Lb}-Q_{Lc})^{-1}+(Q_{Lb}-Q_{La}-Q_{Lc})^{-1}} \end{bmatrix} = \begin{bmatrix} -36.3 \\ 88.2 \\ -23.2 \end{bmatrix}$
C_{PF} Design	[68 μF, −67 μF, 54 μF]$_{max}$ $\leq C_{PF} \leq$ 2026 μF	[84 μF, −36 μF, 129 μF]$_{max}$ $\leq C_{PF} \leq$ 2026 μF
L_{PF} Design	$L_{PF} \leq$ [156 mH, 54 mH, 195 mH]$_{min}$	$L_{PF} \leq$ [134 mH, 52 mH, 324 mH]$_{min}$

$$\begin{bmatrix} P_{La} + jQ_{La} \\ P_{Lb} + jQ_{Lb} \\ P_{Lc} + jQ_{Lc} \end{bmatrix} = \begin{bmatrix} 435 + j432 \\ 571 + j291 \\ 629 + j470 \end{bmatrix} \tag{4.38a}$$

$$\begin{bmatrix} P_{La} + jQ_{La} \\ P_{Lb} + jQ_{Lb} \\ P_{Lc} + jQ_{Lc} \end{bmatrix} = \begin{bmatrix} 185 + j268 \\ 425 - j245 \\ 745 + j214 \end{bmatrix} \tag{4.38b}$$

Based on load power given in (4.38a), the parameter design procedures of the proposed methods are summarized in Tables 4.3 and 4.4.

According to constructed loading harmonic current (I_{Lxn}) in Sects. 4.5 and 4.6, the V_{DC} designed to be 60 V based on (4.36) can be sufficient for nonlinear loads compensations. And according to simulation and experimental settings (f_s = 5 kHz and $\Delta i_{L_c \max}$ = 0.4A), the L_c designed to be 5 mH can satisfy (4.37) (>3.75 mH). From above discussion, Tables 4.3 and 4.4 and matching with the experimental components and settings simultaneously, the final designed parameters for TCLC-HAPF for both loading cases are shown in Table 4.5.

Table 4.5 Design of TCLC-HAPF by applying the balanced design method (Sect. 4.3) and unbalanced design method (Sect. 4.4)

	C_{PF} (μF)	L_{PF} (mH)	V_{DC} (V)(4.36)	L_c (mH)(4.37)
Balanced design method	120	30	60	5
Unbalanced design method	160	30	60	5

By following the parameter designed methods in Tables 4.3 and 4.4 and matching the experimental results simultaneously, the final designed parameters of TCLC for both loading cases are shown in Table 4.5. With the consideration of IEEE standard 519-2014 [21], the Total Demand Distortion (TDD) = 15% and I_{SC}/I_L in 100 < 1000 scale at a typical case, the nominal rate current is assumed to be equal to fundamental load current at the worst case analysis, which results in $THD = TDD = 15\%$. Therefore, this chapter evaluates the compensation performance by setting $THD < 15\%$.

To compensate the two sets of testing loads as shown in (4.38a), Figs. 4.5, 4.6, 4.7 and 4.8 show the simulation compensation results of source voltage and current, source reactive and active power by using (a) TCLC-HAPF designed by balanced design method (Sect. 4.3), (b) TCLC-HAPF designed by unbalanced design method (Sect. 4.4) respectively. Figures 4.9 and 4.10 illustrate the phasor diagrams of source voltage and current. Finally, the corresponding simulation results for two sets of testing loads are summarized in Table 4.6.

After the TCLC-HAPF compensation with balanced design method (Sect. 4.3), the fundamental source reactive power of *phase a* and *c* (unbalanced inductive loads in Fig. 4.5) and *phase c* (mixed inductive and capacitive loads in Fig. 4.7) cannot be compensated to be close to zero, and the fundamental source active power are not balanced for both loadings cases. The source voltage and current are not all in phase with each other after compensations as shown in Figs. 4.9 and 4.10. From Figs. 4.5a, 4.7a and Table 4.6, the worst phase source current total harmonic distortions (THD_{isx}) after TCLC-HAPF (designed by balanced method) compensation are 11.5% for unbalanced inductive loads case and 18.5% for mixed inductive and capacitive loads case, which do not satisfy the IEEE standard ($THD < 15\%$) [21]. The source current unbalanced factor (UBI_{fS}) is 12.0% for unbalanced inductive loads compensation and 24.1% for mixed inductive and capacitive loads compensation, respectively.

With the proposed parameter design method (Sect. 4.4), the TCLC-HAPF can balance the three-phase active power and compensate the reactive power for both loading cases as shown in Figs. 4.6 and 4.8 and Table 4.6. As shown in Figs. 4.9 and 4.10, the source voltage and current are in phase with each other after TCLC-HAPF compensation. Moreover, the THD_{isx} has been reduced to less than 6% for unbalanced inductive loading compensation and 10% for mixed inductive and capacitive loading compensation, which can satisfy the IEEE standard [21]. In Table 4.6 it is shown that the source current unbalanced factor (UBI_{fS}) is less than 3.0% after compensation by TCLC-HAPF with proposed design method.

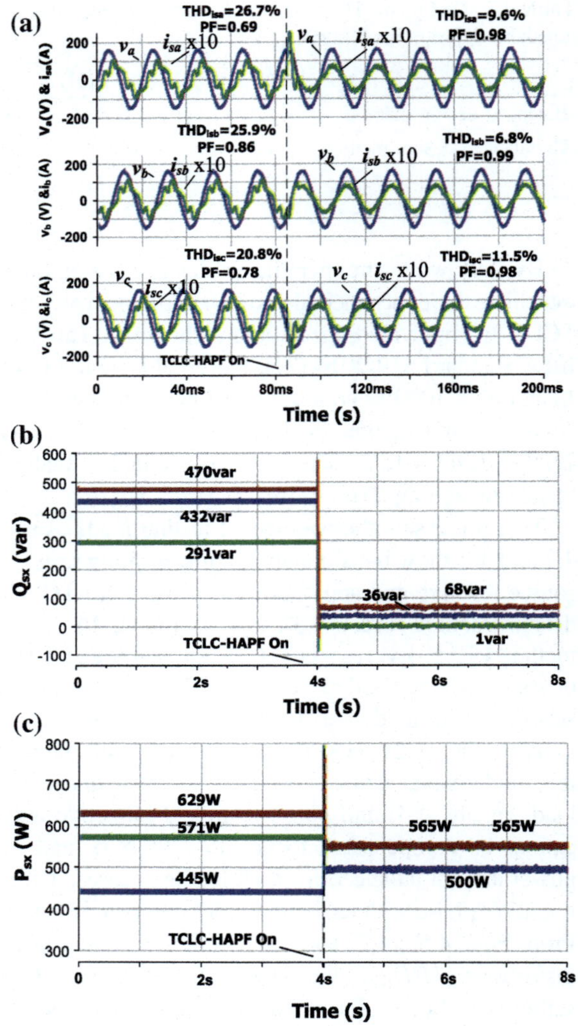

Fig. 4.5 Dynamic performance by using TCLC-HAPF with balanced design method (Sect. 4.3) for unbalanced inductive loads compensation: **a** v_{sx} and i_{sx}, **b** Q_{sx} and **c** P_{sx}

Based on simulation case studies, it can be seen that the TCLC-HAPF with balanced design method cannot fully compensate reactive power, balance active power and achieve acceptable current harmonic compensation results. With the proposed parameter designed method, the TCLC-HAPF can fully compensate reactive power, balance active power and obtain acceptable current harmonic compensation results.

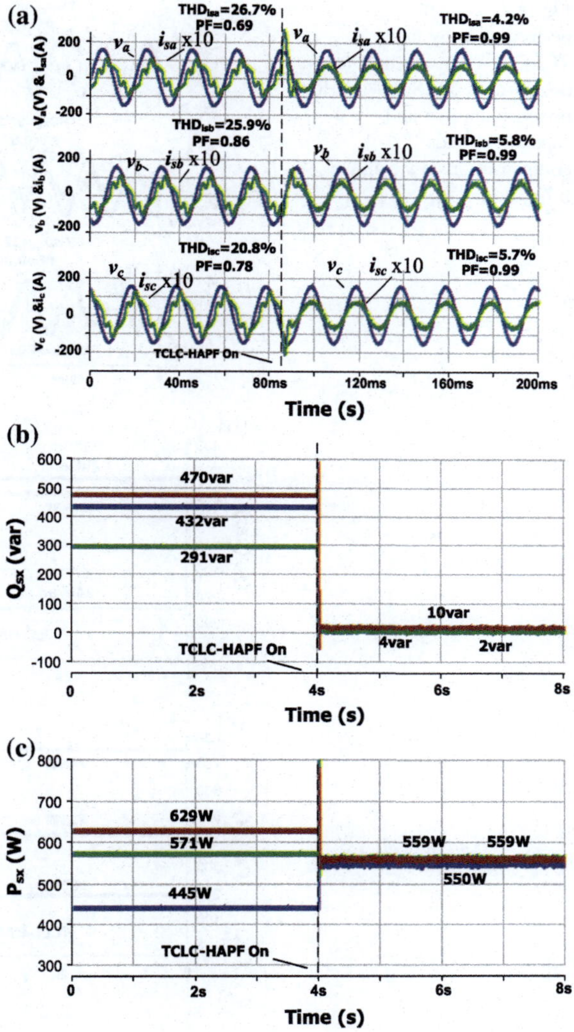

Fig. 4.6 Dynamic performance by using TCLC-HAPF with proposed design method (Sect. 4.4) for unbalanced inductive loads compensation: **a** v_{sx} and i_{sx}, **b** Q_{sx} and **c** P_{sx}

4.6 Experimental Results

In this section, an 110 V–5 kVA experimental prototype of TCLC-HAPF with proposed design method is built in laboratory. Its component values and the testing loads are the same as the simulation case studies.

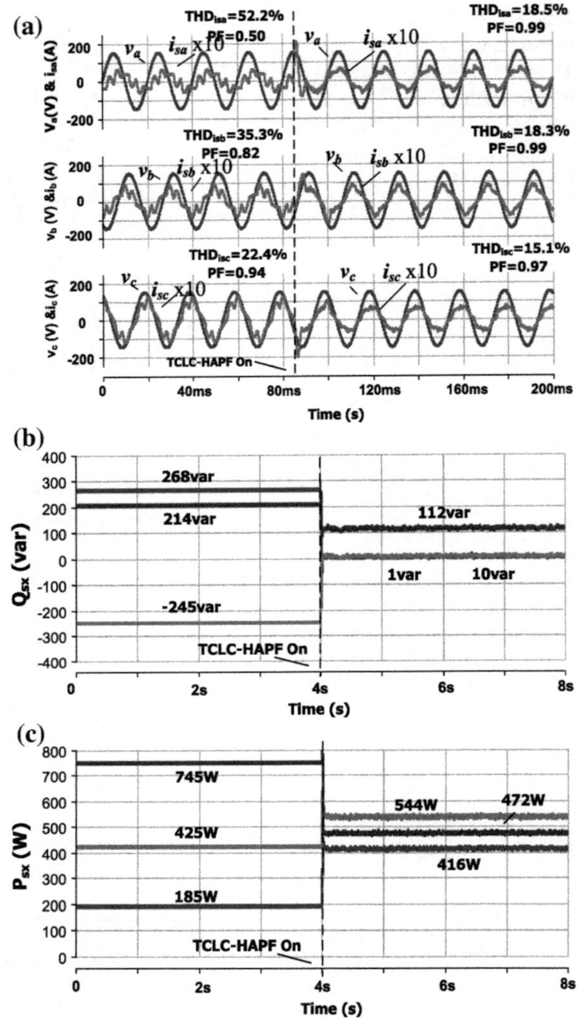

Fig. 4.7 Dynamic performance by using TCLC-HAPF with balanced design method (Sect. 4.3) for mixed inductive and capacitive loads compensation: **a** v_{sx} and i_{sx}, **b** Q_{sx} and **c** P_{sx}

From Figs. 4.11 and 4.12 and Table 4.7, with the proposed parameter design method, the TCLC-HAPF can compensate the fundamental reactive power and balance the active power. From Figs. 4.11a and 4.12a, after TCLC-HAPF compensation, the worst phase PF has been compensated from original 0.56 for

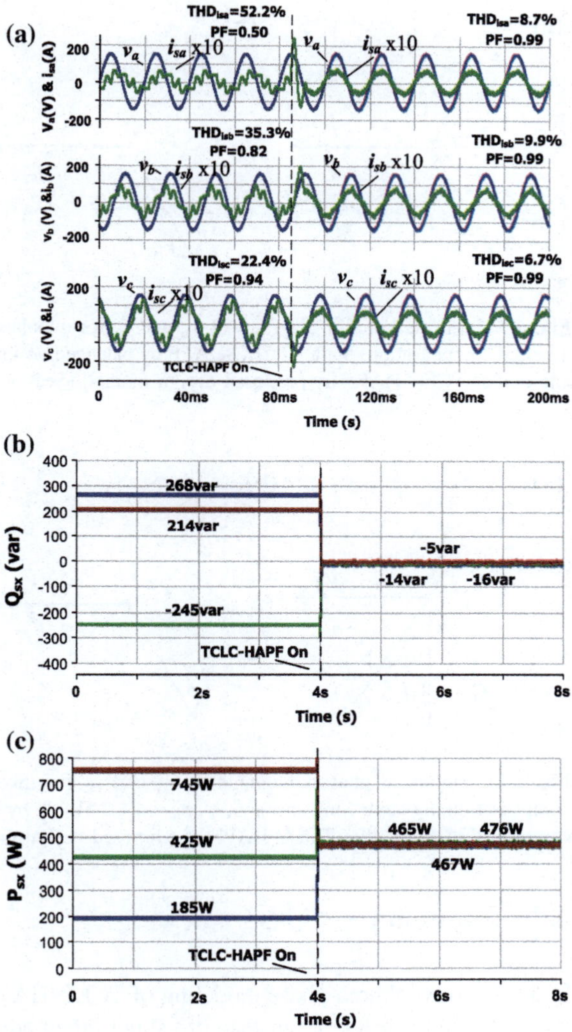

Fig. 4.8 Dynamic performance by using TCLC-HAPF with proposed design method (Sect. 4.4) for mixed inductive and capacitive loads compensation: **a** v_{sx} and i_{sx}, **b** Q_{sx} and **c** P_{sx}

unbalanced inductive loads and 0.60 for mixed inductive and capacitive loads to 0.99. The worst phase THD_{isx} have been compensated to 9.5 and 10.6% from the original 22.7% for unbalanced inductive loads and 38.6% for mixed inductive and capacitive loads, which can meet the IEEE standard (THD < 15%) [21]. As shown in Fig. 4.13, the source voltage and current are in phase with each other after TCLC-HAPF compensation. Moreover, the UBI_{fS} is less than 3.0% after compensation for two sets of loading.

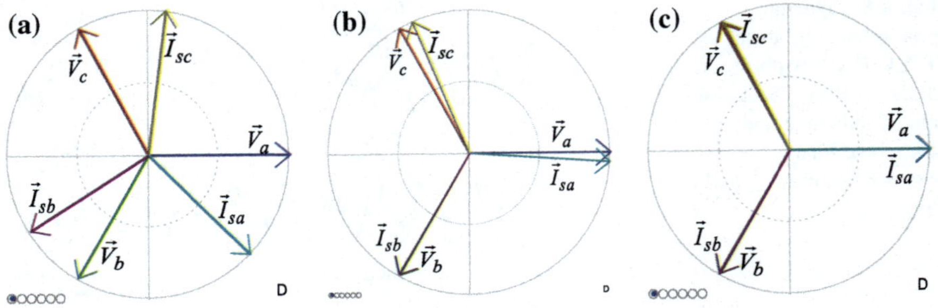

Fig. 4.9 Simulated phasor diagrams of v_{sx} and i_{sx} for unbalanced inductive loads compensation: **a** before compensation, **b** after TCLC-HAPF by balanced design method (Sect. 4.3) compensation and **c** after TCLC-HAPF by proposed design method (Sect. 4.4) compensation

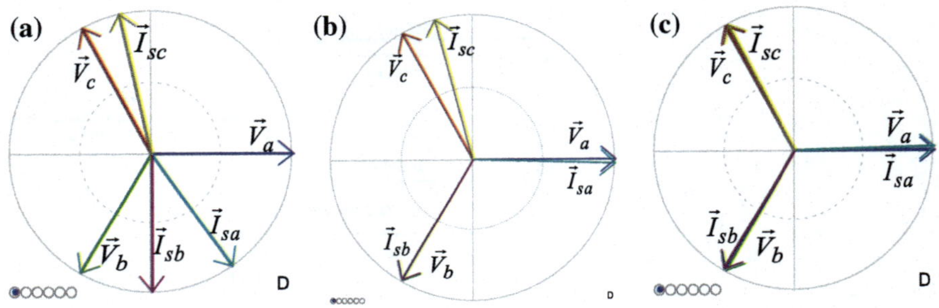

Fig. 4.10 Simulated phasor diagrams of v_{sx} and i_{sx} for mixed inductive and capacitive loads compensation: **a** before compensation, **b** after TCLC-HAPF by balanced design method (Sect. 4.3) compensation and **c** after TCLC-HAPF by proposed design method (Sect. 4.4) compensation

4.7 Summary

In this chapter, a three-phase modeling of TCLC-HAPF under balanced/unbalanced compensation is proposed instead of using of the traditional equivalent single-phase modeling. Based on the three-phase modeling, the parameter design method for TCLC-HAPF is proposed to balance the source active power, compensate loading reactive power and harmonic current. The balanced parameter design method of TCLC-HAPF is proposed based on maximum reactive power compensation among the three phases, while the inner-flowed unbalanced power has not be considered.

Table 4.6 Simulation results for unbalanced loads compensation before and after TCLC-HAPF compensation

			i_{sx}(A)	P_{sx}(W)	Q_{sx}(var)	*PF*	THD_{isx}(%)	UBI_{fS}(%)	V_{DC}(V)
Unbalanced Inductive loads	Before Comp.	A	5.6	445	432	0.69	26.7	25.8	--
		B	5.8	571	291	0.86	25.9		
		C	7.2	629	470	0.78	20.8		
	TCLC-HAPF (Balanced method)	A	4.5	500	36	0.98	9.6	12.0	60
		B	5.1	565	1	0.99	6.8		
		C	5.1	565	68	0.98	11.5		
	TCLC-HAPF (Proposed method)	A	4.8	550	4	0.99	4.2	2.0	60
		B	4.9	559	2	0.99	5.8		
		C	4.9	559	10	0.99	5.7		
Mixed Inductive& Capacitive loads	Before Comp.	A	3.1	185	268	0.50	52.2	76.5	--
		B	4.6	425	-245	0.82	35.3		
		C	6.8	745	214	0.94	22.4		
	TCLC-HAPF (Balanced method)	A	4.0	416	1	0.99	18.5	24.1	60
		B	5.1	544	10	0.99	18.3		
		C	4.6	472	112	0.97	15.1		
	TCLC-HAPF (Proposed method)	A	4.4	465	-5	0.99	8.7	2.2	60
		B	4.5	476	-14	0.99	9.9		
		C	4.5	467	-16	0.99	6.7		

Note The shades areas mean undesirable results

The unbalanced parameter design method has taken consideration of the inner-flowed unbalanced power among each phase under three different types of loading cases: unbalanced inductive loads, unbalanced capacitive loads, and mixed inductive and capacitive loads. Though the simulation and experimental results, it is proved that the proposed unbalanced parameter design method (Sect. 4.4) for TCLC-HAPF can effectively balance the source active power, compensate load reactive power and harmonic current under unbalanced loading, in which the balanced parameter design method (Sect. 4.3) cannot achieve satisfactory compensation during unbalanced loading case.

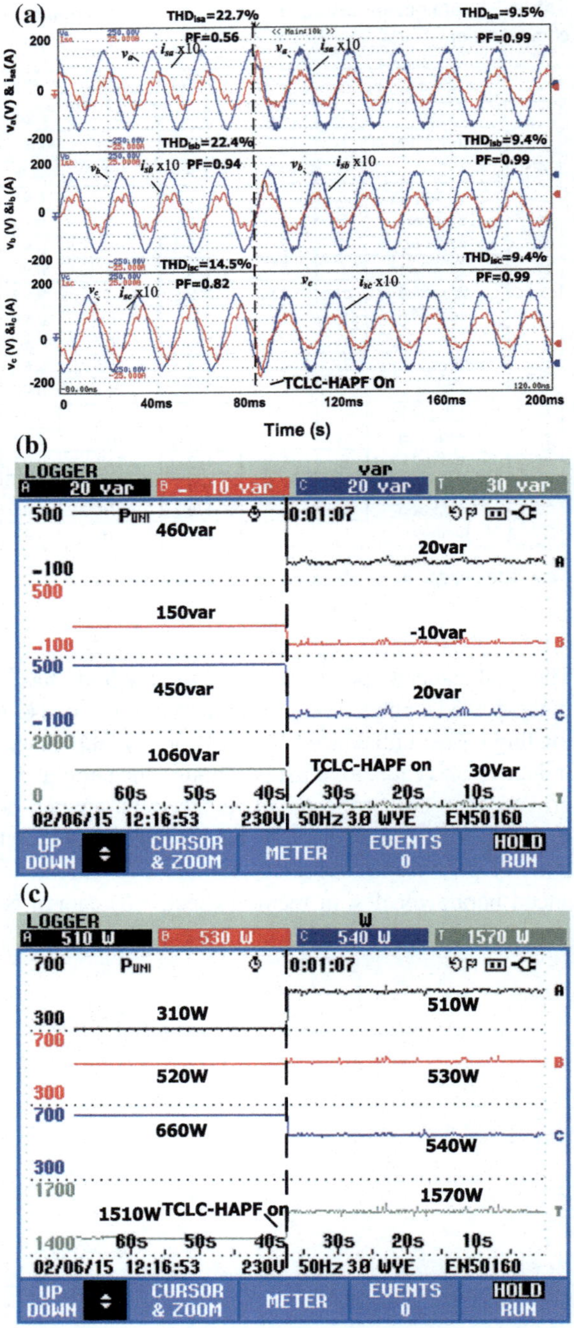

Fig. 4.11 Dynamic unbalanced inductive loads compensation by using TCLC-HAPF with proposed design: **a** v_{sx} and i_{sx}, **b** Q_{sx} and **c** P_{sx}

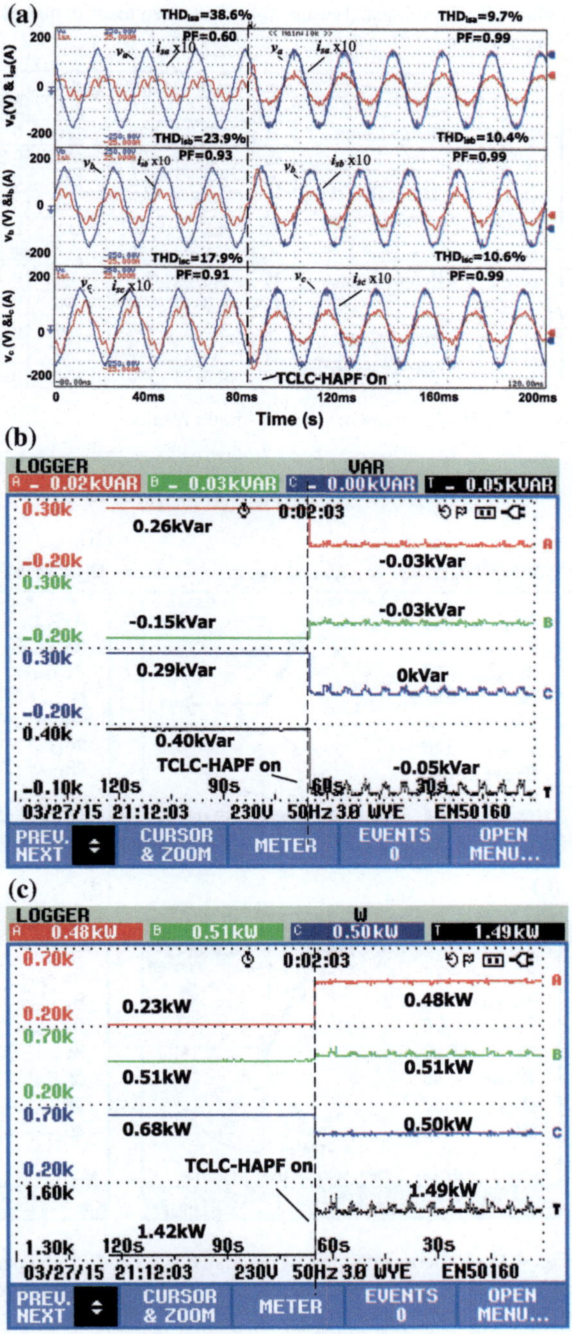

Fig. 4.12 Dynamic mixed inductive and capacitive loads compensation by using TCLC-HAPF with proposed design: **a** v_{sx} and i_{sx}, **b** Q_{sx} and **c** P_{sx}

Table 4.7 Experimental results for unbalanced loads compensation before and after TCLC-HAPF compensation

			i_{sx}(A)	P_{sx}(W)	Q_{sx}(var)	*PF*	THD_{isx}(%)	UBI_{fS}(%)	V_{DC}(V)
Unbalanced Inductive loads	Before Comp.	A	5.1	310	460	0.56	22.7	43.1	--
		B	4.9	520	150	0.94	22.4		
		C	7.4	600	450	0.82	14.5		
	TCLC-HAPF (Proposed method)	A	4.6	510	20	0.99	9.5	4.3	60
		B	4.6	530	-10	0.99	9.4		
		C	4.8	540	20	0.99	9.4		
Mixed Inductive& Capacitive loads	Before Comp.	A	3.1	230	260	0.60	38.6	54.3	--
		B	4.6	510	-150	0.93	23.9		
		C	6.8	680	290	0.91	17.9		
	TCLC-HAPF (Proposed method)	A	4.5	480	-30	0.99	9.7	2.2	60
		B	4.5	510	-30	0.99	10.4		
		C	4.6	500	0	0.99	10.6		

Note The shades areas mean undesirable results

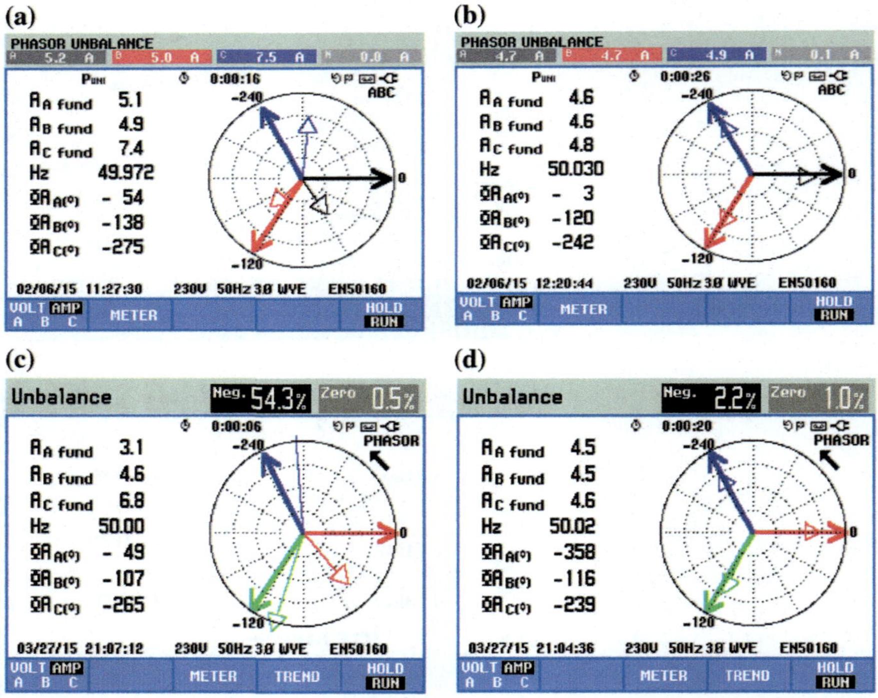

Fig. 4.13 Experimental phasor diagrams of v_{sx} and i_{sx} compensation: **a** before compensation for unbalanced inductive loads, **b** after TCLC-HAPF compensation for unbalanced inductive loads, **c** before compensation for mixed inductive and capacitive loads and **d** after TCLC-HAPF compensation for mixed inductive and capacitive loads

References

1. T. Baldwin, T. Hogans, S. Henry, F. Renovich, P. Latkovic, Reactive power compensation for voltage control at resistance welders. IEEE Trans. Ind. Appl. **41**(6), 1485–1492 (2005)
2. Y.C. Chang, Multi-objective optimal svc installation for power system loading margin improvement. IEEE Trans. Power Systems **27**(2), 984–992 (2012)
3. S. Morello, T.J. Dionise, T.L. Mank, Comprehensive analysis to specify a static var compensator for an electric arc furnace upgrade. IEEE Trans. Ind. Appl. **51**(6), 4840–4852 (2015)
4. N. Daratha, B. Das, J. Sharma, Coordination between OLTC and SVC for voltage regulation in unbalanced distribution system distributed generation. IEEE Trans. Power Syst. **29**(1), 289–299 (2014)
5. H. Nian, Y. Shen, H. Yang, Y. Quan, Flexible grid connection technique of voltage-source inverter under unbalanced grid conditions based on direct power control. IEEE Trans. Ind. Appl. **51**(5), 4041–4050 (2015)
6. P.R. Martinez-Rodriguez, G. Escobar, A.A. Valdez-Fernandez, M. Hernandez-Gomez, J.M. Sosa, Direct power control of a three phase rectifier based on positive sequence detection. IEEE Trans. Ind. Electron. **61**(8), 4084–4092 (2014)
7. Z.X. Zou, K.L. Zhou, Z. Wang, M. Cheng, Frequency-adaptive fractional-order repetitive control of shunt active power filters. Trans. Ind. Electron. **62**(3), 1659–1668 (2015)
8. M. Qasim, P. Kanjiya, V. Khadkikar, Artificial-neural-network-based phase-locking scheme for active power filters. Trans. Ind. Electron. **61**(8), 3857–3866 (2014)
9. H. Hu, Y. Xing, Design considerations and fully digital implementation of 400-Hz active power filter for aircraft applications. IEEE Trans. Ind. Electron. **61**(8), 3823–3834 (2014)
10. J. Dixon, Y. del Valle, M. Orchard, M. Ortuzar, L. Moran, C. Maffrand, A full compensating system for general loads, based on a combination of thyristor binary compensator, and a PWM-IGBT active power filter. IEEE Trans. Ind. Electron. **50**(5), 982–989 (2003)
11. A. Luo, Z. Shuai, W. Zhu, Z. John Shen, Combined system for harmonic suppression and reactive power compensation. IEEE Trans. Ind. Electron. **56**(2), 418–428 (2009)
12. A. Luo, S. Peng, C. Wu, J. Wu, Z. Shuai, Power electronic hybrid system for load balancing compensation and frequency-selective harmonic suppression. IEEE Trans. Ind. Electron. **59** (2), 723–732 (2012)
13. S. Rahmani, A. Hamadi, K. Al-Haddad, A combination of shunt hybrid power filter and thyristor-controlled reactor for power quality. IEEE Trans. Ind. Electron. **61**(5), 2152–2164 (2014)
14. L. Wang, C.S. Lam, M.C. Wong, An unbalanced control strategy for a thyristor controlled LC-coupling hybrid active power filter (SVC-HAPF) in three-phase three-wire systems. IEEE Trans. Power Electron. **32**(2), 1056–1069 (2017)
15. L. Wang, C.S. Lam, M.C. Wong, Hardware and software design of a low dc-link voltage and wide compensation range thyristor controlled LC-coupling hybrid active power filter, in *TENCON 2015 IEEE Region 10 Conference Proceedings* (2015)
16. C.S. Lam, L. Wang, S.I. Ho, M.C. Wong, Adaptive thyristor controlled LC—hybrid active power filter for reactive power and current harmonics compensation with switching loss reduction. IEEE Trans. Power Electron. **32**(10), 7577–7590 (2017)

17. L. Wang, C.S. Lam, M.C. Wong, Selective compensation of distortion, unbalanced and reactive power of a thyristor controlled LC-coupling hybrid active power filter (SVC-HAPF). IEEE Trans. Power Electron. **32**(12), 9065–9077 (2017)
18. L. Wang, C.S. Lam, M.C. Wong, A hybrid-STATCOM with wide compensation range and low dc-link voltage. IEEE Trans. Ind. Electron. **63**(6), 3333–3343 (2016)
19. R.B. Gonzatti et al., Smart impedance: a new way to look at hybrid filters. IEEE Trans. Smart Grid **7**(2), 837–846 (2016)
20. C.H. Silva et al., Smart impedance: expanding the hybrid active series power filter concept, in *Proceedings of Annual Conference IEEE Industrial Electronics Society (IECON)*, Montreal, QC, Canada, (2012), pp. 1416–1421
21. IEEE Standard 519-2014, IEEE recommended practices and requirements for harmonic control in electrical power systems (2014)

Chapter 5
Proposed Unbalanced Control Strategy for Thyristor Controlled LC-Coupling Hybrid Active Power Filter (TCLC-HAPF)

Abstract This chapter proposes a control strategy for a three-phase three-wire thyristor controlled LC-coupling hybrid active power filter (TCLC-HAPF), which can balance active power and compensate reactive power and harmonic current under unbalanced loading. Compared with TCLC-HAPF with conventional control strategy, active power filters (APFs) and hybrid active power filters (HAPFs), which either fail to perform satisfactory compensation or require high rating active inverter part for unbalanced compensation, a control strategy was proposed for TCLC-HAPF to operate with a small rating active inverter part for a variety of loads with satisfactory performance. The control idea is to provide different firing angles for each phase of the thyristor controlled LC-coupling part (TCLC) to balance active power and compensate reactive power, while the active inverter part aims to compensate harmonic current. Firstly, the required different TCLC impedances are deduced. Then, independent firing angles referenced to the phase angle of voltage across TCLC are calculated. After angle transformations, final firing angles referenced to phase angle of load voltage are obtained. In this chapter, a novel controller for TCLC-HAPF under unbalanced loading is proposed. Simulation and experimental results are provided to verify the effectiveness of the proposed controller in comparison with a state-of-the-art controller.

Keywords Active power · Current harmonics · Hybrid active power filter (HAPF) Reactive power · Thyristor controlled LC-coupling hybrid active power filter (TCLC-HAPF) · Unbalanced compensation

5.1 Introduction

When unbalanced nonlinear loads are connected to the three-phase system, a number of current quality problems can come out such as unbalanced current, low power factor, and harmonic pollution. Implementation of power filters like Static Var Compensators (SVCs), active power filters (APFs) hybrid active power filter (HAPF) is one of the solutions for power quality problems. Many control

L. Wang et al., *Adaptive Hybrid Active Power Filters*, Power Systems,
https://doi.org/10.1007/978-981-10-8827-8_5

techniques have been proposed to improve the performance of SVCs, APFs and HAPFs and solve the unbalanced problems [1–17]. The different method can be classified as below and explained as:

(1) Eliminate the oscillating power components though the instantaneous power theories [1–6];
(2) Compensate the negative- and zero-sequence components of the load current [7–15];
(3) Compensate reactive power and balance active power through the deduced active, reactive and apparent power' formulas (or power analysis) [16, 17].

To eliminate the oscillating power, Akagi et al. [1] proposed well-known instantaneous p-q control method. In order to adapt instantaneous p-q control method under different voltage conditions (distorted, unbalanced, etc.), many other control techniques were further developed such as: d-q control method [2–7], p-q-r control method [8–10], Lyapunov function-based control method [11], etc. However, those instantaneous power control methods [1–11] are dedicated to inverter/converter-based structures. Another popular control method for APFs and HAPFs is to balance the system by compensating the negative- and zero-sequence current components (+, − and 0 sequences), as the oscillating power/voltage/current can be analytically expressed as positive-, negative- and zero-sequence components [12]. However, the sequence components introduced by harmonics have not been taken into consideration. To solve this problem, the authors in [13–15] combine the above instantaneous control methods with the +, − and 0 sequence control method, but the computation steps increase a lot, thus significantly increasing the control complexity. Different with the above two types of methods, Czarnecki et al. [16, 17] proposed a power analysis control method based on the theory of current' physical component (CPC) to compensate the reactive power in unbalanced three-phase four-wire system. However, the active power remains unbalanced after this power analysis method compensation, which means the unbalance power cannot be completely eliminated.

With all the above control methods, both APFs and HAPFs can compensate the reactive, harmonic and unbalanced power. However, both APFs and HAPFs probably require high active inverter rating (high initial cost and switching loss) to perform unbalanced power compensation due to the inductive coupling structures of APFs and the narrow compensation range limitations of HAPFs. And, the above control methods [1–17] are all focused on controlling the active inverter part. In 2014, Rahmani et al. [18] proposed the thyristor controlled *LC*-coupling hybrid active power filter (TCLC-HAPF) which can operate with a low rating active inverter part for reactive power and harmonic current compensation. Different with APFs and HAPFs, the TCLC-HAPF contains both controllable thyristor controlled LC-coupling part (TCLC) and active inverter part. To control TCLC-HAPF, a state-of-the-art control method is proposed in [18] to reduce the steady-state error of the TCLC part and improve the performance of current tracking and voltage regulation of the active inverter part. However, the control method proposed in [18] was designed based on the assumption of balanced loading condition. If this control method is applied to TCLC-HAPF for unbalanced loading compensation, it either fails to perform acceptable current quality compensation or requires a high rating active inverter part for compensation, which results in increasing the system initial cost, switching loss and switching noise.

Therefore, this chapter proposes a hybrid unbalanced control method for the TCLC-HAPF, which can balance source side active power and compensate the reactive power and harmonic current with low dc-link voltage characteristic. The control idea is to generate different firing angles to each phase of the TCLC in order to compensate reactive power and balance active power, and the active inverter part aims to compensate the harmonic current. As a result, the voltage rating of the active inverter part can be low, and consequently the system initial cost and switching noise can be significantly reduced. Moreover, the TCLC-HAPF can still maintain a wide compensation range with acceptable performance comparable with the conventional solutions [1–17].

In this chapter, a brief introduction of research background and motivation is covered in Sect. 5.1. In Sect. 5.2, the circuit configuration of the TCLC-HAPF is presented and discussed. In Sect. 5.3, the hybrid controller for the TCLC-HAPF under unbalanced loading is proposed and explained. Afterwards, simulation and experimental results are given to verify the effectiveness of the proposed unbalanced control strategy for the TCLC-HAPF in comparison with the state-of-the-art control method [18] in Sect. 5.4. Finally, summary is drawn in Sect. 5.5.

5.2 Circuit Configuration of Three-Phase Three-Wire TCLC-HAPF

The circuit configuration of a three-phase three-wire TCLC-HAPF is given in Fig. 5.1, where the subscript 'x' denotes phase $x = a, b, c$. v_{sx}, v_x and v_{invx} are the system voltage, load voltage and inverter output voltage respectively; L_s is the system inductance; i_{sx}, i_{Lx} and i_{cx} are the source, load and compensated current respectively. C_{DC} and V_{DC} are DC-link capacitor and dc-link voltage; L_c is the coupling inductor; L_{PF} and C_{PF} are the TCLC part inductor and capacitor.

In this topology, the TCLC part and the active inverter part can complement each other's disadvantages. As the TCLC part offers the reactive power compensation range and provides a large voltage drop between load voltage and inverter voltage, the voltage rating of the active inverter part can be significantly reduced. On the other hand, the active inverter part can solve the inherent problems of using TCLC alone, such as inrush current, resonance problem, noise of thyristors turning on/off, mistuning of firing angles and low harmonic compensation ability. Based on the circuit configuration as shown in Fig. 5.1, the unbalanced control strategy for the TCLC-HAPF will be proposed in next section.

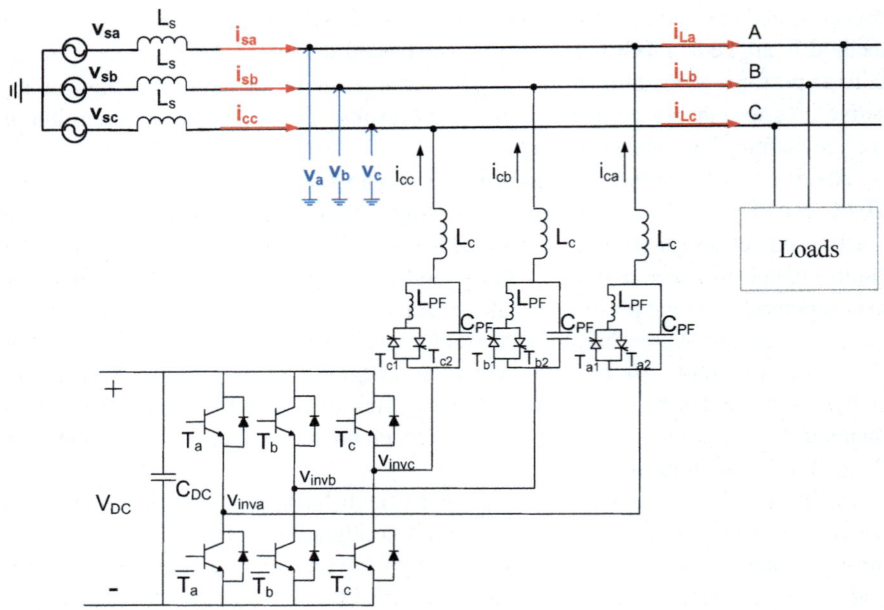

Fig. 5.1 Circuit configuration of a three-phase three-wire TCLC-HAPF

5.3 Proposed Unbalanced Control Strategy for TCLC-HAPF

The purposes of the proposed unbalanced control strategy can be described as follows: the TCLC part is controlled to balance active power and compensate reactive power, while the active inverter part aims to compensate harmonic current. The equivalent fundamental circuit models of the TCLC-HAPF for power analysis are illustrated in Fig. 5.2, where the subscripts '*f*' denotes the fundamental frequency component. In this chapter, V_{sxf} and V_{xf} are assumed to be pure sinusoidal without harmonic components ($V_x = V_{sxf} = V_{xf}$) for simplicity.

Figure 5.2 is used to calculate the required impedances and the corresponding firing angles for each phase of the TCLC part in order to balance and compensate active and reactive power. In Fig. 5.2a, the active inverter can be treated as a controlled voltage source, and the required fundamental inverter voltage (V_{invxf}) depends on the TCLC impedance [19–22]. If the TCLC impedance is perfectly matched with the load impedance, then the required V_{invxf} can be equal to zero. In this chapter, it is assumed that the TCLC is controlled to be perfectly matched with the loading to simplify the following analysis, thus $V_{invxf} = 0$, then the required TCLC impedance can be calculated based on Fig. 5.2b. In the following, the proposed hybrid control strategy for the TCLC-HAPF under unbalanced loading compensation will be presented and explained in three parts: Sect. 5.3.1: TCLC part

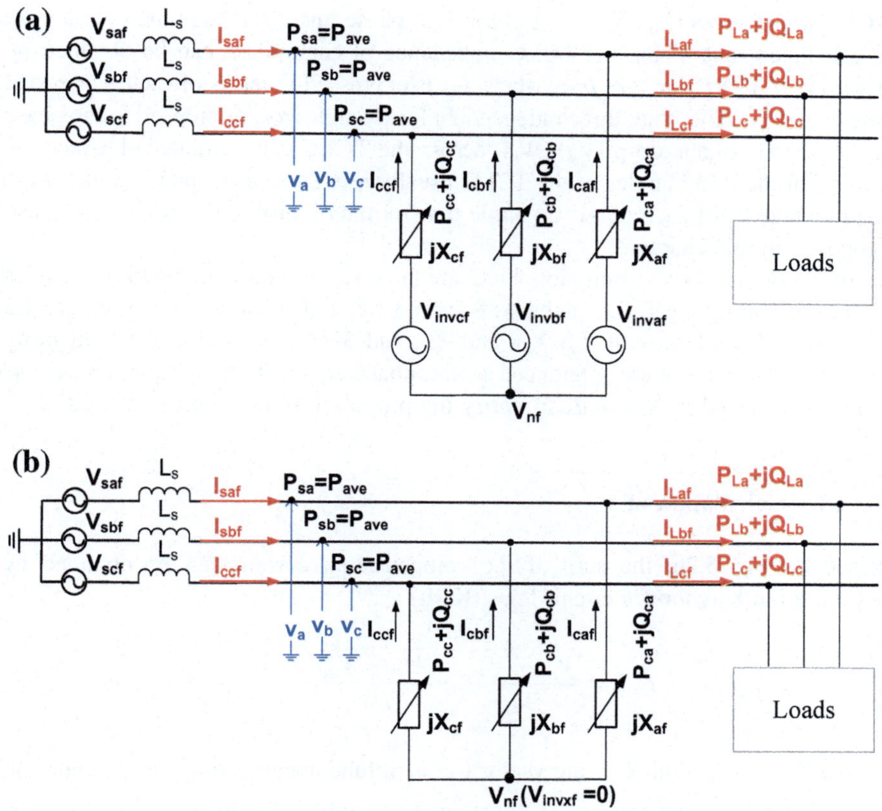

Fig. 5.2 Equivalent fundamental circuit models of the TCLC-HAPF when: **a** $V_{invxf} \neq 0$ and **b** $V_{invxf} = 0$

control strategy, which is based on the fundamental model in Fig. 5.2b, Sect. 5.3.2: Active inverter part control strategy and Sect. 5.3.3: The overall hybrid controller for TCLC-HAPF.

5.3.1 TCLC Part Control Strategy

From the mathematical analysis as shown in Appendix, it can be clearly illustrated that the three-phase source active power can become balanced once the reactive power is compensated. With this idea, the control target of the TCLC part is dedicated to balance and compensate the fundamental active and reactive power via the calculated required firing angles in this chapter. The firing angles are determined by the required TCLC impedances (X_{af}, X_{bf} and X_{cf}) for compensation which can be deduced by Ohm's law. Referring to Fig. 5.2b, in order to obtain the required

TCLC impedances (X_{af}, X_{bf} and X_{cf}) for each phase, the virtual common point (V_{nf}) is calculated first. Then, the TCLC impedance of each phase can be obtained by $X_{xf} = (V_x - V_{nf})/I_{cxf}$, $x = a, b, c$, where I_{cxf} is expressed in terms of each phase load power and V_x. After that, three independent firing angles $\alpha_{0,x}$ referenced to the phase angle of the voltage drop ($V_x - V_{nf}$) across the TCLC can be obtained. Since the control of the firing angles of the TCLC usually reference to the phase angle of the load voltage V_x of each phase, an angle transformation process is required and also proposed in this chapter.

Based on the above discussion, there are three steps to find the final firing angles α_x for controlling the TCLC, namely: Sect. 5.3.1.1: Calculation of V_{nf}, Sect. 5.3.1.2 Obtain the impedances of X_{af}, X_{bf}, and X_{cf}, and Sect. 5.3.1.3 find the final firing angles α_x for each phase referenced to the phase angle of V_x. In addition, a case study is provided in *Appendix* to verify the proposed TCLC control method.

5.3.1.1 Calculation of V_{nf}

Based on Fig. 5.2b, the sum of the compensating current can be obtained by applying the Kirchhoff's circuit laws (KCL) as:

$$\vec{I}_{caf} + \vec{I}_{cbf} + \vec{I}_{ccf} = \frac{\vec{V}_a - \vec{V}_{nf}}{jX_{af}} + \frac{\vec{V}_b - \vec{V}_{nf}}{jX_{bf}} + \frac{\vec{V}_c - \vec{V}_{nf}}{jX_{cf}} = 0 \tag{5.1}$$

where $\vec{I}_{caf}$, $\vec{I}_{cbf}$ and $\vec{I}_{ccf}$ are vector forms of fundamental compensating current. $\vec{V}_a$, $\vec{V}_b$ and $\vec{V}_c$ are the vector forms of load voltage. From (5.1), $\vec{V}_{nf}$ can be obtained as:

$$\begin{aligned}\vec{V}_{nf} = {} & \frac{X_{bf} \cdot X_{cf}}{X_{af}X_{bf} + X_{bf}X_{cf} + X_{cf}X_{af}} \cdot \vec{V}_a + \frac{X_{cf} \cdot X_{af}}{X_{af}X_{bf} + X_{bf}X_{cf} + X_{cf}X_{af}} \cdot \vec{V}_b \\ & + \frac{X_{af} \cdot X_{bf}}{X_{af}X_{bf} + X_{bf}X_{cf} + X_{cf}X_{af}} \cdot \vec{V}_c\end{aligned} \tag{5.2}$$

With the expression of $\vec{V}_{nf}$, the TCLC impedances can be obtained by Ohm's law as discussed in the next section.

5.3.1.2 Obtain the Impedance of X_{af}, X_{bf} and X_{cf}

From (5.1) and (5.2), the relationship among the phase compensating current $\vec{I}_{cxf}$, load voltage $\vec{V}_x$ and coupling impedances X_{xf} can be expressed as:

$$\begin{bmatrix} \vec{I}_{caf} \\ \vec{I}_{cbf} \\ \vec{I}_{ccf} \end{bmatrix} = \begin{bmatrix} (jX_{af})^{-1} & 0 & 0 \\ 0 & (jX_{bf})^{-1} & 0 \\ 0 & 0 & (jX_{cf})^{-1} \end{bmatrix} \cdot \begin{bmatrix} \vec{V}_a - \vec{V}_{nf} \\ \vec{V}_b - \vec{V}_{nf} \\ \vec{V}_c - \vec{V}_{nf} \end{bmatrix}$$
$$= \begin{bmatrix} -j\left(\frac{X_b + X_c}{m}\right) & j\frac{X_{cf}}{m} & j\frac{X_{bf}}{m} \\ j\frac{X_{cf}}{m} & -j\left(\frac{X_{af} + X_{cf}}{m}\right) & j\frac{X_{af}}{m} \\ j\frac{X_b}{m} & j\frac{X_{af}}{m} & -j\left(\frac{X_{af} + X_{bf}}{m}\right) \end{bmatrix} \cdot \begin{bmatrix} \vec{V}_a \\ \vec{V}_b \\ \vec{V}_c \end{bmatrix} \quad (5.3)$$

where $\vec{V}_{nf}$ can be obtained from (5.2) and $m = X_{af}X_{bf} + X_{bf}X_{cf} + X_{cf}X_{af}$.

The compensating apparent power S_{cx} can be defined as: $S_{cx} = P_{cx} + jQ_{cx} = \vec{V}_x \cdot \vec{I}^{\,*}_{cxf}$ where $\vec{I}^{\,*}_{cxf}$ is the conjugate of $\vec{I}_{cxf}$. Modifying S_{cx} equation, the expression of the phase compensating current can be given as:

$$\begin{bmatrix} \vec{I}_{caf} \\ \vec{I}_{cbf} \\ \vec{I}_{ccf} \end{bmatrix} = \begin{bmatrix} \left((P_{ca} + jQ_{ca})/\vec{V}_a\right)^* \\ \left((P_{cb} + jQ_{cb})/\vec{V}_b\right)^* \\ \left((P_{cc} + jQ_{cc})/\vec{V}_c\right)^* \end{bmatrix} \quad (5.4)$$

where the note "*" denotes the conjugate complex numbers. $\vec{V}_a$ is set to be the reference phasor, so $\vec{V}_a = \bar{V}_x\angle 0°$, $\vec{V}_b = \bar{V}_x\angle -120°$ and $\vec{V}_c = \bar{V}_x\angle 120°$, where $1\angle -120° = -1/2 - j\sqrt{3}/2$ and $1\angle 120° = -1/2 + j\sqrt{3}/2$, $\bar{V}_x$ is the root mean square (RMS) value of load voltage which can be obtained in real time as $\bar{V}_x = \|v_L\|/\sqrt{3} = \sqrt{v_a^2 + v_b^2 + v_c^2}\Big/\sqrt{3}$ [1]. Q_{cx} and P_{cx} are the compensating reactive power and active power. Theoretically, after compensation, the source reactive power should become zero and the source active power should all be equal to $(P_{La} + P_{Lb} + P_{Lc})/3$. Therefore, the reactive power and active power generated/absorbed by the TCLC part can be expressed as:

$$Q_{cx} = -Q_{Lx} \text{ and } P_{cx} = -\left(P_{Lx} - \frac{P_{La} + P_{Lb} + P_{Lc}}{3}\right) \quad (5.5)$$

where $x = a, b, c$, Q_{cx} and Q_{Lx} are the compensating and load reactive power, and P_{cx} and P_{Lx} are the compensating and load active power. Combining (5.1)–(5.5), the relationship between TCLC impedances and load power can be deduced as:

$$\begin{bmatrix} X_{af}/m \\ X_{bf}/m \\ X_{cf}/m \end{bmatrix} = \begin{bmatrix} (Q_{ca} - Q_{cb} - Q_{cc})/3\bar{V}_x^2 \\ (Q_{cb} - Q_{ca} - Q_{cc})/3\bar{V}_x^2 \\ (Q_{cc} - Q_{cb} - Q_{ca})/3\bar{V}_x^2 \end{bmatrix} \quad (5.6)$$

where $m = X_{af}X_{bf} + X_{bf}X_{cf} + X_{cf}X_{af}$. Further simplifying (5.6), the coupling impedances X_{af}, X_{bf} and X_{cf} can be obtained as:

$$\begin{bmatrix} X_{af} \\ X_{bf} \\ X_{cf} \end{bmatrix} = \begin{bmatrix} \frac{3\cdot \bar{V}_x^2 \cdot (Q_{Lc}-Q_{Lb}-Q_{La})^{-1}\cdot (Q_{Lb}-Q_{La}-Q_{Lc})^{-1}}{(Q_{Lc}-Q_{Lb}-Q_{La})^{-1}+(Q_{La}-Q_{Lb}-Q_{Lc})^{-1}+(Q_{Lb}-Q_{La}-Q_{Lc})^{-1}} \\ \frac{3\cdot \bar{V}_x^2 \cdot (Q_{Lc}-Q_{Lb}-Q_{La})^{-1}\cdot (Q_{La}-Q_{Lb}-Q_{Lc})^{-1}}{(Q_{Lc}-Q_{Lb}-Q_{La})^{-1}+(Q_{La}-Q_{Lb}-Q_{Lc})^{-1}+(Q_{Lb}-Q_{La}-Q_{Lc})^{-1}} \\ \frac{3\cdot \bar{V}_x^2 \cdot (Q_{La}-Q_{Lb}-Q_{Lc})^{-1}\cdot (Q_{Lb}-Q_{La}-Q_{Lc})^{-1}}{(Q_{Lc}-Q_{Lb}-Q_{La})^{-1}+(Q_{La}-Q_{Lb}-Q_{Lc})^{-1}+(Q_{Lb}-Q_{La}-Q_{Lc})^{-1}} \end{bmatrix} \tag{5.7}$$

From (5.6), the TCLC impedances are expressed in terms of the phase load reactive power Q_{Lx} and RMS voltage of phase load voltage $\bar{V}_x$. Then, the TCLC impedances are used to find the firing angles to control the TCLC part, which will be discussed in the following part.

5.3.1.3 Find the Final Firing Angles α_x Referenced to the Phase Angle of V_x

The impedance of the TCLC part can be considered as the X_{L_c} connected in series with the combination of $X_{C_{PF}}$ in parallel with a thyristor controlled reactance (TCR) $X_{TCR}(\alpha_{0,x})$ $(= \frac{\pi}{2\pi - 2\alpha_{0,x} + \sin 2\alpha_{0,x}} \cdot X_{L_{PF}})$. Therefore, the expression of the TCLC impedances (X_{af}, X_{bf} and X_{cf}) in terms of passive components and firing angles ($\alpha_{0,x}$) referenced to the phase angle of voltage across TCLC ($V_x - V_{nf}$) is shown as:

$$\begin{bmatrix} X_{af}(\alpha_{0,a}) \\ X_{bf}(\alpha_{0,b}) \\ X_{cf}(\alpha_{0,c}) \end{bmatrix} = \begin{bmatrix} \frac{\pi X_{L_{PF}} X_{C_{PF}}}{X_{C_{PF}}[2\pi - 2\alpha_{0,a} + \sin(2\alpha_{0,a})] - \pi X_{L_{PF}}} + X_{L_c} \\ \frac{\pi X_{L_{PF}} X_{C_{PF}}}{X_{C_{PF}}[2\pi - 2\alpha_{0,b} + \sin(2\alpha_{0,b})] - \pi X_{L_{PF}}} + X_{L_c} \\ \frac{\pi X_{L_{PF}} X_{C_{PF}}}{X_{C_{PF}}[2\pi - 2\alpha_{0,c} + \sin(2\alpha_{0,c})] - \pi X_{L_{PF}}} + X_{L_c} \end{bmatrix} \tag{5.8}$$

where X_{L_c}, $X_{C_{PF}}$ and $X_{L_{PF}}$ are the impedances of the coupling inductor, the TCLC capacitor and inductor, respectively (5.8) has a term of $-2\alpha_{0,x} + \sin(2\alpha_{0,x})$ and it does not have a closed-form solution. Therefore, a look up table (LUT) has been implemented into the controller for performing simulations and experiments. Once the required coupling impedance $X_{xf}(\alpha_{0,x})$ is obtained from (5.7), the pre-calculated firing angle $\alpha_{0,x}$ can be extracted based on the LUT of (5.8). As the control of the firing angles of the TCLC are usually referenced to the phase angle of the load voltage V_x of each phase, an angle transformation process is required and proposed in the following.

Figure 5.3 shows the angle relationship between the phase angle of the load voltage $\vec{V}_x$ and the phase angle of the voltage across TCLC ($V_x - V_{nf}$). From Fig. 5.3, there will be an angle difference for each phase between that of the load voltage V_x and that of voltage across TCLC ($V_x - V_{nf}$), thus the above calculated firing angles $\alpha_{0,x}$ should have angle transformation if the control of the TCLC firing angles is referenced to the load voltage.

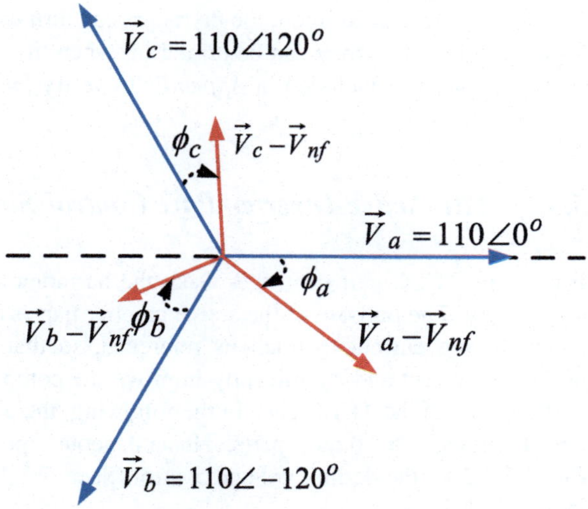

Fig. 5.3 Phasor diagram of load voltage $\vec{V}_x$ and the TCLC impedance voltage $\vec{V}_x - \vec{V}_{nf}$

To keep the same impedance X_x in part 5.3.1.2, an angle transformation process is required in order to obtain the final firing angle α_x referenced to the phase angle of the load voltage V_x. The final firing angles α_x can be obtained as:

$$\begin{bmatrix} \alpha_a \\ \alpha_b \\ \alpha_c \end{bmatrix} = \begin{bmatrix} \alpha_{0,a} \\ \alpha_{0,b} \\ \alpha_{0,c} \end{bmatrix} - \begin{bmatrix} \varphi_a \\ \varphi_b \\ \varphi_c \end{bmatrix} \tag{5.9}$$

where the phase angle φ_x is the phase angle of the voltage across TCLC. And φ_x can be calculated from the expression of $\overline{V}_{nf}$ in (5.2) as:

$$\begin{bmatrix} \varphi_a \\ \varphi_b \\ \varphi_c \end{bmatrix} = \begin{bmatrix} \tan^{-1}\left(\frac{X_{cf}-X_{bf}}{\sqrt{3}(X_{bf}+X_{cf})}\right) \\ \tan^{-1}\left(\frac{X_{af}-X_{cf}}{\sqrt{3}(X_{af}+X_{cf})}\right) \\ \tan^{-1}\left(\frac{X_{bf}-X_{af}}{\sqrt{3}(X_{bf}+X_{af})}\right) \end{bmatrix}_{\tan^{-1}\theta\in[-90^\circ,90^\circ]} \tag{5.10}$$

Based on (5.7)–(5.10), the final firing angle α_x can be obtained in terms of phase load reactive power, load voltage and the passive component values of the TCLC. The phase load reactive power can be obtained with the help of single phase *pq* theory [23] and low pass filters (LPFs).

The trigger signals to control the TCLC part are generated by comparing the calculated α_x in (5.9) with the phase angle of the load voltage θ_x, which is instantaneously measured by the phase lock loop (PLL). The TCLC part has two back-to-back connected thyristors T_{x1} and T_{x2} for each phase, and they are triggered alternately in every half cycle. When $\theta_x > \alpha_x$, the gate pulse for thyristor T_{x1} can be generated. On the other hand, when $\theta_x < 180^o - \alpha_x$, the gate pulse for thyristor T_{x2}

can be generated. In addition, the thyristors can turn on immediately when there is a trigger signal, while they can be turned off when the inductor current goes to zero.

A case study is included in *Appendix* to verify the above analysis.

5.3.2 The Active Inverter Part Control Strategy

If only the TCLC part control is used, the harmonic current cannot be eliminated satisfactory. The purpose of the active inverter part is to instantaneously control the compensating current to track its reference, so that it can compensate the load harmonic current and significantly improve the compensation ability and dynamic performance of the TCLC part. In the following, the active inverter part control can be discussed in three parts: Instantaneous power compensation control (Sect. 5.3.2.1), the dc-link voltage control (Sect. 5.3.2.2) and current PWM control (Sect. 5.3.2.3).

5.3.2.1 Instantaneous Power Compensation Control

As the instantaneous power theory [1] is valid for unbalanced three-phase systems, it is chosen to calculate the reference compensating current i_{cxref}. The calculated i_{cxref} contains the harmonics, reactive power, unbalanced power and the DC-link voltage regulating components. By controlling the compensating current i_{cx} to its reference i_{cxref}, the active inverter part can compensate the load harmonic current, improve the reactive power compensation ability and dynamic performance of the TCLC part and also regulate the DC-link voltage to its reference value. The i_{cxref} can be calculated through the well-known instantaneous p-q theory [1] as:

$$\begin{bmatrix} i_{caref} \\ i_{cbref} \\ i_{ccref} \end{bmatrix} = \sqrt{\frac{2}{3}} \cdot \frac{1}{v_{\alpha}^2 + v_{\beta}^2} \begin{bmatrix} 1 & 0 \\ -1/2 & \sqrt{3}/2 \\ -1/2 & -\sqrt{3}/2 \end{bmatrix} \cdot \begin{bmatrix} v_{\alpha} & -v_{\beta} \\ v_{\beta} & v_{\alpha} \end{bmatrix} \cdot \begin{bmatrix} \tilde{p}_{\alpha\beta} + dc_p \\ q_{\alpha\beta} + dc_q \end{bmatrix} \tag{5.11}$$

where the dc_p and dc_q are the active and reactive components for DC-link voltage regulation. And the discussions of dc_p and dc_q are provided in Sect. 5.3.2.2. $p_{\alpha\beta}$ and $q_{\alpha\beta}$ is the instantaneous active and reactive power which include dc components $\bar{p}_{\alpha\beta}$ and $\bar{q}_{\alpha\beta}$, and ac components $\tilde{p}_{\alpha\beta}$ and $\tilde{q}_{\alpha\beta}$. $\bar{p}_{\alpha\beta}$ and $\bar{q}_{\alpha\beta}$ contain the fundamental active and reactive current components respectively, while $\tilde{p}_{\alpha\beta}$ and $\tilde{q}_{\alpha\beta}$ contain harmonic current and negative sequence components. $\tilde{p}_{\alpha\beta}$ is obtained by passing $p_{\alpha\beta}$ through a high pass filter (HPF). $p_{\alpha\beta}$ and $q_{\alpha\beta}$ can be obtained as:

$$\begin{bmatrix} p_{\alpha\beta} \\ q_{\alpha\beta} \end{bmatrix} = \begin{bmatrix} v_{\alpha} & v_{\beta} \\ -v_{\beta} & v_{\alpha} \end{bmatrix} \cdot \begin{bmatrix} i_{\alpha} \\ i_{\beta} \end{bmatrix} \tag{5.12}$$

In (5.11) and (5.12), the voltage (v_α and v_β) and current (i_α and i_β) in $\alpha - \beta$ plane are transformed from *a–b–c* frames by:

$$\begin{bmatrix} v_\alpha \\ v_\beta \end{bmatrix} = \begin{bmatrix} 1 & -1/2 & -1/2 \\ 0 & \sqrt{3}/2 & -\sqrt{3}/2 \end{bmatrix} \cdot \begin{bmatrix} v_a \\ v_b \\ v_c \end{bmatrix} \text{ and}$$
$$\begin{bmatrix} i_\alpha \\ i_\beta \end{bmatrix} = \begin{bmatrix} 1 & -1/2 & -1/2 \\ 0 & \sqrt{3}/2 & -\sqrt{3}/2 \end{bmatrix} \cdot \begin{bmatrix} i_{La} \\ i_{Lb} \\ i_{Lc} \end{bmatrix} \quad (5.13)$$

where v_x and i_{Lx} are load voltage and current signals.

5.3.2.2 The DC-Link Voltage Control

The active inverter part can effectively control the DC-link voltage by feeding back the DC voltage controlled signals which include both reactive and active components (dc_q, dc_p) [19–22] as shown in Fig. 5.4.

$$dc_q = -k_q \cdot (V_{DCref} - V_{DC}) \quad (5.14)$$

$$dc_p = k_p \cdot (V_{DCref} - V_{DC}) \quad (5.15)$$

In (5.14) and (5.15), k_q and k_p are the proportional gains of the DC-link voltage controller. Through the directly hysteresis band control, dc_q can be used to step change the DC-link voltage during the start-up process, while dc_p can be used to maintain the DC-link voltage level due to system loss. With the help of the three-phase instantaneous *p-q* theory [1] and the terms of dc_q and dc_p, the DC-link voltage V_{DC} can track its reference V_{DCref} by transforming the DC voltage control signals into reference compensating current through the above (5.11). Notice that the DC-link control block requires a small amount of active power to maintain DC-link voltage at its reference value. The active power for maintaining DC-link voltage is taken equally from three phases, and it is relatively small comparing with the compensating one, so that it will not affect the unbalanced loading compensation of the TCLC part.

5.3.2.3 Current PWM Control

The final reference and actual compensating current $i_{cxfinal}$ and i_{cx} will be sent to the PWM control, and then PWM trigger signals for controlling the active inverter switching devices can be generated. The PWM method such as: triangular carrier-based sinusoid PWM or hysteresis PWM can be applied. In this chapter the hysteresis current PWM [24] is selected due to its advantages of fast response, ease of implementation and good current limiting capability, etc.

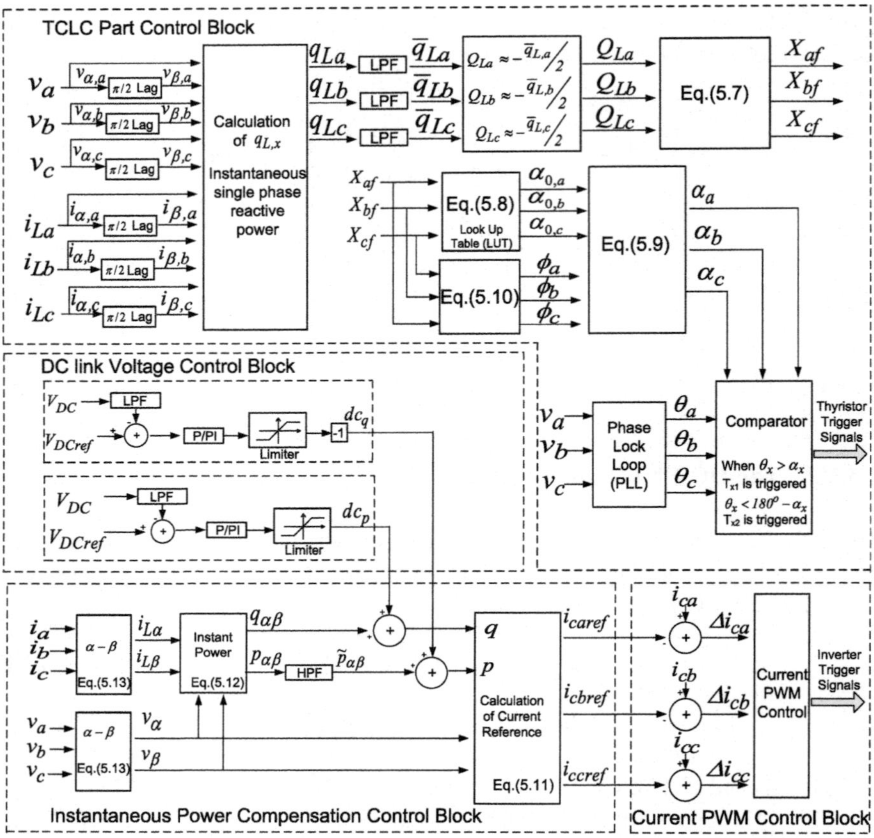

Fig. 5.4 Proposed hybrid control block diagram for the TCLC-HAPF under unbalanced loads compensation

5.3.3 *The Proposed Hybrid Controller for TCLC-HAPF*

Based on the above discussions, the proposed hybrid control block diagram for the TCLC-HAPF under unbalanced loads compensation is shown in Fig. 5.4. It consists of five main control blocks: the TCLC part control block (discussed in Sect. 5.3.1), the instantaneous power compensation control block (discussed in Sect. 5.3.2.1), the dc-link control block (discussed in Sect. 5.3.2.2) and the current PWM control block (discussed in Sect. 5.3.2.3).

In the following, simulation and experimental results of the proposed control strategy will be presented in comparison with the state-of-the-art control strategy in [18] in order to verify its effectiveness and superior compensating performances under unbalanced loading condition.

5.4 Simulation and Experimental Results

In this section, simulation and experimental results of the proposed TCLC-HAPF control strategy under unbalanced loading conditions will be presented and discussed in comparison with the results of the state-of-the-art control method in [18], in which the same DC-link voltage is applied to both of them. An 110 V–5 kVA three-phase three-wire TCLC-HAPF experimental prototype is designed and constructed in the laboratory. The simulations are carried out by using PSCAD/EMTDC. In addition, with reference to the IEEE standard 519-2014 [26], the Total Demand Distortion (TDD) = 15% with I_{SC}/I_L is in 100 < 1000 scale (for 110 V–5 kVA experimental prototype). The nominal rate current is assumed to be equal to the fundamental load current in the worst case, which results in $THD = TDD = 15\%$. Therefore, this chapter focuses on a *THD* below 15%.

Figures 5.5 and 5.6 (simulation results) and Figs. 5.7 and 5.8 (experimental results) illustrate the waveforms of source current, load current, compensating current, capacitor (C_{PF}) current, inductor (L_{PF}) current and dc-link voltage, source reactive and active power before and after compensation using the state-of-the-art control method [18] and the proposed control method. Figures 5.9 and 5.10 (simulation results) and Figs. 5.11 and 5.12 (experimental results) demonstrate the source current spectrums and phasor diagrams of source voltage and current before and after the state-of-the-art control method [18] and the proposed control method. For each harmonic order of the current spectrum as in Fig. 5.11, the three bars from left to right represent phase *a*, *b* and *c*, respectively. Moreover, Tables 5.1 and 5.2 summarize all the simulation and experimental results shown in the aforementioned figures.

By using the control method in [18], the simulated and experimental system source power factors (*PFs*) have been compensated to 0.93 and 0.95 (showing worst phase) respectively as shown in Figs. 5.5 and 5.7, Tables 5.1 and 5.2. From Figs. 5.5b–c and 5.7b–c, the source reactive power are compensated to 160, 9 and 237 var in the simulations and 130, 70 and 120 var in the experiments. Besides, the simulated source active power (483, 662 and 698 W) and experimental source active power (510, 680 and 690 W) are not balanced after compensations. Moreover, from Figs. 5.9b and 5.11b, Tables 5.1 and 5.2, the simulated and experimental source current *THD* are compensated to 22.5 and 18.9% (showing worst phase) respectively. From Figs. 5.10b and 5.12b, it can be seen that source voltage and current are not in phase (especially for phases *a* and *c*). In addition, as shown in Tables 5.1 and 5.2, the simulated and experimental source current unbalanced factor (UBI_{fS}) is 33.9 and 32.4% respectively after compensation. From Figs. 5.5 and 5.7, Figs. 5.9, 5.10, 5.11 and 5.12, Tables 5.1 and 5.2, it is proved that the state-of-the-art control method [18] cannot balance the source current and provide good compensation performance with the small rating active inverter part under unbalanced loading condition.

By applying the proposed control method, the simulated and experimental system source power factors (*PFs*) have been compensated to 0.99 or above in both

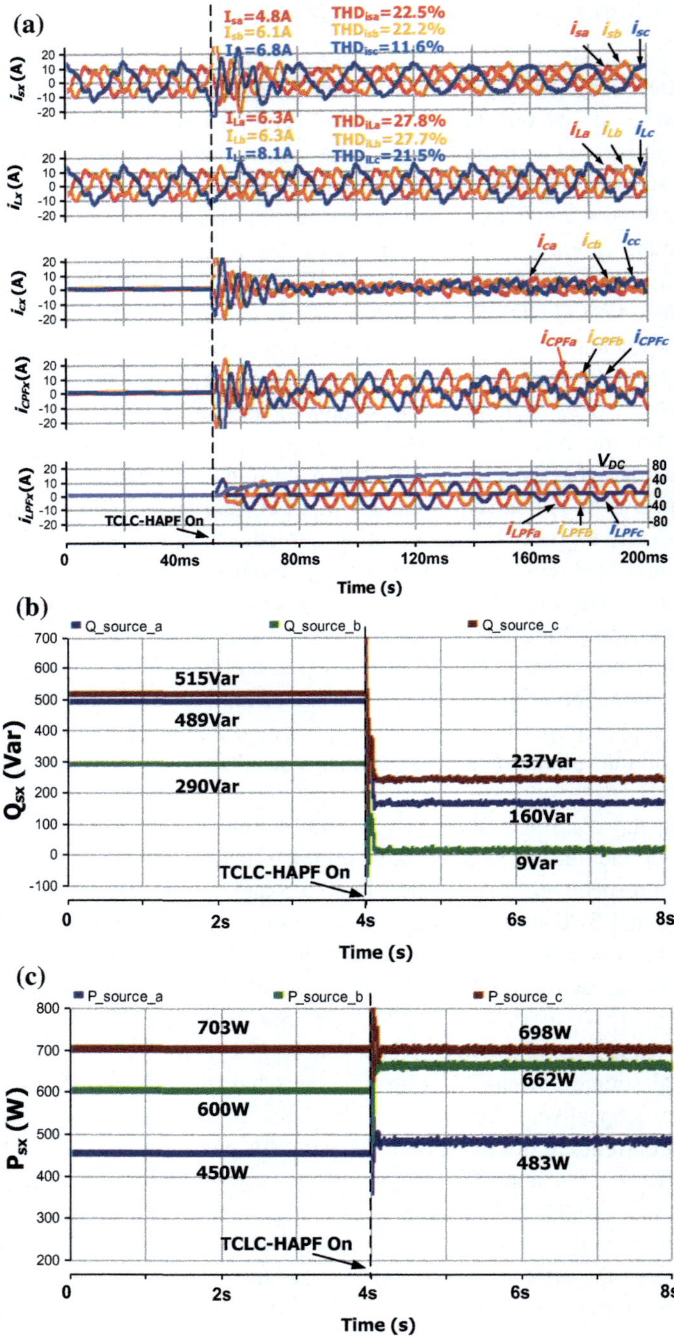

Fig. 5.5 Simulation results of dynamic performance by using the state-of-the-art control method [18] before and after compensation: **a** source current, compensating current, capacitor (C_{PF}) current, inductor (L_{PF}) current and dc-link voltage, **b** source reactive power and **c** source active power

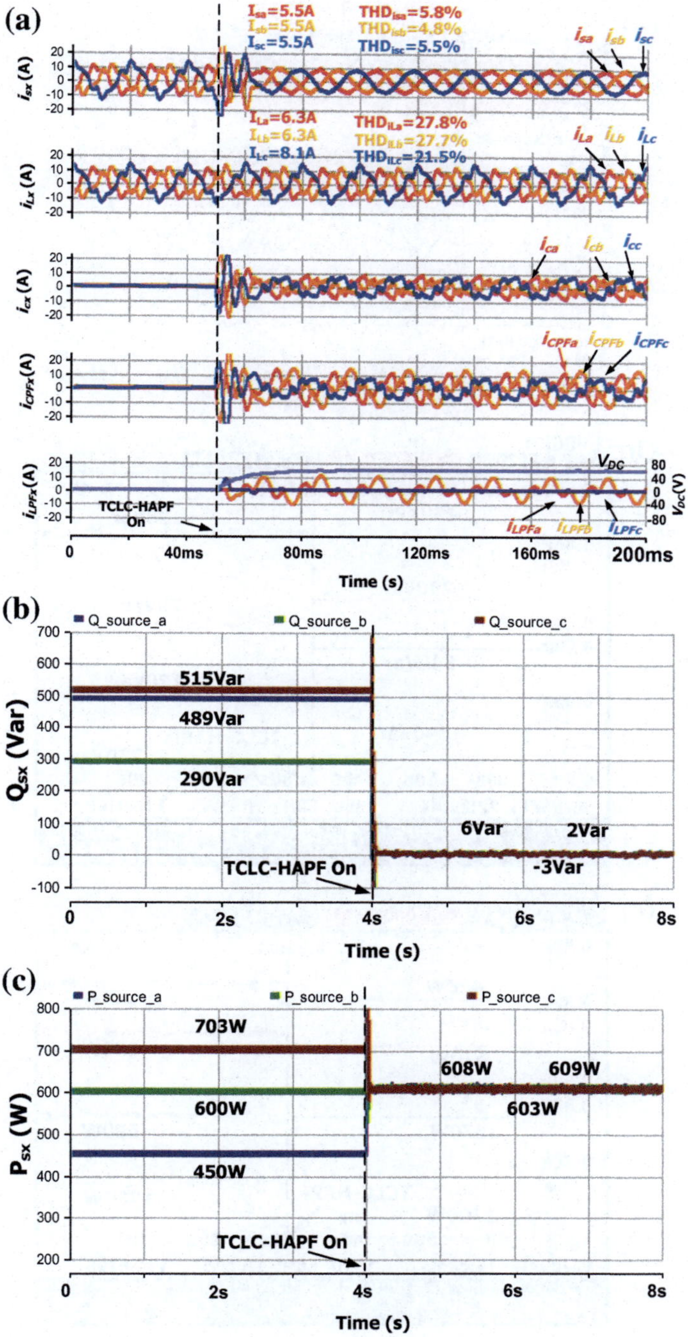

Fig. 5.6 Simulation results of dynamic performance by using the proposed control method before and after compensation: **a** source current, compensating current, capacitor (C_{PF}) current, inductor (L_{PF})current and dc-link voltage, **b** source reactive power and **c** source active power

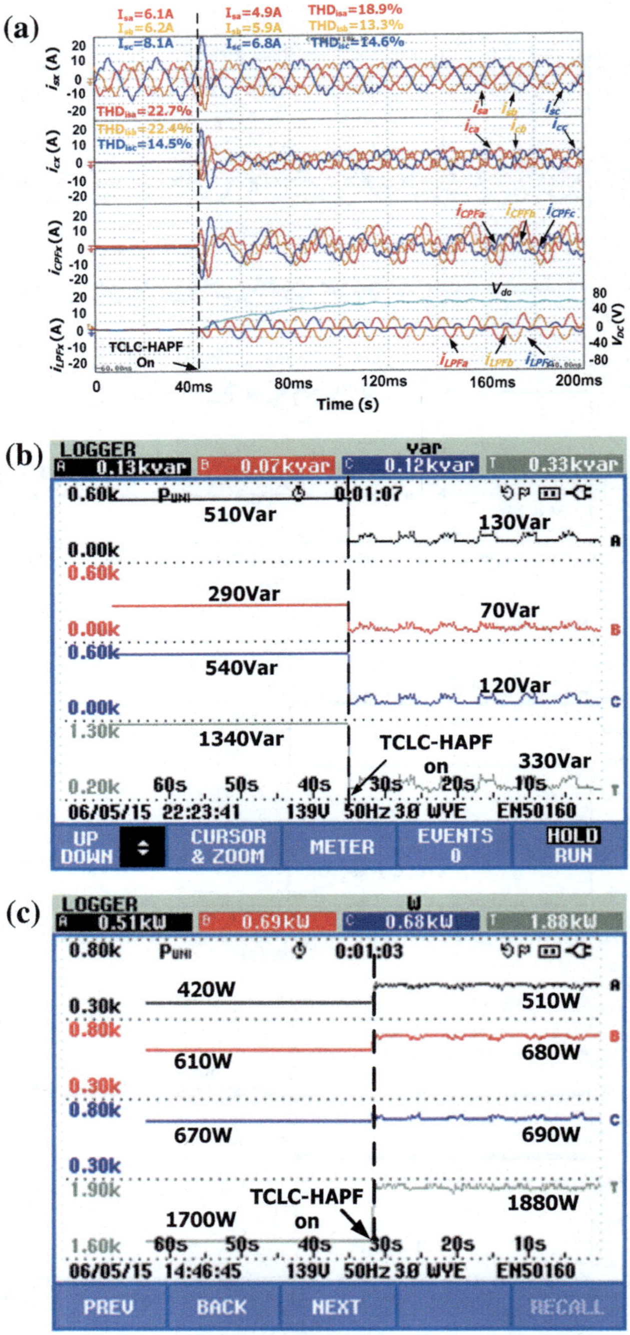

Fig. 5.7 Experimental results of dynamic performance by using the state-of-the-art control method [18] before and after compensation: **a** source current, compensating current, capacitor (C_{PF}) current, inductor (L_{PF}) current and dc-link voltage, **b** source reactive power and **c** source active power

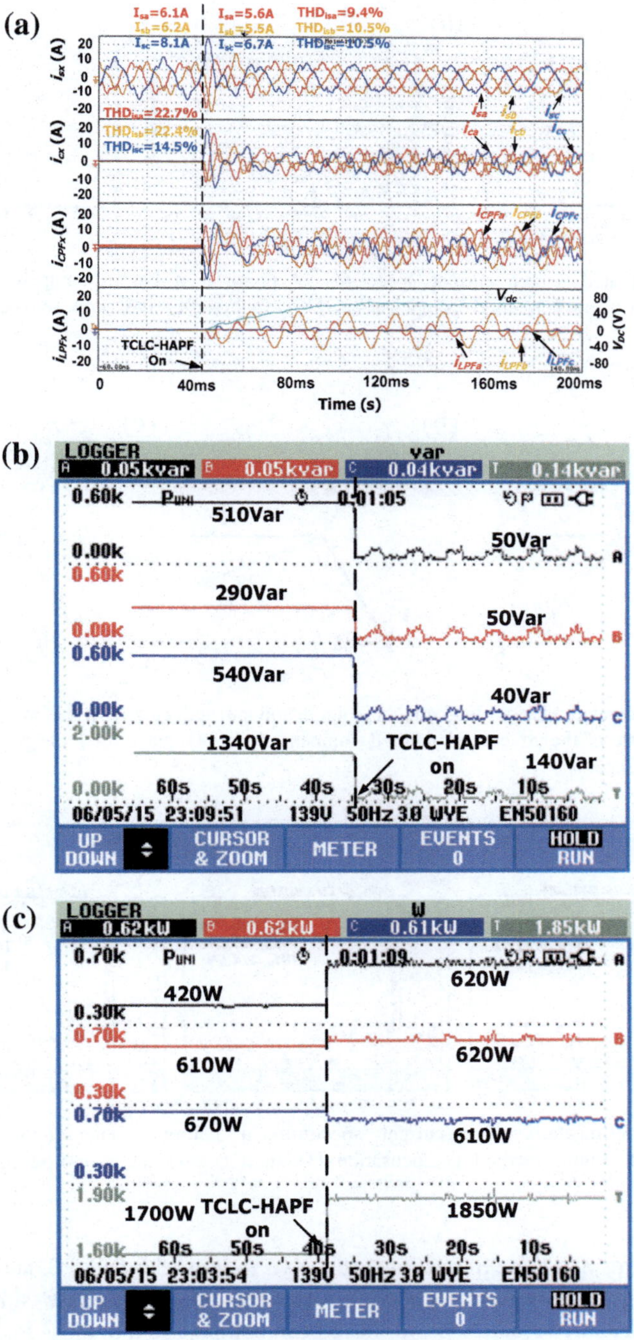

Fig. 5.8 Experimental results of dynamic performance by using the proposed control method before and after compensation: **a** source current, compensating current, capacitor (C_{PF}) current, inductor (L_{PF}) current and dc-link voltage, **b** source reactive power and **c** source active power

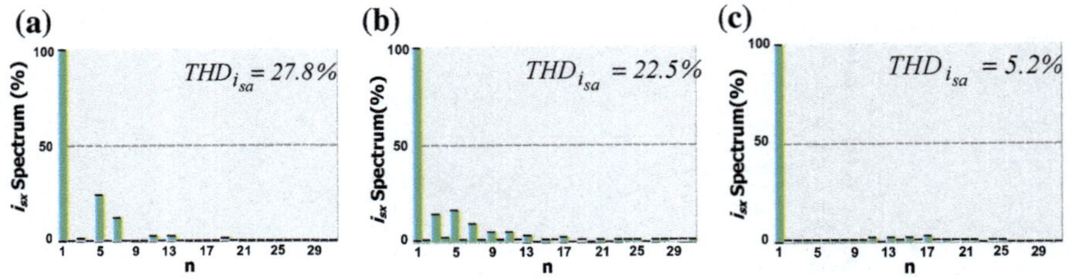

Fig. 5.9 Simulated source current spectrums of phase *a* **a** before compensation, **b** after state-of-the-art control method compensation [18] and **c** after the proposed control method compensation

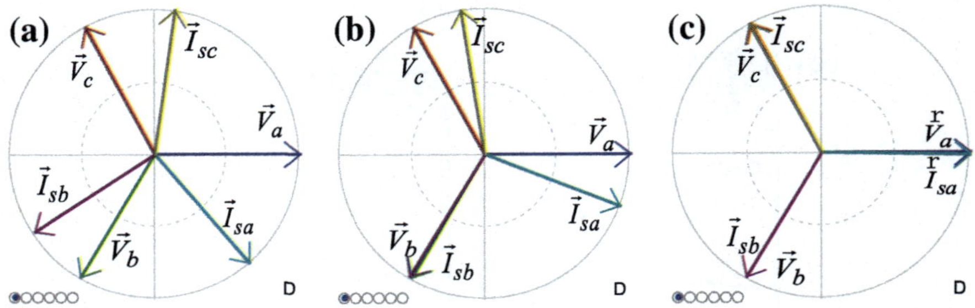

Fig. 5.10 Simulated phasor diagrams of source voltage and current **a** before compensation, **b** after the state-of-the-art control method compensation [18] and **c** after the proposed control method compensation

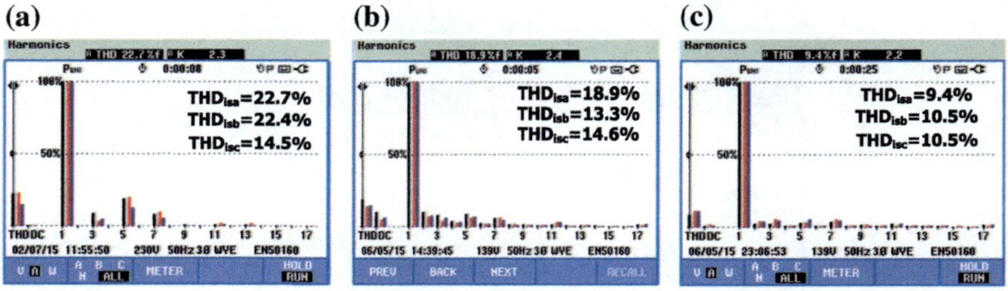

Fig. 5.11 Experimental source current spectrums **a** before compensation, **b** after the state-of-the-art control method compensation [18] and **c** after the proposed control method compensation

the simulations and the experiments, as shown in Figs. 5.6 and 5.8, and Tables 5.1 and 5.2, Figs. 5.6b–c and 5.8b–c illustrate that the three-phase simulated and experimental source reactive power have been compensated to close to zero and the source active power are approximately balanced after compensation. Moreover, from Fig. 5.9c and Fig. 5.11c and Tables 5.1 and 5.2, the simulated and experimental source current *THD* have been compensated to 5.5 and 10.5% (showing

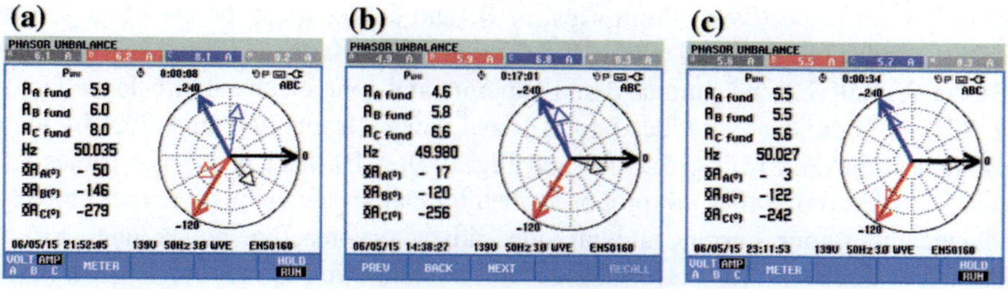

Fig. 5.12 Experimental phasor diagrams of source voltage and current **a** before compensation, **b** after the state-of-the-art control method compensation [18] and **c** after the proposed control method compensation

Table 5.1 Simulation results before and after TCLC-HAPF compensation by using the state-of-the-art control method [18] and the proposed method

		i_{sxf} (A)	Q_{sx} (Var)	P_{sx} (W)	PF	THD_{isx} (%)	UBI_{fS} (%)	V_{DC} (V)
Before Comp.	A	6.3	489	450	0.68	27.8	26.0	–
	B	6.3	290	600	0.90	27.7		
	C	8.1	515	703	0.80	21.5		
Control method [18]	A	4.8	160	483	0.94	22.5	33.9	60
	B	6.1	9	662	0.99	22.2		
	C	6.8	237	698	0.94	11.6		
Proposed method	A	5.5	6	608	0.99	5.2	0.1	60
	B	5.5	2	603	0.99	4.8		
	C	5.5	−5	609	0.99	5.5		

Table 5.2 Experimental results before and after TCLC-HAPF compensation by using the state-of-the-art control method [18] and the proposed method

		i_{sxf} (A)	Q_{sx} (Var)	P_{sx} (W)	PF	THD_{isx} (%)	UBI_{fS} (%)	V_{DC} (V)
Before Comp.	A	6.1	510	420	0.63	22.7	29.4	–
	B	6.2	290	610	0.89	22.4		
	C	8.1	540	670	0.78	14.5		
Control method [18]	A	4.9	130	510	0.95	18.9	32.4	60
	B	5.9	70	680	0.98	13.3		
	C	6.8	120	690	0.97	14.6		
Proposed method	A	5.6	50	620	0.99	9.4	3.6	60
	B	5.5	50	610	0.99	10.5		
	C	5.7	40	610	0.99	10.5		

worst phase) respectively. From Figs. 5.10c and 5.12c, it can be seen that source voltage and current are in phase for all three phases. In addition, as shown in Tables 5.1 and 5.2, the simulated and experimental source current unbalance factor (UBI_{fS}) is significantly reduced to 0.1 and 3.6% (from 26.0 and 29.4%) after compensation. From Figs. 5.5, 5.7 and Figs. 5.9, 5.10, 5.11 and 5.12, Tables 5.1 and 5.2, they prove that the proposed method can simultaneously and effectively balance the source current and provide satisfactory reactive power and current harmonics compensation performances with a small rating active inverter part under unbalanced loading condition.

In addition, the simulation results are consistent with experimental results, which clearly verify the effectiveness and viability of the proposed unbalanced control strategy in comparison with the state-of-the-art one [18] for unbalanced loading compensation.

5.5 Summary

In this chapter, a novel control strategy for a three-phase three-wire TCLC-HAPF is proposed, which can maintain it operating with a small rating active inverter part and at the same time it can balance the active power and compensate the reactive power and harmonic current under unbalanced loading compensation. The design idea and operation steps of the proposed hybrid controller for the TCLC-HAPF under unbalanced loading is presented and discussed in details. Finally, simulation and experimental results are given to verify the proposed control method in comparison with the state-of-the-art control method [18], which shows its superior compensating performances under the unbalanced loading condition.

5.6 Appendix: Balancing Three-phase Fundamental Active Power by Reactive power compensation

The mathematical analysis is provided to show how the active power can be controlled to be balanced when the reactive power is compensated. Figure 5.13 shows the power flow analysis before TCLC-HAPF compensation.

The phase fundamental apparent power is defined as $S_{sx} = P_{sx} + jQ_{sx} = \vec{V}_{sx} \cdot \vec{I}^{*}_{sxf}$ where the note "*" denotes the complex conjugate. In Fig. 5.13, the sum of $\vec{I}^{*}_{sxf}$ can be expressed as [25]:

$$\vec{I}^{*}_{saf} + \vec{I}^{*}_{sbf} + \vec{I}^{*}_{scf} = \frac{P_{sa} + jQ_{sa}}{\vec{V}_{saf}} + \frac{P_{sb} + jQ_{sb}}{\vec{V}_{sbf}} + \frac{P_{sc} + jQ_{sc}}{\vec{V}_{scf}} = 0 \qquad (5.16)$$

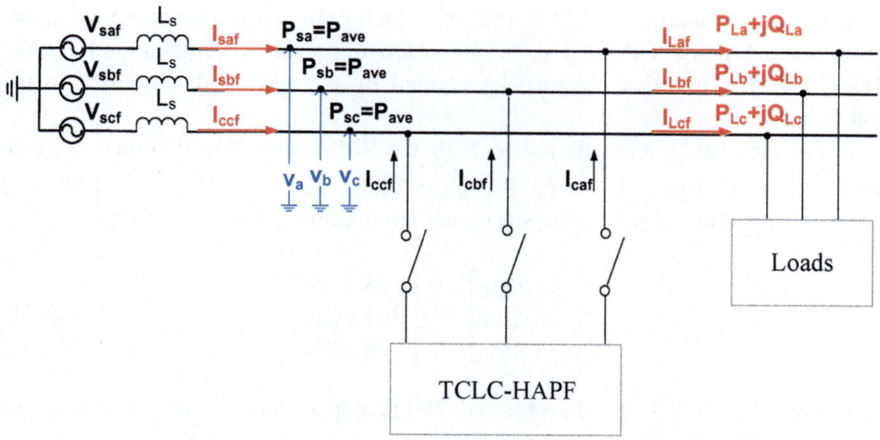

Fig. 5.13 Power flow analysis before compensation

where $\vec{I}_{saf}$, $\vec{I}_{sbf}$ and $\vec{I}_{scf}$ are the fundamental source current phasors. $\vec{V}_{saf}$, $\vec{V}_{sbf}$ and $\vec{V}_{scf}$ are the fundamental load voltage phasors. $\vec{V}_{saf}$ is set to be the reference phasor, so $\vec{V}_{saf} = \bar{V}_{xf}\angle 0°$, $\vec{V}_{sbf} = \bar{V}_{xf}\angle -120°$ and $\vec{V}_{scf} = \bar{V}_{xf}\angle 120°$, where $1\angle -120° = -1/2 - j\sqrt{3}/2$ and $1\angle 120° = -1/2 + j\sqrt{3}/2$, $\bar{V}_{xf}$ is the root mean square (rms) value of the source voltage. Simplifying (5.16), one can get:

$$\begin{aligned}&\left(2P_{sa} - P_{sb} - \sqrt{3}Q_{sb} - P_{sc} + \sqrt{3}Q_{sc}\right)\\&+j\left(2Q_{sa} - Q_{sb} + \sqrt{3}P_{sb} - Q_{sc} - \sqrt{3}P_{sc}\right) = 0\end{aligned} \tag{5.17}$$

In (5.17), both real part and imaginary part needs to be zero. Thus,

$$\begin{aligned}2P_{sa} - P_{sb} - \sqrt{3}Q_{sb} - P_{sc} + \sqrt{3}Q_{sc} = 0\\2Q_{sa} - Q_{sb} + \sqrt{3}P_{sb} - Q_{sc} - \sqrt{3}P_{sc} = 0\end{aligned} \tag{5.18}$$

If the source reactive power is compensated to be zero by the TCLC-HAPF as:

$$Q_{sa} = Q_{sb} = Q_{sc} = 0 \tag{5.19}$$

By substituting (5.19) into (5.18), the relationship of the three-phase source active power can be obtained as:

$$P_{sa} = P_{sb} = P_{sc} \tag{5.20}$$

Based on the analysis of (5.17)–(5.20), it can be concluded that the source active power can become balanced once the reactive power is compensated by the TCLC-HAPF. This idea can also be verified by the case study provided in the following.

In this case study, referring to Fig. 5.2b, the RMS value of load voltage is given as: $\bar{V}_x = 110$ V, where $\vec{V}_a = \bar{V}_x\angle 0°$, $\vec{V}_b = \bar{V}_x\angle -120°$ and $\vec{V}_c = \bar{V}_x\angle 120°$, and the unbalanced phase load apparent power are assumed to be:

$$\begin{bmatrix} P_{La}+jQ_{La} \\ P_{Lb}+jQ_{Lb} \\ P_{Lc}+jQ_{Lc} \end{bmatrix} = \begin{bmatrix} 233+j438 \\ 363+j203 \\ 498+j429 \end{bmatrix} \tag{5.21}$$

From (5.7) and (5.21), the required TCLC impedances for compensating the above unbalanced loading can be obtained as:

$$\begin{bmatrix} X_{af} \\ X_{bf} \\ X_{cf} \end{bmatrix} = \begin{bmatrix} \frac{3\cdot\bar{V}_x^2\cdot(Q_{Lc}-Q_{Lb}-Q_{La})^{-1}\cdot(Q_{Lb}-Q_{La}-Q_{Lc})^{-1}}{(Q_{Lc}-Q_{Lb}-Q_{La})^{-1}+(Q_{La}-Q_{Lb}-Q_{Lc})^{-1}+(Q_{Lb}-Q_{La}-Q_{Lc})^{-1}} \\ \frac{3\cdot\bar{V}_x^2\cdot(Q_{Lc}-Q_{Lb}-Q_{La})^{-1}\cdot(Q_{La}-Q_{Lb}-Q_{Lc})^{-1}}{(Q_{Lc}-Q_{Lb}-Q_{La})^{-1}+(Q_{La}-Q_{Lb}-Q_{Lc})^{-1}+(Q_{Lb}-Q_{La}-Q_{Lc})^{-1}} \\ \frac{3\cdot\bar{V}_x^2\cdot(Q_{La}-Q_{Lb}-Q_{Lc})^{-1}\cdot(Q_{Lb}-Q_{La}-Q_{Lc})^{-1}}{(Q_{Lc}-Q_{Lb}-Q_{La})^{-1}+(Q_{La}-Q_{Lb}-Q_{Lc})^{-1}+(Q_{Lb}-Q_{La}-Q_{Lc})^{-1}} \end{bmatrix} = \begin{bmatrix} -22.75 \\ -77.58 \\ -24.62 \end{bmatrix} \tag{5.22}$$

With the obtained TCLC impedances above, the final firing angles can be found through (5.8)–(5.10) as:

$$\begin{bmatrix} \alpha_a \\ \alpha_b \\ \alpha_c \end{bmatrix} = \begin{bmatrix} \alpha_{0,a} \\ \alpha_{0,b} \\ \alpha_{0,c} \end{bmatrix} - \begin{bmatrix} \varphi_a \\ \varphi_b \\ \varphi_c \end{bmatrix} = \begin{bmatrix} 145.4° \\ 122.3° \\ 141.8° \end{bmatrix} - \begin{bmatrix} -16.6° \\ -1.7° \\ 17.6° \end{bmatrix} = \begin{bmatrix} 162.0° \\ 124.0° \\ 124.2° \end{bmatrix} \tag{5.23}$$

Then the corresponding source active power P_{sx} and reactive power Q_{sx} after TCLC compensation can be calculated by using the above X_{af}, X_{bf} and X_{cf} values or simulated by using α_a, α_b and α_c, which will be discussed in the following. If the conditions: $P_{sa}= P_{sb}= P_{sc}$ and $Q_{sa}= Q_{sb}= Q_{sc}= 0$ are satisfied, it means that TCLC part can balance the active power and compensate reactive power of the unbalanced loading in (5.21). Thus, the validity of the proposed method can be verified.

With the required TCLC impedances in (5.22), load voltage $\vec{V}_x$ and the help of (5.3), the compensating current $\vec{I}_{cxf}$ can be calculated as:

$$\begin{bmatrix} \vec{I}_{caf} \\ \vec{I}_{cbf} \\ \vec{I}_{ccf} \end{bmatrix} = \begin{bmatrix} -j\left(\frac{X_b+X_c}{m}\right) & j\frac{X_{cf}}{m} & j\frac{X_{bf}}{m} \\ j\frac{X_{cf}}{m} & -j\left(\frac{X_{af}+X_{cf}}{m}\right) & j\frac{X_{af}}{m} \\ j\frac{X_b}{m} & j\frac{X_{af}}{m} & -j\left(\frac{X_{af}+X_{bf}}{m}\right) \end{bmatrix} \cdot \begin{bmatrix} \vec{V}_a \\ \vec{V}_b \\ \vec{V}_c \end{bmatrix}$$
$$= \begin{bmatrix} 4.16\angle 72.98^\circ \\ 1.88\angle -30.72^\circ \\ 4.14\angle -133.35^\circ \end{bmatrix} \tag{5.24}$$

where $\begin{bmatrix} X_a/m \\ X_b/m \\ X_c/m \end{bmatrix} = \begin{bmatrix} \frac{X_{af}}{X_{af}X_{bf}+X_{bf}X_{cf}+X_{cf}X_{af}} \\ \frac{X_{bf}}{X_{af}X_{bf}+X_{bf}X_{cf}+X_{cf}X_{af}} \\ \frac{X_{cf}}{X_{af}X_{bf}+X_{bf}X_{cf}+X_{cf}X_{af}} \end{bmatrix} = \begin{bmatrix} 5.37\times 10^{-3} \\ 18.32\times 10^{-3} \\ 5.81\times 10^{-3} \end{bmatrix}$

The phase load current $\vec{I}_{Lxf}$ can be calculated as:

$$\begin{bmatrix} \vec{I}_{Laf} \\ \vec{I}_{Lbf} \\ \vec{I}_{Lcf} \end{bmatrix} = \begin{bmatrix} \left[(P_{La}+jQ_{La})/\vec{V}_a\right]^* \\ \left[(P_{Lb}+jQ_{Lb})/\vec{V}_b\right]^* \\ \left[(P_{Lc}+jQ_{Lc})/\vec{V}_c\right]^* \end{bmatrix} = \begin{bmatrix} 4.51\angle -61.93^\circ \\ 3.76\angle -149.50^\circ \\ 5.99\angle 79.14^\circ \end{bmatrix} \tag{5.25}$$

From (5.24) and (5.25), the source current after TCLC compensation can be calculated as:

$$\begin{bmatrix} \vec{I}_{saf} \\ \vec{I}_{sbf} \\ \vec{I}_{scf} \end{bmatrix} = \begin{bmatrix} \vec{I}_{Laf} \\ \vec{I}_{Lbf} \\ \vec{I}_{Lcf} \end{bmatrix} + \begin{bmatrix} \vec{I}_{caf} \\ \vec{I}_{cbf} \\ \vec{I}_{ccf} \end{bmatrix} = \begin{bmatrix} 3.34\angle 0^\circ \\ 3.34\angle -120^\circ \\ 3.34\angle 120^\circ \end{bmatrix} \tag{5.26}$$

Moreover, the active power and reactive power at source side can be calculated as:

$$\begin{bmatrix} P_{sa}+jQ_{sa} \\ P_{sb}+jQ_{sb} \\ P_{sc}+jQ_{sc} \end{bmatrix} = \begin{bmatrix} \vec{V}_a \cdot \vec{I}^{\,*}_{saf} \\ \vec{V}_b \cdot \vec{I}^{\,*}_{sbf} \\ \vec{V}_c \cdot \vec{I}^{\,*}_{scf} \end{bmatrix} = \begin{bmatrix} 367+j0 \\ 367+j0 \\ 367+j0 \end{bmatrix} \tag{5.27}$$

In (5.27), as $P_{sa}= P_{sb}= P_{sc}$ and $Q_{sa}= Q_{sb}= Q_{sc}= 0$, it can be clearly shown that the calculated TCLC impedances in Fig. 5.14 can balance and compensate the active and reactive power. Moreover, under unbalanced loading as shown in (5.21), Fig. 5.14 shows the simulated Q_{sx} and P_{sx} before and after compensation by applying the proposed TCLC control method with the firing angles α_x obtained in (5.24). In Fig. 5.14, Q_{sx} have been compensated to zero while P_{sx} are balanced after the TCLC compensation. Besides, the simulation results are consistent with the

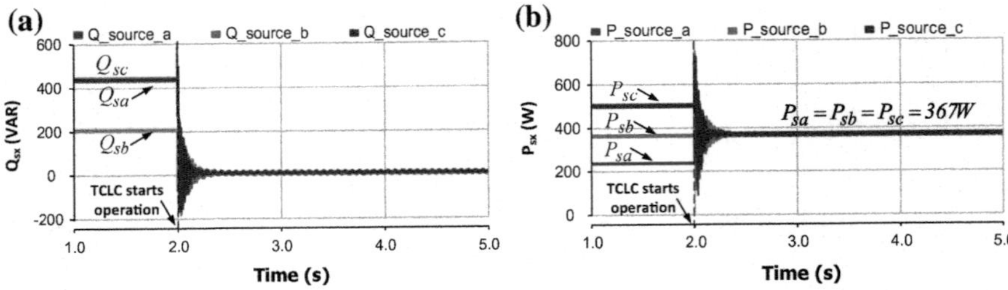

Fig. 5.14 Simulated source reactive power Q_{sx} and active power P_{sx} before and after TCLC compensation: **a** Q_{sx} and **b** P_{sx}

theoretical results. Therefore, from the results shown in (5.27) and Fig. 5.14 of the case study, it can be verified that the proposed TCLC control method can balance active power and compensate reactive power.

References

1. H. Akagi, Y. Kanazawa, A. Nabae, Instantaneous reactive power compensators comprising switching devices without energy storage components. IEEE Trans. Ind. Appl. **IA-20**(3), 625–630 (1984)
2. H. Haibing, X. Yan, Design considerations and fully digital implementation of 400-Hz active power filter for aircraft applications. IEEE Trans. Ind. Electron. **61**(8), 3823–3834 (2014)
3. Y. Hu, Z. Zhu, K. Liu, Current control for dual three-phase permanent magnet synchronous motors accounting for current unbalance and harmonics. IEEE Trans. Emerg. Sel. Topics Power Electron. **2**(2), 272–284 (2014)
4. P. Salmeron, S.P. Litran, A control strategy for hybrid power filter to compensate four-wires three-phase systems. IEEE Trans. Power Electron. **25**(7), 1923–1931 (2010)
5. W.C. Lee, T.K. Lee, D.S. Hyun, A three-phase parallel active power filter operating with PCC voltage compensation with consideration for an unbalanced load. IEEE Trans. Power Electron. **17**(5), 807–814 (2002)
6. S. Senini, P.J. Wolfs, Hybrid active filter for harmonically unbalanced three phase three wire railway traction loads. IEEE Trans. Power Electron. **15**(4), 702–710 (2000)
7. S. Rahmani, K. Al-Haddad, F. Fnaiech, A three phase shunt hybrid power filter adopted a general algorithm to compensate harmonics, reactive power and unbalanced load under nonideal mains voltage, in *Proceedings of IEEE International Conference on Industrial Technology. IEEE ICIT04* (2004), pp. 651–656
8. M. Aredes, H. Akagi, E.H. Watanabe, E. Vergara Salgado, L.F. Encarnacao, Comparisons between the p–q and p–q–r theories in three-phase four-wire systems. IEEE Trans. Power Electron. **24**(4), 924–933 (2009)
9. B. Wen, D. Boroyevich, R. Burgos, P. Mattavelli, Z. Shen, Analysis of D-Q small-signal impedance of grid-tied inverters. IEEE Trans. Power Electron. **31**(1), 675–687 (2016)
10. S. Srianthumrong, H. Akagi, A medium-voltage transformerless AC/DC Power conversion system consisting of a diode rectifier and a shunt hybrid filter. IEEE Trans. Ind. Appl. **39**, 874–882 (2003)
11. S. Rahmani, A. Hamadi, K. Al-Haddad, A Lyapunov-function-based control for a three-phase shunt hybrid active filter. IEEE Trans. Ind. Electron. **59**(3), 1418–1429 (2012)

12. L. Shaohua, W. Xiuli, Y. Zhiqing, L. Tai, P. Zhong, Circulating current suppressing strategy for MMC-HVDC based on nonideal proportional resonant controllers under unbalanced grid conditions. IEEE Trans. Power Electron. **30**(1), 387–397 (2015)
13. X. Guo, W. Liu, X. Zhang, X. Sun, Z. Lu, J.M. Guerrero, Flexible control strategy for grid-connected inverter under unbalanced grid faults without PLL. IEEE Trans. Power Electron. **30**(4), 1773–1778 (2015)
14. K. Ma, W. Chen, M. Liserre, F. Blaabjerg, Power controllability of a three-phase converter with an unbalanced AC source. IEEE Trans. Power Electron. **30**(3), 1591–1604 (2015)
15. M. Castilla, J. Miret, A. Camacho, L. Garcia de Vicuna, J. Matas, Modeling and design of voltage support control schemes for three-phase inverters operating under unbalanced grid conditions. IEEE Trans. Power Electron. **29**(11), 6139–6150 (2014)
16. L.S. Czarnecki, S.E. Pearce, Compensation objectives and current' physical components-based generation of reference signals for shunt switching compensator control. IET Power Electron. **2**(1), 33–41 (2009)
17. L.S. Czarnecki, P.M. Haley, Unbalanced power in four-wire systems and its reactive compensation. IEEE Trans. Power Del. **30**(1), 53–63 (2015)
18. S. Rahmani, A. Hamadi, K. Al-Haddad, A combination of shunt hybrid power filter and thyristor-controlled reactor for power quality. IEEE Trans. Ind. Electron. **61**(5), 2152–2164 (2014)
19. C.S. Lam, X.X. Cui, W.H. Choi, M.C. Wong, Y.D. Han, Minimum inverter capacity design for three-phase four-wire LC-hybrid active power filters. IET, Power Electron. **5**(7), 956–968 (2012)
20. C.-S. Lam, W.-H. Choi, M.-C. Wong, Y.-D. Han, Adaptive dc-link voltage controlled hybrid active power filters for reactive power compensation. IEEE Trans. Power Electron. **27**(4), 1758–1772 (2012)
21. C.-S. Lam, M.-C. Wong, W.-H. Choi, X.-X. Cui, H.-M. Mei, J.-Z. Liu, Design and performance of an adaptive low-dc-voltage-controlled LC-Hybrid active power filter with a neutral inductor in three-phase four-wire power systems. IEEE Trans. Ind. Electron. **61**(6), 2635–2647 (2014)
22. W.-H. Choi, C.-S. Lam, M.-C. Wong, Y.-D. Han, Analysis of dc-link voltage controls in three-phase four-wire hybrid active power filters. IEEE Trans. Power Electron. **28**(5), 2180–2191 (2013)
23. V. Khadkikar, A. Chandra, B.N. Singh, Generalized single-phase p-q theory for active power filtering: simulation and DSP-based experimental investigation. IET Power Electron. **2**, 67–78 (2009)
24. L. Wang et al., Non-linear adaptive hysteresis band pulse width modulation control for hybrid active power filters to reduce switching loss. IET Power Electron. **8**(11), 2156–2167 (2015)
25. F.R. Quintela, J.M.G. Arevalo, R.C. Redondo, Power analysis of static var compensators. Electr. Power Syst. Res. **30**(6), 376–382 (2008)
26. IEEE Standard 519-2014, IEEE recommended practices and requirements for harmonic control in electrical power systems (2014)

Chapter 6
Minimizing Inverter Capacity Design and Comparative Performance Evaluation of Static Var Compensator Coupling Hybrid Active Power Filters (SVC-HAPFs)

Abstract The static var compensator (SVC) coupling hybrid active power filters (SVC-HAPFs) which consist of a SVC part and an active inverter part have been proposed for power quality compensation recently. The cost of SVC part is much lower than that of active inverter part, thus the reduction of power rating ratio between the active inverter part and SVC part can lead to a decrease in the total cost of SVC-HAPF. However, the SVC-HAPF with too low inverter capacity may fail to perform satisfactory compensation. In this chapter, minimizing inverter capacity design and comparative performance evaluation study of different SVC-HAPFs are proposed. At first, the power rating ratio analysis between the active inverter part and SVC part is proposed. With different SVC parts consideration, the required active inverter capacity of SVC-HAPFs is deduced and analyzed. Finally, simulation and laboratory-scaled experimental results are presented to show and verify the performance and the deduced required active inverter capacity of different SVC-HAPFs. Through minimizing inverter capacity design, simulation and experimental result evaluations, a SVC-HAPF structure is recommended for better compensation performance with low inverter capacity.

Keywords Power quality · Static var compensator (SVC) · Hybrid active power filter (HAPF) · Compensation range · Low DC-link voltage · Inverter capacity

6.1 Introduction

The fixed capacitors and thyristor-controlled reactor (FC-TCR) and thyristor controlled LC (TCLC) as two types of static var compensator (SVC) structures can be used in SVC coupling hybrid active power filters (SVC-HAPFs) for power quality compensation. Thus, the SVC-HAPFs are named as FC-TCR-HAPF and TCLC-HAPF, respectively for further study in this chapter. The cost of SVC part is much lower than that of active inverter part (shown in Fig. 2.4 in Sect. 2.4.2), thus, the reduction of power rating ratio between the active inverter part and SVC part

L. Wang et al., *Adaptive Hybrid Active Power Filters*, Power Systems,
https://doi.org/10.1007/978-981-10-8827-8_6

can lead to a decrease in the total cost of SVC-HAPF. However, the SVC-HAPF with low inverter capacity may fall to perform satisfactory results. Therefore, the minimizing inverter capacity design and comparative performance evaluation of FC-TCR-HAPF and TCLC-HAPF have been proposed and discussed in this chapter.

As the cost of active inverter part is usually much higher than passive part and SVC part [1], researchers have proposed different structures and design methods to reduce the power rating of the active inverter part [1–6]. Lam et al. [2] deduced and analyzed minimum DC-link voltage of active power filters (APFs) for reactive power and current harmonics compensation. However, due to the inherent limitation of APFs, the required minimum DC-link voltage still need to maintain at high voltage level (larger than source voltage). The similar minimum DC-link voltage analysis idea in [2] has also implemented in LC-coupling hybrid active power filters (LC-HAPFs) to develop an adaptive DC voltage control strategy [3, 4]. Beside of those DC-link voltage analysis methods, researches in [4] also suggest to add a tuned coupling neutral inductor to further reduce DC-link voltage of the LC-HAPF. Moreover, in high-speed co-phase traction power supply system, the parameter design procedures of hybrid power quality conditioner (HPQC) has been explored in [5] for minimum DC-link voltage operation. Unfortunately, even though minimizing the DC-link operating voltage can lower the system initial cost, switching loss and switching noise, this action for both LC-HAPFs and HPQC leads to further reduce the originally narrow compensation range. To keep a wide compensation range and low active inverter rating simultaneously, the combined systems of SVC in parallel with APF (SVC//APF) [7], SVC in parallel with HAPF (SVC//HAPF) [8, 10], SVC-HAPF [11, 12] have been introduced. However, the minimizing inverter capacity design and the design of power rating ratio between the active inverter part and SVC part have not been presented in [7–12]. As the cost and control complexity of the SVC-HAPFs [11, 12] are the lowest and simplest among SVC//APF and SVC//HAPF, the SVC-HAPF has large potential to further develop.

In the existing literatures [11, 12], the required DC-link voltage for harmonic compensation has not been discussed and included, in which both harmonic current portions injected by SVC and nonlinear loading should actually be taken into consideration. Due to the limitations among the existing literatures [11, 12], this chapter aims to:

- reduce the overall SVC-HAPF cost by appropriately design the power rating ratio between active inverter part and SVC part through mathematical analysis;
- propose a minimizing DC-link voltage (inverter capacity) design for SVC-HAPFs reactive power and harmonic current compensation generated by nonlinear load and SVC part in comparison among different SVC parts;
- provide representative simulation (10 kV–5 MVA) and laboratory-scaled experimental (110 V–5 kVA) results to evaluate and verify the performance and the deduced active inverter capacity of different SVC-HAPFs.

In this chapter, the system configuration and modeling of the SVC-HAPFs are introduced in Sect. 6.2. In Sect. 6.3, the parameter design and characteristics of different coupling SVC parts of SVC-HAPFs are studied and analyzed. Then, the power rating ratio ($R_{\_tot}$) between active inverter part and SVC part, and also the required minimum DC-link voltage (V_{DC_tot}) of different SVC-HAPFs are deduced. Finally, simulation (Sect. 6.4) and experimental results (Sect. 6.5) are provided to evaluate and verify the performance and the deduced active inverter capacity of different SVC-HAPFs. And, a SVC-HAPF structure is also recommended for better compensation performance with low inverter capacity.

6.2 Circuit Configuration and Modeling of SVC-HAPFs

The three-phase three-wire SVC-HAPFs are given in Fig. 6.1. For the below analysis, the subscript 'x' denotes phase $x = a, b, c$. v_{sx}, v_{Lx} and v_{invx} are the system voltage, load voltage and inverter output voltage respectively. L_s is the system inductance. i_{sx}, i_{Lx} and i_{cx} are the source, load and compensating current respectively. C_{DC} and V_{DC_tot} are DC-link capacitor and DC-link voltage. The SVC part can be FC-TCR or thyristor controlled-LC (TCLC). Thus, the SVC-HAPFs are named as FC-TCR-HAPF and TCLC-HAPF to differentiate them in the following discussions. For the FC-TCR, L_{TCR} and C_{FC} are the inductor of TCR and capacitor of FC. For TCLC part, L_c is the coupling inductor, L_{PF} is the TCLC inductor and C_{PF} is the TCLC capacitor.

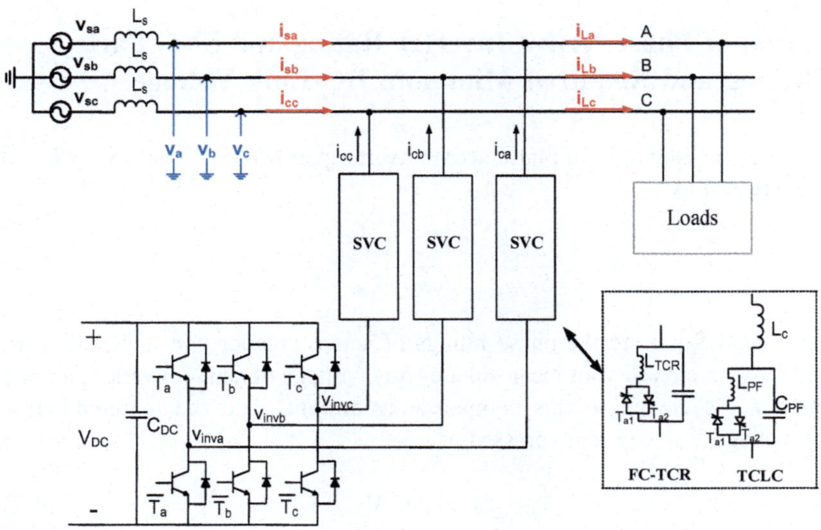

Fig. 6.1 The three-phase three-wire SVC-HAPFs

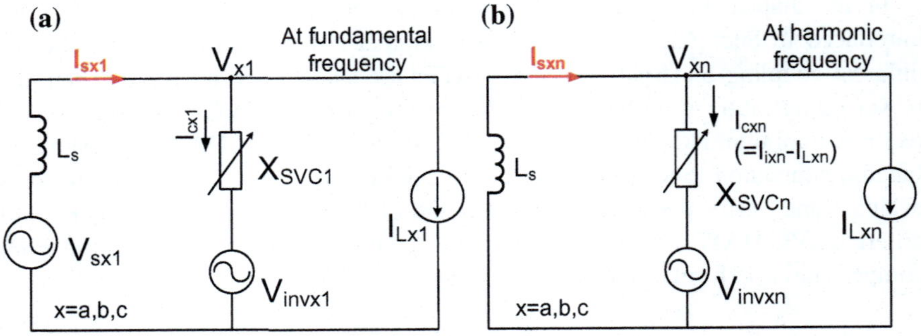

Fig. 6.2 Equivalent circuit models of the SVC-HAPFs: **a** fundamental frequency model and **b** harmonic frequency model

Among them, the one with both the low inverter capacity and better compensation performance will be selected as the recommended SVC-HAPF structure. The equivalent fundamental and harmonic frequency circuit models of SVC-HAPF are shown in Fig. 6.2. In fundamental frequency model (Fig. 6.2a), the purpose of SVC part (FC-TCR or TCLC) of SVC-HAPF is used to dynamically compensate the fundamental reactive power of the loading by controlling their firing angle. In harmonic frequency model (Fig. 6.2b), the small rating active inverter part is used to enlarge the reactive power compensation range and compensate the harmonic current which includes the self-harmonic injection current by SVC (I_{ixn}) and the loading harmonic current (I_{Lxn}).

6.3 Ratio of Phase Active Inverter Rating and SVC Part Rating and Required Minimum DC-Link Voltage

Based on Fig. 6.1, the ratio of phase active inverter part and SVC part (S_{invx}/S_{SVCx}) can be expressed as:

$$R_{tot} = \frac{S_{invx}}{S_{SVCx}} = \frac{V_{invx} \cdot I_{cx}}{V_{SVCx} \cdot I_{cx}} = \frac{V_{invx}}{V_{SVCx}} \tag{6.1}$$

where S_{invx} and S_{SVCx} are the phase ratings of active inverter part and SVC part, V_{invx} and V_{SVCx} are phase root mean square (rms) voltage of active inverter part and SVC part, I_{cx} is the phase rms compensating current. And the required phase DC-link voltage V_{DC} can be expressed as:

$$V_{DC\_tot} = \sqrt{6} \cdot V_{invx} \tag{6.2}$$

Based on the models in Fig. 6.2a and b, the current passing through the SVC-HAPFs include both fundamental and harmonic components. Therefore, the V_{SVCx} and V_{invx} in (6.1) and (6.2) can be expressed as:

$$V_{SVCx} = \sqrt{V_{SVCx1}^2 + \sum_{n=2}^{\infty} V_{SVCxn}^2} = \sqrt{(X_{SVC1} \cdot I_{cx1})^2 + \sum_{n=2}^{\infty} (X_{SVCn} \cdot I_{cxn})^2} \quad (6.3)$$

$$V_{invx} = \sqrt{V_{invx1}^2 + \sum_{n=2}^{\infty} V_{invxn}^2} = \sqrt{V_{invx1}^2 + \sum_{n=2}^{\infty} (X_{SVCn} \cdot I_{cxn})^2} \quad (6.4)$$

Based on (6.3) and (6.4), the R_{tot} and V_{DC_tot} also can be expressed in fundamental and harmonic components as:

$$R_{tot} = \sqrt{\frac{V_{invx1}^2}{V_{SVCx1}^2 + \sum_{n=2}^{\infty} V_{SVCxn}^2} + \frac{\sum_{n=2}^{\infty} V_{invxn}^2}{V_{SVCx1}^2 + \sum_{n=2}^{\infty} V_{SVCxn}^2}} = \sqrt{R_{tot1}^2 + \sum_{n=2}^{\infty} R_{totn}^2} \quad (6.5)$$

$$V_{DC\_tot} = \sqrt{6 \cdot V_{invx1}^2 + 6 \cdot \sum_{n=2}^{\infty} V_{invxn}^2} = \sqrt{V_{DC\_tot1}^2 + \sum_{n=2}^{\infty} V_{DC\_totn}^2} \quad (6.6)$$

In above (6.1)–(6.6), the subscript 'n' stands for the fundamental ($n = 1$) and harmonic ($n \geq 2$) frequency components respectively. X_{SVC1} and X_{SVCn} are the impedance of SVC part at fundamental and each harmonic frequency order. V_{invx1} and V_{invxn} are the fundamental and each harmonic inverter output voltage, I_{cx1} and I_{cxn} are the fundamental and each harmonic compensating current, where I_{cxn} includes two parts: I_{ixn} represents the self-harmonic injection current components by SVC and I_{Lxn} represents the loading harmonic current components. R_{tot1} and V_{DC_tot1}, and R_{totn} and V_{DC_totn} are the fundamental and harmonic components of power rating ratio R_{tot} and DC-link voltage of V_{DC_tot}, respectively. In the following analysis, the R_{tot} in (6.5) and V_{DC_tot} in (6.6) will be deduced, compared and discussed in following parts:

- In Sect. 6.3.1, the parameter design and characteristics of FC-TCR and TCLC (X_{SVC1} and X_{SVCn}) will be proposed and discussed;
- In Sect. 6.3.2, the fundamental components (R_{tot1} and V_{DC_tot1}) of R_{tot} and V_{DC_tot} are deduced;
- In Sect. 6.3.3, the harmonic components (R_{totn} and V_{DC_totn}) of R_{tot} and V_{DC_tot} are obtained;
- In Sect. 6.3.4, the required R_{tot} and V_{DC_tot} are deduced and discussed;
- In Sect. 6.3.5, a summary is drawn.

6.3.1 The Parameter Design and Characteristics of FC-TCR and TCLC

Referred to Fig. 6.1, the expressions of the impedance of FC-TCR ($X_{FC\text{-}TCRn}$) and TCLC (X_{TCLCn}) at fundamental (n = 1) and harmonic frequency order (n = 2, 3, 4 …) can be given as:

$$X_{FC-TCRn}(\alpha) = \left| \frac{\pi(n\omega L_{FC})}{(2\pi - 2\alpha + \sin 2\alpha) - \pi(n\omega)^2 \cdot L_{TCR}C_{FC}} \right| \tag{6.7}$$

$$X_{TCLCn}(\alpha) = \left| \frac{\pi(n\omega L_{PF})}{(2\pi - 2\alpha + \sin 2\alpha) - \pi(n\omega)^2 \cdot L_{PF}C_{PF}} + n\omega L_c \right| \tag{6.8}$$

where α is the firing angle, $\omega = 2\pi f$, f is the fundamental frequency. In the following, FC-TCR and two different designs of TCLC with zero harmonic impedance point at around 3.7th harmonic order ($X_{TCLC}(n \approx 3.7) = 0$) and the dominant 5th harmonic order ($X_{TCLC}(n \approx 5) = 0$) will be discussed.

At the fundamental frequency as shown in Fig. 6.2a, the compensating reactive power of SVC part can be obtained with assuming the $V_{inv1} = 0$, and the deduction of V_{inv1} are included in Sect. 6.3.2. Therefore, the Q_{cx_SVC} can be expressed as:

$$Q_{cx\_SVC} = \frac{V_{x1}^2}{X_{SVC1}(\alpha)} \tag{6.9}$$

where V_{x1} is rms value of fundamental load voltage, X_{SVC1} for FC-TCR and TCLC can be obtained from (6.7) and (6.8) with n = 1. From Fig. 6.1, the SVC parts (FC-TCR and TCLC) have two back-to-back connected thyristors T_{x1} and T_{x2} for each phase, and they are triggered alternately in every half cycle. When both thyristors T_{x1} and T_{x2} for each phase are turned off for the whole fundamental period (firing angle α = 180°), the SVC part provides the maximum capacitive compensating reactive power for load inductive reactive power compensation $Q_{cx\_SVC(MaxCap)}$ ($= -Q_{Lx(MaxInd)}$). On the other hand, when one of the thyristors (T_{x1} and T_{x2}) is turned on for half of the fundamental period (firing angle α = 90°), the SVC part provides the maximum inductive reactive power $Q_{cx\_SVC(MaxInd)}$ ($= -Q_{Lx(MaxCap)}$). The L_{TCR} and C_{FC} of FC-TCR part, and L_{PF} and C_{PF} of TCLC part are mainly used to cover the entire reactive power compensation range. Therefore, based on (6.7)–(6.9) the L_{TCR} and C_{FC} for FC-TCR part, L_{PF} and C_{PF} of TCLC part can be designed as:

$$C_{FC} = -\frac{Q_{Lx(MaxInd)}}{\omega_1 \cdot V_{x1}^2} \tag{6.10}$$

$$L_{TCR} = \frac{V_{x1}^2}{\omega_1 \cdot Q_{Lx(MaxCap)} + \omega_1^2 \cdot V_{x1}^2 \cdot C_{PF}} \tag{6.11}$$

$$C_{PF} = \frac{Q_{Lx(MaxInd)}}{\omega_1^2 \cdot Q_{Lx(MaxInd)} \cdot L_c - \omega_1 \cdot V_{x1}^2} \tag{6.12}$$

$$L_{PF} = \frac{V_{x1}^2 - \omega_1 L_c Q_{Lx(MaxCap)}}{\omega_1 \cdot Q_{Lx(MaxCap)} - \omega_1^3 \cdot L_c \cdot C_{PF} \cdot Q_{Lx(MaxCap)} + \omega_1 \cdot V_{x1}^2 \cdot C_{PF}} \tag{6.13}$$

where $Q_{Lx(MaxInd)}$ and $Q_{Lx(MaxCap)}$ are the load maximum inductive and capacitive reactive power, respectively.

The purposes of L_c are (a) to filter out the current ripple caused by the power switches of active inverter part and (b) to determine the zero harmonic impedance point ($X_{TCLC}(n) = 0$), which can easily affect the required R_{tot} in (6.5) and V_{DC_tot} in (6.6). For the (a) purpose, the value of the L_c can be designed as:

$$L_c \geq \frac{V_{DC\_tot}}{8 \cdot f_s \cdot \Delta i_{L_c\max}} \tag{6.14}$$

where f_s is the switching frequency of active inverter, $\Delta i_{L_c\max}$ is the maximum allowed output current ripple value. In the following, two different designs of TCLC part with zero harmonic impedance point around 3.7th harmonic order ($X_{TCLC}(n \approx 3.7) = 0$) and the dominant harmonic order 5th harmonic order ($X_{TCLC}(n \approx 5) = 0$) will be discussed. Based on (6.7) and (6.8) and the system parameters as shown in Table 6.1, the impedances of FC-TCR part and TCLC part versus with harmonic order and firing angle can be plotted as shown in Fig. 6.3.

The load harmonic spectrum usually has the characteristic of high harmonic current amplitude at low harmonic order, and decreasing amplitude with increasing harmonic order. From (6.2) and (6.4), the required inverter voltage V_{invx} or DC-link voltage V_{DC_tot} of SVC-HAPFs is proportional to the SVC part harmonic impedance X_{SVCn}.

According to Fig. 6.3a, (6.2) and (6.3), the $X_{SVCn} = X_{FC\text{-}TCRn}$ can provide low impedance for harmonic current, which can reduce the required V_{DC_tot}. However, the FC-TCR does not have an L_c to filter out the switching noise generated by active inverter part, so that the current tracking ability can be low and the switching noise can easily inject into the power grid to deteriorate the compensation performance.

Table 6.1 System parameters of power grid and SVC-HAPFs

	Parameters	Physical values
Power grid	v_x, f, L_s	10 kV, 50 Hz, 0.5 mH
FC-TCR-HAPF	L_{TCR}, C_{FC}	30 mH, 160 μF
TCLC-HAPF ($X_{TCLC}(n \approx 3.7) = 0$)	L_c, L_{PF}, C_{PF}	5 mH, 30 mH, 160 μF
TCLC-HAPF ($X_{TCLC}(n \approx 5) = 0$)	L_c, L_{PF}, C_{PF}	2.5 mH, 30 mH, 160 μF

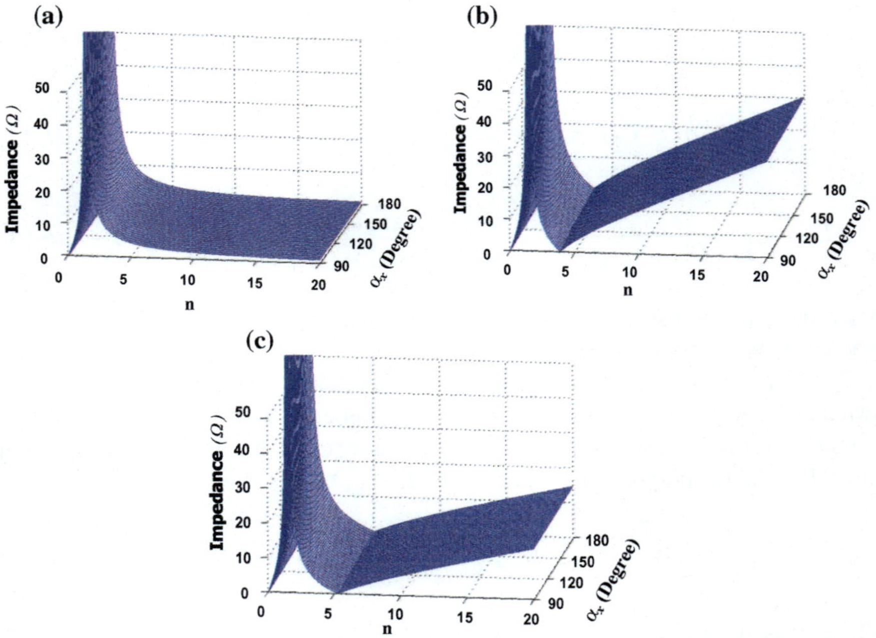

Fig. 6.3 The impedance of SVC part of SVC-HAPFs **a** FC-TCR impedance **b** TCLC impedance with ($X_{TCLC}(n \approx 3.7) = 0$) and **c** TCLC impedance with ($X_{TCLC}(n \approx 5) = 0$)

Figure 6.3b shows the TCLC impedance with $X_{TCLC}(n \approx 3.7) = 0$, which is diverging from the possible current harmonic orders ($6k \pm 1$ with $k = 1, 2, 3\ldots$) of three-phase three-wire power system. Therefore, the self-harmonic current injection by TCLC can be very small (will be discussed in Sect. 6.3.3). But, the required V_{DC_tot} at the high harmonic frequency order is high due to a larger $X_{SVCn} = X_{TCLCn}$ according to (6.2) and (6.3).

Figure 6.3c shows the TCLC part impedance with $X_{TCLC}(n \approx 5) = 0$. With this TCLC design, the required V_{DC_tot} for compensating 5th order loading harmonic current can be minimized. The required V_{DC_tot} at high harmonic frequency order will be lower than $X_{TCLC}(n \approx 3.7) = 0$ case due to smaller X_{SVCn}.

6.3.2 Fundamental Frequency Analysis of R_{tot1} and V_{DC_tot1}

From Fig. 6.2a, the fundamental inverter voltage V_{invx1} can be expressed as:

$$V_{invx1} = V_{x1} - X_{\mathrm{SVC1}}(\alpha) \cdot I_{cx1} \tag{6.15}$$

where V_{invx1}, I_{cx1} and X_{SVC1} is the fundamental load voltage and compensating current and SVC impedance, respectively. The loading reactive power can be expressed as:

$$Q_{Lx} = V_{x1} \cdot I_{Lxq1} = V_{x1} \cdot (-I_{cx1}) \tag{6.16}$$

where Q_{Lx} is the loading reactive power, the I_{Lxq1} is the load fundamental reactive current which is equal to $-i_{cx1}$ after compensation. Combining (6.3), (6.9), (6.15), (6.16), the R_{tot1} and V_{DC_tot1} can be expressed as:

$$R_{tot1} = \frac{V_{invx1}}{V_{SVC1}} = \left| \frac{|Q_{Lx}| - |Q_{cx\_SVC}|}{Q_{Lx}} \right| \tag{6.17}$$

$$V_{DC\_tot1} = \sqrt{6} \cdot V_{invx1} = \sqrt{6} \cdot V_{x1} \left| \frac{|Q_{Lx}| - |Q_{cx\_SVC}|}{Q_{cx\_SVC}} \right| \tag{6.18}$$

In (6.17), the Q_{cx_SVC} has the opposite direction with Q_{Lx}. If the Q_{Lx} is varying within range of $Q_{Lx} \in [-|Q_{cx\_SVC}(\alpha = 90°)|, \ |Q_{cx\_SVC}(\alpha = 180°)|]$, the $|Q_{Lx}| = |Q_{cx\_SVC}|$ and $V_{DC1} = 0$, otherwise, the $R_{tot1} > 0$ and $V_{DC\_tot1} > 0$.

6.3.3 Harmonic Frequency Analysis of R_{totn} and V_{DC_totn}

As shown in Fig. 6.2b, the harmonic compensating current can be expressed as:

$$I_{cxn} = I_{ixn} + (-I_{Lxn}) \tag{6.19}$$

where I_{ixn} is the self-harmonic current injection by SVC and I_{Lxn} is the loading harmonic current.

In this part, the required R_{totn} and V_{DC_totn} will be separately discussed in two parts. In Sect. 6.3.3.1, the required R_{totn} and V_{DC_totn} to compensate i_{ixn} (injected by SVC part) are deduced under assumptions of $I_{cxn} = I_{ixn}$ and $I_{Lxn} = 0$ (linear loading). In Sect. 6.3.3.2, the total required R_n and V_{DCn} are obtained to compensate nonlinear loading with the current $I_{cxn} = I_{ixn} + (-I_{Lxn})$.

6.3.3.1 The R_{totn} and V_{DC_totn} for Compensating i_{ixn} Only

In this part, the required R_{totn} and V_{DC_totn} can be obtained to compensate i_{ixn} injected by SVC part. In order to clearly observe the i_{ixn} problem, the loading is assumed to be linear ($i_{Lxn} = 0$).The equivalent single-phase self-harmonic current rejection analysis model is provided in Fig. 6.4.

In Fig. 6.4, the thyristors (T_{x1} and T_{x2}) in each phase of the SVCs (FC-TCR or TCLC) can be considered as a pair of bidirectional switches, which would generate

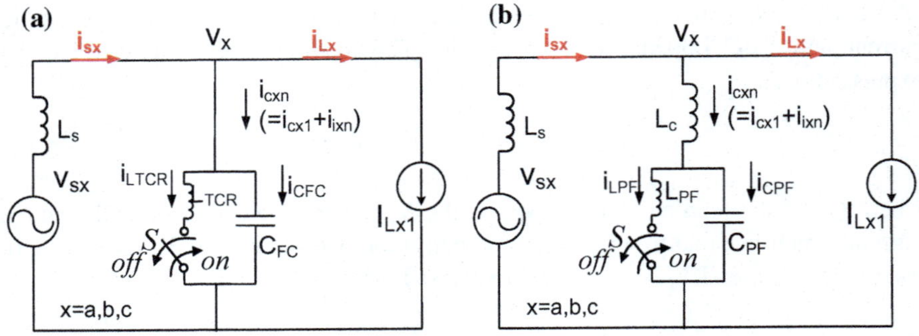

Fig. 6.4 SVC equivalent single phase model for harmonic current rejection analysis **a** FC-TCR and **b** TCLC

low order harmonic current i_{ixn}. During the switch S turns on and off, the two differential equations of $i_{cx}(t)$ (i_{cx_off} and i_{cx_on}) can be obtained as:

$$i_{cx\_off} = i_{cx1} + i_{ixn} = \underset{Fundamental}{A_1 \sin(\omega_1 t - \alpha)} + \underset{Harmonic}{K_1 \sin(\omega_{n1} t + \phi_{n1})} \tag{6.20}$$

$$i_{cx\_on} = i_{cx1} + i_{ixn} = \underset{Fundamental}{A_2 \sin(\omega_1 t + \alpha)} + \underset{Harmonic}{K_2 \cos(\omega_{n2} t + \phi_{n2}) + K_3} \tag{6.21}$$

where A_1 and A_2 are the peak values of the fundamental compensating current during each turn off and turn on; K_1, K_2, K_3, ϕ_{n1} and ϕ_{n2} are constants during each switching cycle, and they depend on the initial conditions of the compensating current and capacitor voltage; ω_{n1} and ω_{n2} are the harmonic angular frequencies. For FC-TCR, the ω_{n1} and ω_{n2} can be obtained as:

$$\omega_{n1} = \frac{1}{\sqrt{L_s C_{FC}}} \left(n_1 = \frac{\omega_1}{2\pi f} = \frac{1}{2\pi f \sqrt{L_s C_{FC}}} \right) \tag{6.22}$$

$$\omega_{n2} = \sqrt{\frac{L_s + L_{TCR}}{L_s L_{TCR} C_{FC}}} \left(n_2 = \frac{\omega_2}{2\pi f} = \frac{1}{2\pi f} \sqrt{\frac{L_s + L_{TCR}}{L_s L_{TCR} C_{FC}}} \right) \tag{6.23}$$

For TCLC part, the ω_{n1} and ω_{n2} can be obtained as:

$$\omega_{n1} = \frac{1}{\sqrt{(L_s + L_c) \cdot C_{PF}}} \left(n_1 = \frac{\omega_1}{2\pi f} = \frac{1}{2\pi f \sqrt{(L_s + L_c) \cdot C_{PF}}} \right) \tag{6.24}$$

$$\omega_{n2} = \sqrt{\frac{L_s + L_c + L_{PF}}{(L_s + L_c) \cdot L_{PF} C_{PF}}} \left(n_2 = \frac{\omega_2}{2\pi f} = \frac{1}{2\pi f} \sqrt{\frac{L_s + L_c + L_{PF}}{(L_s + L_c) \cdot L_{PF} C_{PF}}} \right) \tag{6.25}$$

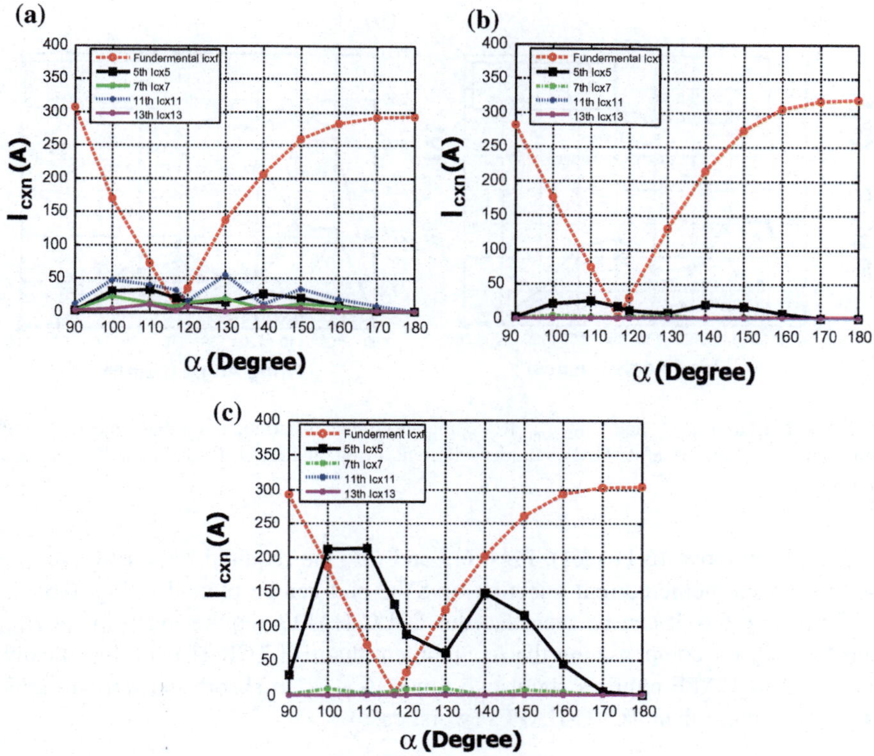

Fig. 6.5 Simulated fundamental and harmonic compensating current under different firing angles with **a** FC-TCR impedance **b** TCLC impedance with ($X_{TCLC}(n \approx 3.7) = 0$) **c** TCLC impedance with ($X_{TCLC}(n \approx 5) = 0$)

where L_s is the system inductor which is usually a small value. For a three-phase three-wire system, the common harmonic orders are $6n \pm 1$ (≥ 5) with $n = 1, 2, 3\ldots$ Based on (6.22) and (6.23), the FC-TCR part may generate low order harmonic current. However, for TCLC part with $X_{TCLC}(n \approx 3.7) = 0$, the possible generated harmonic orders n_1 and n_2 are far from the $6n \pm 1$th order, thus the generated harmonic current can be significantly reduced. For TCLC part with $X_{TCLC}(n \approx 5) = 0$, significant 5th order harmonic current can be generated as the n_1 and n_2 are designed closed to 5.

Based on above analysis, the fundamental and harmonic compensating current under different firing angles are provided in Fig. 6.5.

From Fig. 6.5a, when the FC-TCR is applied, it can be observed that many low order harmonic current are injected into the power grid. On the other hand, when the TCLC ($X_{TCLC}(n \approx 3.7) = 0$) is applied, the injected harmonic current can be significantly reduced as shown in Fig. 6.5b. When TCLC ($X_{TCLC}(n \approx 5) = 0$) is applied as shown in Fig. 6.5c, significant 5th order harmonic current is generated.

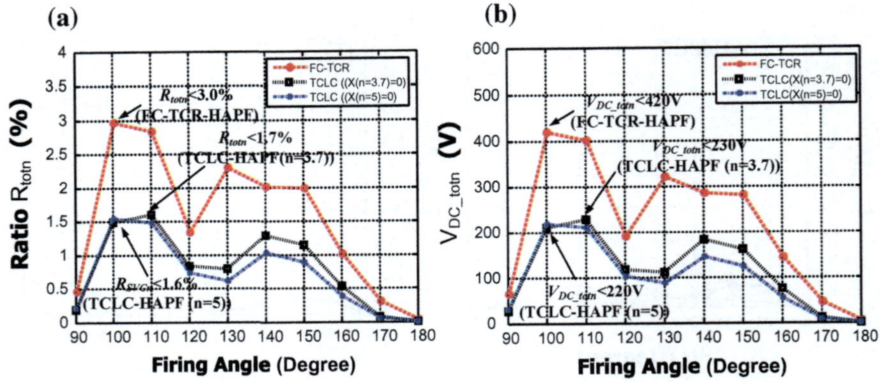

Fig. 6.6 Required R_{totn} and V_{DC_totn} for compensating self-harmonic current injection with different SVC part of SVC-HAPFs under different firing angles: **a** $R_{totn}(= S_{invxn}/S_{SVCxn})$ and **b** V_{DC_totn}

Based on above (6.1)–(6.4), Figs. 6.3 and 6.5, the required R_{SVC} and V_{DC_SVC} due to self-harmonic current injection by SVC part can be plotted in Fig. 6.6.

From Fig. 6.6, it can be seen that the TCLC-HAPFs require much lower R_{totn} and V_{DC_totn} for compensating the i_{Lxn} problem than FC-TCR-HAPF. Specifically, the FC-TCR-HAPF requires about two times V_{DC_totn} to absorb the self-harmonic current injection than TCLC-HAPFs (worst case).

6.3.3.2 The R_n and V_{DCn} for Compensating Both i_{ixn} and i_{Lxn}

Since the SVC has poor harmonic compensation ability, the active inverter part of SVC-HAPF is designed to compensate load harmonic current i_{Lxn} and absorb the SVC self-generated i_{ixn} simultaneously.

The six-pulse rectifier nonlinear loads are commonly used in industrial applications such as AC-DC converters, speed controlled DC motors, and steel hardening machines. In this chapter, the six-pulse rectifier nonlinear loads are considered as the worst case of the three-phase three-wire harmonic loads [13–16]. If the designed V_{DCn} can compensate this kind of nonlinear loads, the SVC-HAPFs can probably compensate other industrial loads. Considering the rating of a nonlinear loading is 10 kV–500 A, the total required $R_{\_tot}$ and V_{DC-tot} of SVC-HAPFs can be plotted as Fig. 6.7 with the help of (6.1)–(6.19).

Figure 6.7 shows that the FC-TCR-HAPF requires the lowest required R_{totn} and V_{DC_totn} among three SVC-HAPFs. Comparing the two designs of TCLC-HAPFs, the required R_{totn} and V_{DC_totn} of TCLC with $X_{TCLC}(n \approx 3.7) = 0$ case is much higher than TCLC with $X_{TCLC}(n \approx 5) = 0$ case. Based on Fig. 6.7, to compensate the commonly used industrial loads with considering i_{ixn} problem as well, the

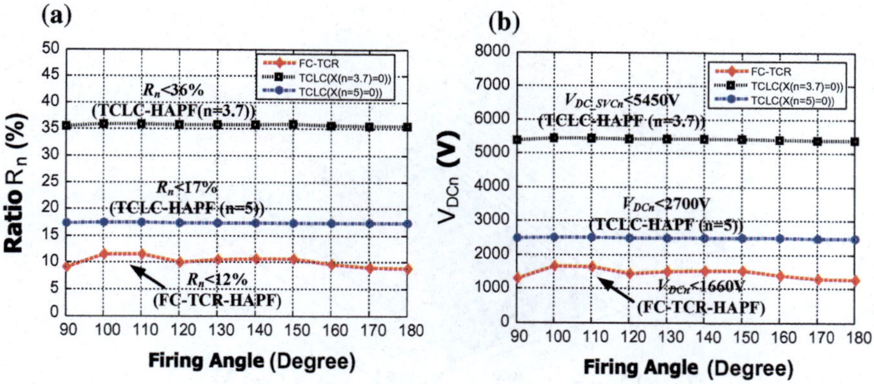

Fig. 6.7 Required R_{totn} and V_{DC_totn} for compensating nonlinear loads and self-harmonic current injection with different SVC part of SVC-HAPFs under different firing angles: **a** $R_{totn}(= S_{invxn}/S_{SVCxn})$ and **b** V_{DC_totn}

V_{DC_totn} are set to be 1660 V (R_{totn} = 12%) for FC-TCR-HAPF, 5450 V (R_{totn} = 36%) for TCLC-HAPF with $X_{TCLC}(n \approx 3.7) = 0$ and 2500 V (R_{totn} = 17%) for TCLC-HAPF with $X_{TCLC}(n \approx 5) = 0$ under 10 kV–500 A nonlinear load consideration.

6.3.4 The Minimizing Inverter Capacity Design of Total R_{tot} and V_{DC_Tot}

In this part, the minimizing inverter capacity design of SVC-HAPFs is discussed which included the fundamental component (discussed in Sect. 6.3.2) and harmonic components (discussed in Sect. 6.3.3). Figures 6.8 and 6.9 show the ratio of R_{tot} and $V_{DC\text{–}tot}$ in terms of Q_{Lx} and I_{Lxn} (also included I_{ixn}) for different SVC-HAPFs.

From Figs. 6.8 and 6.9, at $I_{Lxn} = 0$, it can be seen that the SVC-HAPFs still require the R_{totn} and $V_{DC\text{–}tot} = V_{DC\_totn}$ to absorb the self-harmonic injection current by SVC which is agreed with Fig. 6.6 in Sect. 6.3.3.1. With the I_{Lxn} increasing, the required R_{tot} and $V_{DC\text{–}tot}$ are increasing. To compensated 10 kV–500 A rectifier nonlinear loading with harmonic current I_{Lxn} = 150 A, the required R_{tot} and $V_{DC\text{–}tot}$ in Figs. 6.8 and 6.9 can also be found in Fig. 6.7 of Sect. 6.3.3.2. Moreover, when the Q_{Lx} is outside of the designed Q_{cx_SVC} range, the designed R_{tot} and V_{DC_tot} can further enlarge the reactive power compensation range of SVC parts which is agreed with analysis in Sect. 6.3.2.

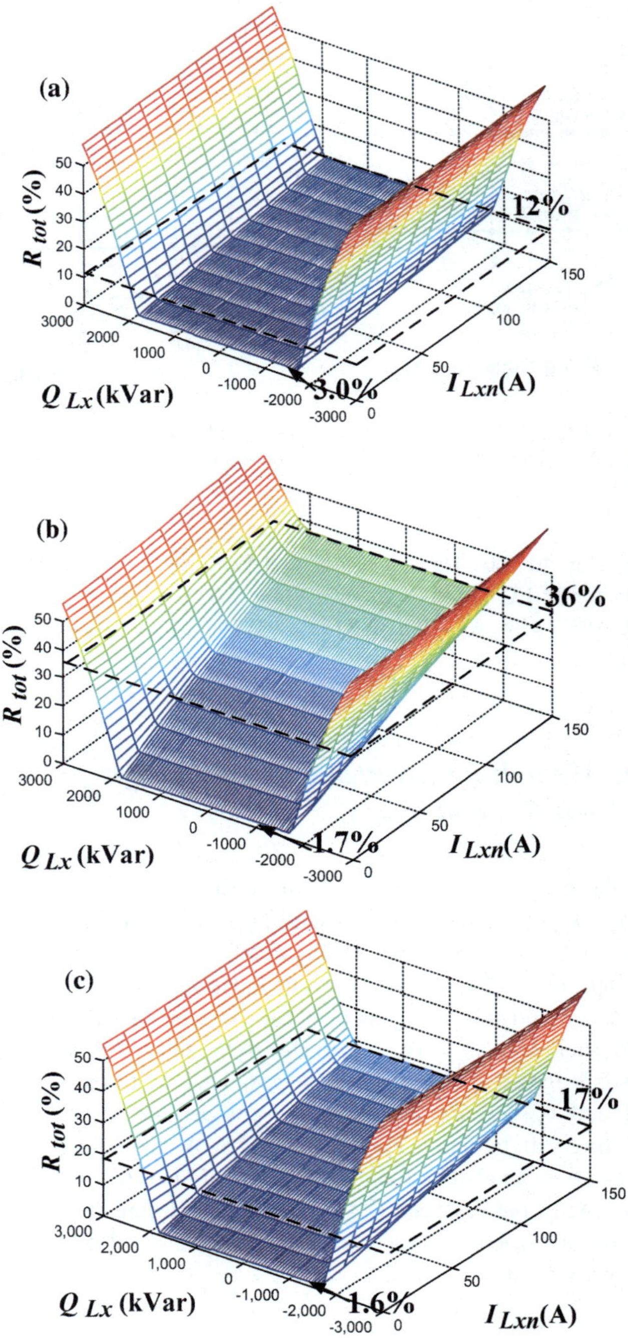

Fig. 6.8 The ratio of R_{tot} (= S_{invx}/S_{SVCx}) in terms of Q_{Lx} and I_{Lxn} for **a** FC-TCR-HAPF **b** TCLC-HAPF with ($X_{TCLC}(n \approx 3.7) = 0$) **c** TCLC-HAPF with ($X_{TCLC}(n \approx 5) = 0$)

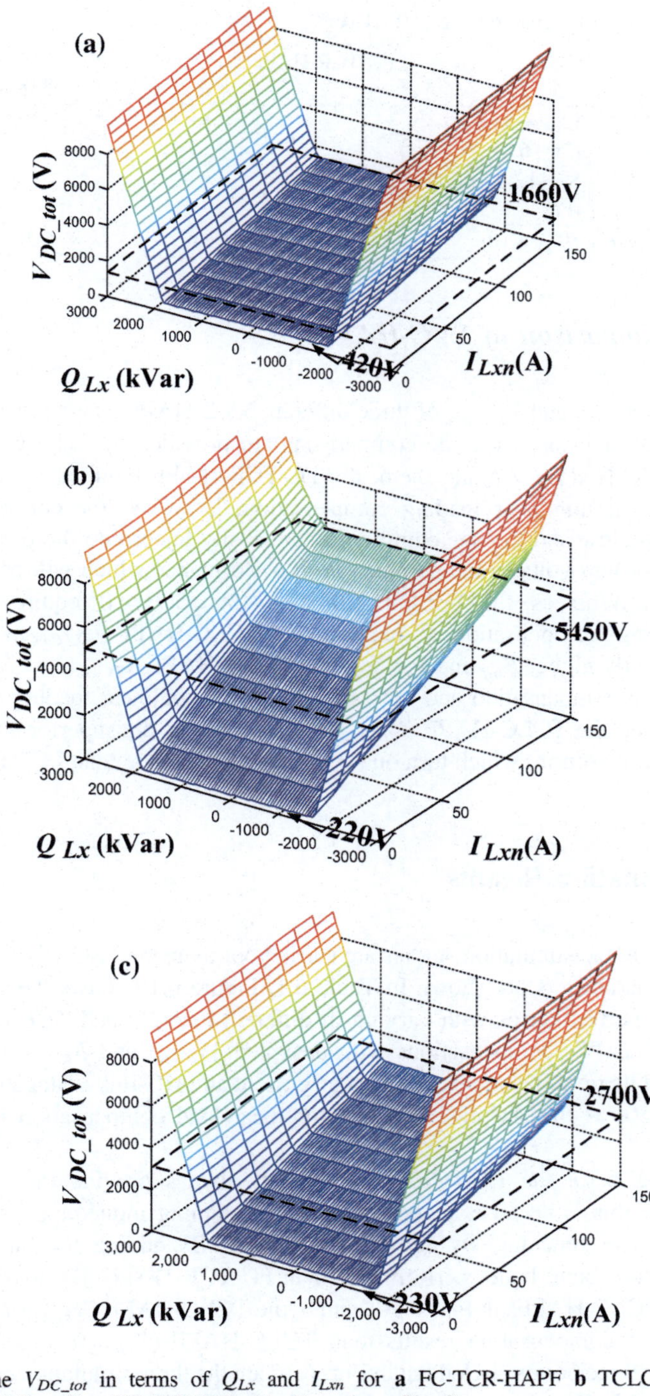

Fig. 6.9 The V_{DC_tot} in terms of Q_{Lx} and I_{Lxn} for **a** FC-TCR-HAPF **b** TCLC-HAPF with ($X_{TCLC}(n \approx 3.7) = 0$) **c** TCLC-HAPF with ($X_{TCLC}(n \approx 5) = 0$)

Table 6.2 Comparison among three SVC-HAPFs

		FC-TCR-HAPF	TCLC-HAPF ($n \approx 3.7$)	TCLC-HAPF ($n \approx 5$)
Required V_{DC_tot} and R_{tot}	Linear load (Fig. 6.6)	*	***	***
	Non-linear load (Fig. 6.7)	***	*	**
Comp. performance (Fig. 6.10)		*	**	***

6.3.5 *Comparison of SVC-HAPFs*

Until now, the R_{tot} and V_{DC_tot} of three different SVC-HAPFs are deduced through mathematical analysis. And, the comparisons are provided in Table 6.2 for three different SVC-HAPFs. Among them, the FC-TCR-HAPF requires the lowest R_{tot} and V_{DC_tot} for nonlinear loading compensation. However, the current tracking ability can be low and the switching noise can easily inject into the power grid to deteriorate system voltage and compensation performance which will be proved in next section. Whereas, the TCLC-HAPF $X_{TCLC}(n \approx 3.7) = 0$ requires lower for linear loading compensation. Whereas, the TCLC-HAPF ($X_{TCLC}(n \approx 5) = 0$) requires slightly higher R_{tot} and V_{DC_tot} than TCLC-HAPF $X_{TCLC}(n \approx 3.7) = 0$ for linear loading compensation and FC-TCR-HAPF for nonlinear loading compensation. However, the TCLC-HAPF ($X_{TCLC}(n \approx 5) = 0$) has the superior performance (proved in nest section) which turn out to be a better structure than FC-TCR-HAPF.

6.4 Simulation Results

In this section, the simulation studies are carried out by using PSCAD/EMTDC and the system parameters are shown in Table 6.1. Figure 6.10 shows the source current THD_{isx} performances with varying of different load PF and THD_{iLx} after different SVC-HAPFs compensations with designed V_{DC_tot} and R_{tot}. And the load voltage (and its ripple), source current, load current and DC-link voltage waveforms for 10 kV–500 A the six-pulse rectifier nonlinear load compensation is given in Fig. 6.11.

From Fig. 6.10, the FC-TCR-HAPF cannot always provide the satisfactory performance this is because it does not have a coupling inductor L_c to filter out switching noise generated by active inverter part. By adding the tuned L_c, the TCLC-HAPFs obtain better performance than FC-TCR-HAPF. By comparing two designs of TCLC-HAPFs in Fig. 6.10a and b, the TCLC-HAPF ($X_{TCLC}(n \approx 5) = 0$) obtains better compensation results than TCLC-HAPF ($X_{TCLC}(n \approx 3.7) = 0$). In addition, the three points circled out in Fig. 6.10 are the three simulation cases which are illustrated in Fig. 6.11.

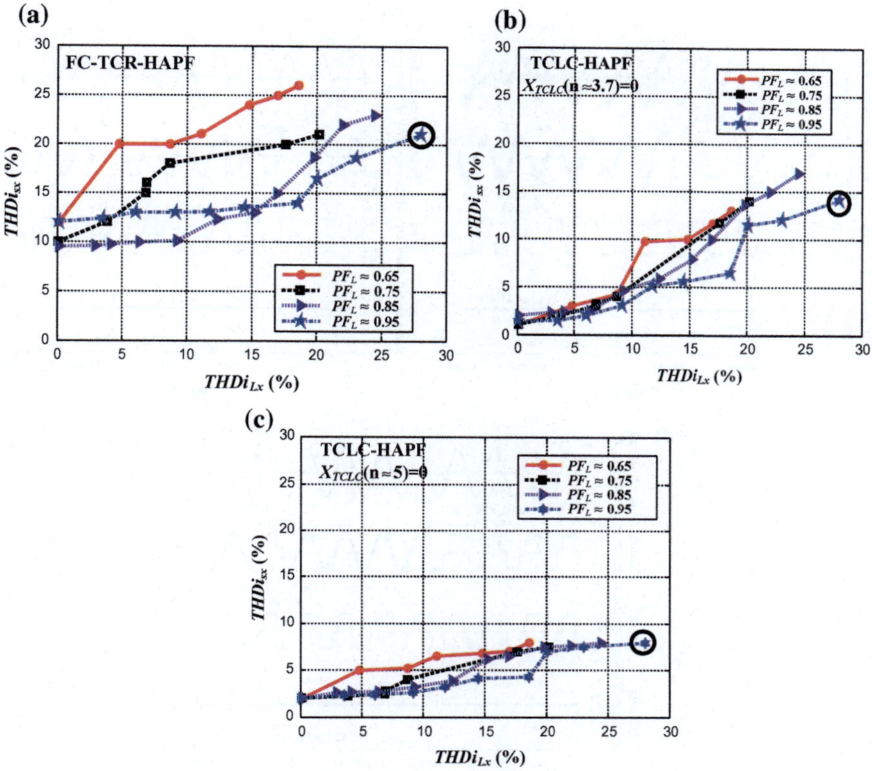

Fig. 6.10 Simulated source current THD_{isx} with varying of different load PF_L and THD_{iLx} with V_{DC_tot} = 2500 V **a** after FC-TCR-HAPF compensation, **b** after TCLC-HAPF (X_{TCLC} ($n \approx 3.7$) = 0) compensation and **c** after TCLC-HAPF ($X_{TCLC}(n \approx 5) = 0$) compensation

From Fig. 6.11, after the FC-TCR-HAPF compensation, the peak voltage ripple of load voltage $V_{rip(p)}$ is about 1.0 kV. Meanwhile, the source current THD_{isx} have been compensated to 21.0% from 28.0% for the worst case loading. In addition, the power factors of the source current (PFs) becomes 0.97 after compensation. After TCLC-HAPF compensations, it can be seen that the source current THD_{isx} are improved to 14.1% for the design of TCLC-HAPF ($X_{TCLC}(n \approx 3.7) = 0$) case and less than 8% for the design of $X_{TCLC}(n \approx 5) = 0$ case. And the peak voltage ripple of load voltage $V_{rip(p)}$ is just about 0.1 kV for both TCLC-HAPF ($X_{TCLC}(n \approx 3.7) = 0$) and ($X_{TCLC}(n \approx 5) = 0$) cases. Besides, the PF is improved to be equal to 0.99. Moreover, it is worth noting that the TCLC-HAPF ($X_{TCLC}(n \approx 5) = 0$) obtains the superior compensating performance.

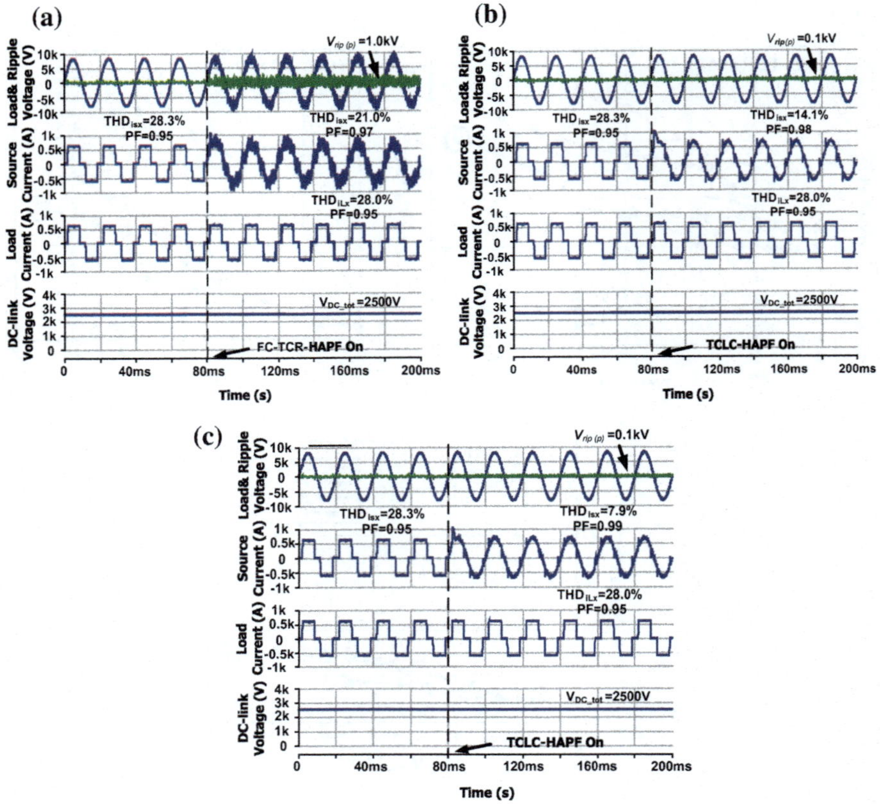

Fig. 6.11 Simulated load voltage (and its ripple), source current, load current and DC-link voltage with $V_{DC\_tot} = 2500$ V after: **a** FC-TCR-HAPF compensation, **b** TCLC-HAPF ($X_{TCLC}(n \approx 3.7) = 0$) compensation and **c** TCLC-HAPF ($X_{TCLC}(n \approx 5) = 0$) compensation

6.5 Experimental Results

Two 110 V–5 kVA experimental prototypes of three-phase three-wire TCLC-HAPFs with $X_{TCLC}(n \approx 3.7) = 0$ and of $X_{TCLC}(n \approx 5) = 0$ are constructed in the laboratory. The control system has a sampling frequency of 25 kHz. The switching devices for the active inverter part and TCLC part are Mitsubishi IGBTs PM300DSA060 and SanRex thyristors PK110FG160.

Figure 6.12 shows the simulated dynamic waveforms of load voltage, source current, load current and DC-link voltage waveforms by applying different SVC-HAPFs with deduced DC-link voltage for different loads compensation.

From Fig. 6.12 and Table 6.3, after the TCLC-HAPF with $X_{TCLC}(n \approx 3.7) = 0$ and $V_{DC\_tot} = 95$ V ($R_{tot} = 36\%$) compensation, the source current $THDi_{sx}$ have been compensated to 5.8% for the worst case loading, 5.3% for 1st loading compensation and 5.0% for 1st and 2nd loading compensation. Meanwhile, the power

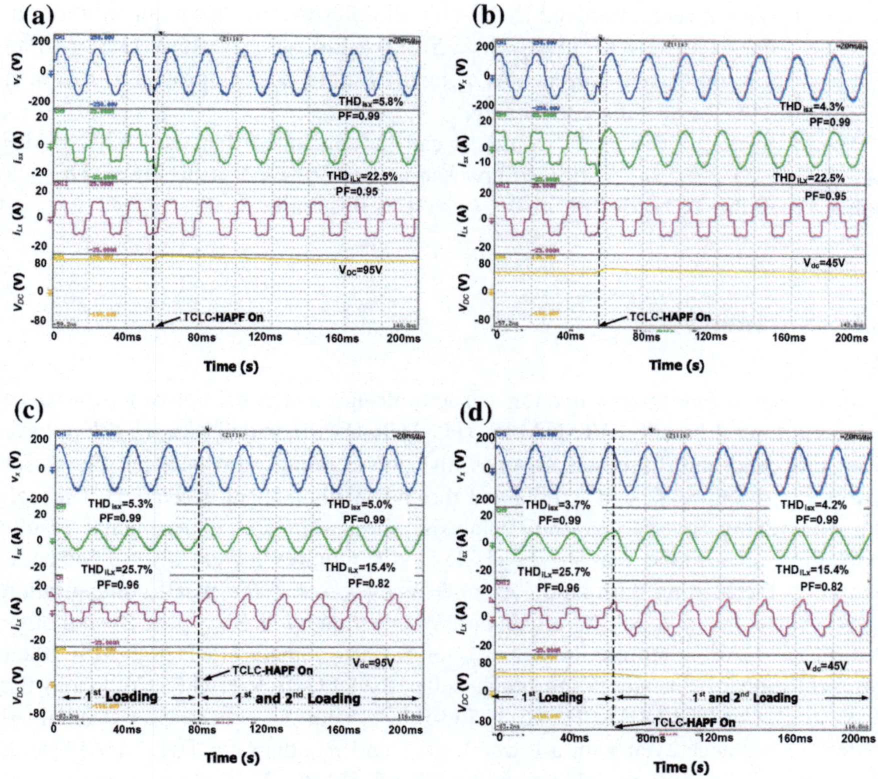

Fig. 6.12 The experimental load voltage (and its ripple), source current, load current and DC-link voltage waveforms for different loading compensation **a** TCLC-HAPF ($X_{TCLC}(n \approx 3.7) = 0$) with $V_{DC\_tot} = 95$ V ($R_{\_tot} = 36\%$) for worst case loading compensation; **b** TCLC-HAPF with $X_{TCLC}(n \approx 5) = 0$ and $V_{DC\_tot} = 45$ V ($R_{\_tot} = 17\%$) for worst case loading compensation; **c** TCLC-HAPF ($X_{TCLC}(n \approx 3.7) = 0$) with $V_{DC\_tot} = 95$ V ($R_{\_tot} = 36\%$) compensation with loading dynamic changing; **d** TCLC-HAPF ($X_{TCLC}(n \approx 5) = 0$) with $V_{DC\_tot} = 45$ V ($R_{\_tot} = 17\%$) compensation with loading dynamic changing

Table 6.3 Compensation results for different loads by applying TCLC-HAPFs

	Rectifier nonlinear load			1st load			1st and 2nd loads		
	PF	$THDi_{sx}$ (%)	V_{dc_tot} (V)	*PF*	$THDi_{sx}$ (%)	V_{dc_tot} (V)	*PF*	$THDi_{sx}$ (%)	V_{dc_tot} (V)
Bef. comp.	0.95	22.5	–	0.96	25.7	–	0.82	15.4	–
TCLC-HAPF ($n \approx 3.7$)	0.99	5.8	95	0.99	5.3	95	0.99	5.0	95
TCLC-HAPF ($n \approx 5$)	0.99	4.3	45	0.99	3.7	45	0.99	4.2	45

factors (PF) have been improved to 0.99 for all different loadings compensation. In contrast, after TCLC-HAPF ($X_{TCLC}(n \approx 5) = 0$ with $V_{DC\_tot} = 45$ V ($R_{tot} = 17\%$) compensation, the source current $THDi_{sx}$ and PF have been improved to less than 4.5% and 0.99 for all different loading.

Based on the experimental results, it can be concluded that the TCLC-HAPF ($X_{TCLC}(n \approx 5) = 0$) has better compensation performance with lower DC-link voltage than the TCLC-HAPF ($X_{TCLC}(n \approx 3.7) = 0$) case.

6.6 Summary

In this chapter, minimizing inverter capacity design and comparative performance evaluation of different SVC-HAPFs (FC-TCR-HAPF and TCLC-HAPFs) have been proposed and discussed. Besides, the power rating ratio between the active inverter part and SVC part is obtained through mathematical derivation. Through the minimizing inverter capacity analysis, simulation and experimental results evaluation, it is found that the FC-TCR-HAPF requires lowest V_{DC_tot} for compensating the loading harmonic current. However, it has the worst compensation performance due to its self-harmonic current injection by its SVC part. Even though the TCLC-HAPFs require higher R_{tot} and V_{DC_tot} than FC-TCR-HAPF, their compensation performances are much better than FC-TCR-HAPF. Compared the two designs of TCLC-HAPF, the TCLC-HAPF with $X_{TCLC}(n \approx 5) = 0$ obtained better performance even with a lower V_{DC_tot} and R_{tot} than the TCLC-HAPF with $X_{TCLC}(n \approx 3.7) = 0$ case. Therefore, the TCLC-HAPF ($X_{TCLC}(n \approx 5) = 0$) turns out to be the superior structure among three SVC-HAPF structures. And the power rating ratio between the active inverter part and SVC part of TCLC-HAPF with $X_{TCLC}(n \approx 5) = 0$ case can be chosen at around 17% for six-pulse nonlinear rectifier load.

References

1. K.W. Lao, M.C. Wong, N. Dai, C.K. Wong, C.S. Lam, A systematic approach to hybrid railway power conditioner design with harmonic compensation for high-speed railway. IEEE Trans. Ind. Electron. **62**(2), 930–942 (2015)
2. C.S. Lam, M.C. Wong, N.Y. Dai, W.H. Choi, X.X. Cui, C.Y. Chung, Switching-loss reduction technique in active power filters without auxiliary circuits. IET Power Electron. **9**(4), 728–742 (2016)
3. C.S. Lam, X.X. Cui, W.H Choi, M.C. Wong, Y.D. Han, Minimum inverter capacity design for three-phase four-wire LC-hybrid active power filters. IET Power Electron. **5**(7), 956–968 (2012)
4. C.S. Lam, M.C. Wong, W.H. Choi et al., Design and performance of an adaptive low-DC-voltage-controlled LC-hybrid active power filter with a neutral inductor in three-phase four-wire power systems. IEEE Trans. Ind. Electron. **61**(6), 2635–2647 (2014)

5. K.W. Lao, M.C. Wong, N.Y. Dai, C.K. Wong, C.S. Lam, Analysis of DC-link operation voltage of a hybrid railway power quality conditioner and its PQ compensation capability in high-speed cophase traction power supply. IEEE Trans. Power Electron. **31**(2), 1643–1656 (2016)
6. L. Wang, C.S. Lam, M.C. Wong et al., Non-linear adaptive hysteresis band pulse-width modulation control for hybrid active power filters to reduce switching loss. IET Power Electron. **8**(11), 2156–2167 (2015)
7. J. Dixon, Y. del Valle, M. Orchard, M. Ortuzar, L. Moran, C. Maffrand, A full compensating system for general loads, based on a combination of thyristor binary compensator, and a PWM-IGBT active power filter. IEEE Trans. Ind. Electron. **50**(5), 982–989 (2003)
8. A. Luo, S. Peng, C. Wu, J. Wu, Z. Shuai, Power electronic hybrid system for load balancing compensation and frequency-selective harmonic suppression. IEEE Trans. Ind. Electron. **59** (2), 723–732 (2012)
9. A. Luo, Z. Shuai, W. Zhu, Z.J. Shen, Combined system for harmonic suppression and reactive power compensation. IEEE Trans. Ind. Electron. **56**(2), 418–428 (2009)
10. B. Chen, C. Zhang, C. Tian, J. Wang, J. Yuan, A hybrid electrical magnetic power quality compensation system with minimum active compensation capacity for V/V cophase railway power supply system. IEEE Trans. Power Electron. **31**(6), 4159–4170 (2016)
11. S. Rahmani, A. Hamadi, K. Al-Haddad, A combination of shunt hybrid power filter and thyristor-controlled reactor for power quality. IEEE Trans. Ind. Electron. **61**(5), 2152–2164 (2014)
12. L. Wang, C.S. Lam, M.C. Wong, A hybrid-STATCOM with wide compensation range and low DC-link voltage. IEEE Trans. Ind. Electron. **63**(6), 3333–3343 (2016)
13. H. Haibing, X. Yan, Design considerations and fully digital implementation of 400-Hz active power filter for aircraft applications. IEEE Trans. Ind. Electron. **61**(8), 3823–3834 (2014)
14. D.E. Rice, A detailed analysis of six-pulse converter harmonic current. IEEE Trans. Ind. Appl. **30**(2), 294–304 (1994)
15. D. Basic, V.S. Ramsden, P.K. Muttik, Harmonic filtering of high power 12-pulse rectifier loads with a selective hybrid filter system. IEEE Trans. Ind. Electron. **48**(6), 1118–1127 (2001)
16. J.C. Brown, J. Allan, B. Mellitt, Six-pulse three phase rectifier bridge model for calculating close up and remote short circuit transients on dc supplied railways. Proc. Inst. Elect. Eng. **138**(pt. B, no. 6) (1991)

Chapter 7
Adaptive DC-Link Voltage Control of Thyristor Controlled LC-Coupling Hybrid Active Power Filter (TCLC-HAPF)

Abstract In this chapter, an adaptive DC-link voltage controlled thyristor controlled LC-coupling hybrid active power filter (TCLC-HAPF) is proposed for reducing switching loss, switching noise and enhancing the compensating performance. Unfortunately, the TCLC-HAPF has both controllable active TCLC part and active inverter part, thus the conventional minimum DC-link voltage calculation methods for active power filter (APF) and *LC*-coupling hybrid active power filter (LC-HAPF) cannot be directly applied to the TCLC-HAPF. Moreover, the aforementioned DC-link voltage calculation methods were developed based on the Fast Fourier Transform (FFT), which makes the calculation complex. This chapter also presents a simplified minimum DC-link voltage calculation method for TCLC-HAPF reactive power and current harmonics compensation, which can significantly reduce the large amount of the calculation steps by using the FFT method. After that, an adaptive DC-link voltage controller for the TCLC-HAPF is developed to dynamically keep its operating at its minimum DC-link voltage level to reducing its switching loss and switching noise. Finally, representative simulation and experimental results are given to verify the proposed simplified DC-link voltage calculation method and the adaptive DC-link voltage control method of TCLC-HAPF.

Keywords Thyristor controlled LC-hybrid active power filter (TCLC-HAPF) Reactive power · Current harmonics · Adaptive DC-link voltage control

7.1 Introduction

With the proliferation and increased use of power electronics devices (nonlinear loads) and motor loadings, such as converters, adjustable speed drives (ASDs), arc furnaces, bulk rectifiers, power supplies, computers, fluorescent lamps, elevators, escalators, large air conditioning systems, compressors, etc. in distribution power systems, the power quality (PQ) problems become more serious, especially for

L. Wang et al., *Adaptive Hybrid Active Power Filters*, Power Systems,
https://doi.org/10.1007/978-981-10-8827-8_7

lower power factor and harmonic pollution [1–11]. To solve the above PQ issues, different PQ compensators have been developed.

The active power filters (APFs) have been widely used for dynamically compensate reactive power and harmonic current. However, the APFs require high DC-link voltage (V_{DC}) for performing compensation, so their initial and operating costs are high [5–11]. Afterwards, different hybrid active power filter (HAPF) topologies composed of APF and passive power filter (PPF) in series and/or parallel have been proposed [4–9], aiming to improve the compensation characteristics of PPFs and reduce the voltage and/or current ratings (costs) of the APFs. *LC*-coupled hybrid active power filters (LC-HAPFs) [12–16] can be considered as a good tradeoff between the system cost and compensation performance, which aims to reduce the DC-link operating voltage. However, the LC-HAPF has a narrow reactive power variation range, which may require a high V_{DC} when it is operating outside its compensation range, thus losing its low V_{DC} characteristic. To enlarge the compensation range and keep at a low rating of active inverter part simultaneously, a thyristor controlled LC-coupling-hybrid active power filters (TCLC-HAPFs) were proposed in 2014 and 2016 for distribution [17, 18] and transmission power systems [19], respectively. The TCLC-HAPF can provide a much wider reactive power compensation range than the LC-HAPF and keeps the low DC-link operating voltage characteristics as the LC-HAPF.

In practical case, there is always a minimum DC-link voltage for PQ compensators performing load reactive power and harmonic current compensation. As the switching loss is directly proportional to the V_{DC} [14, 16, 20], the PQ compensators have higher switching loss if V_{DC} is higher, and vice versa. On the other hand, a sufficient V_{DC} can ensure satisfactory compensation performance. Thus, it is necessary to obtain an appropriate V_{DC} to achieve satisfactory compensation performance with low switching loss and switching noise.

Different minimum V_{DC} design methods for different PQ compensators have been reported among the existing literatures [14–16, 21–24]. In [22–24], the V_{DC} of the APFs are designed to be equal or larger than certain voltage levels, such as line to line voltage peak value [22], $2\sqrt{2}$ times the fundamental output voltage of the active inverter [23], $2\sqrt{2}$ times the root mean square (rms) value of the system source voltage [24]. However, the V_{DC} calculation methods in [22–24] are not related to the loading situation, so that the calculated V_{DC} are not accurate for compensation. To obtain a more accurate V_{DC} requirement for APF, the authors in [21] proposed a detailed deduction analysis of the minimum V_{DC} value for both fundamental reactive power and current harmonic compensation based on the complicated Fast Fourier Transform (FFT). Later on, the minimum V_{DC} calculation method of APF has been extended to the LC-HAPF systems [14–16]. Even through the V_{DC} calculation method for the adaptive V_{DC} controlled LC-HAPF in [14] is based on the single-phase instantaneous p-q theory [25], the work in [14] aims to compensate dynamic reactive power problem only without current harmonics consideration. When the LC-HAPFs are designed to compensate both reactive power and current harmonics simultaneously, similar as APF case, the V_{DC}

calculation still requires the FFT [15, 16], which makes the V_{DC} calculation complex. As the TCLC-HAPF has controllable active TCLC part and active inverter part, it should have different V_{DC} calculation equations in comparison to the APFs and the LC-HAPFs as they have fixed passive L/LC part.

Even though TCLC-HAPF was first proposed in 2014 [17], the derivation of its V_{DC} design was reported in [18]. However, the V_{DC} in [18] was proposed to compensate the fundamental reactive power left by the TCLC part, while the current harmonic components and the loading reactive power over the TCLC part compensation range situation have not taken into consideration. If the V_{DC} design in [18] is directly applied to compensate time-varying nonlinear loads, the TCLC-HAPF may fail to perform satisfactory current quality compensation.

Moreover, the TCLC-HAPF is always operating at a fixed V_{DC} level [17–19], the TCLC-HAPF will obtain a larger switching loss if a higher V_{DC} is used, and vice versa. Therefore, if the V_{DC} can be adaptively changed according to different loading situations, the TCLC-HAPF can achieve better performances and operational flexibility. Besides, if the V_{DC} calculation for the TCLC-HAPF is based on the instantaneous p-q theory instead of using the complicated FFT as in [15, 16, 21], the number of V_{DC} calculation steps and its corresponding processing time can be significantly reduced, which can relax the original complex V_{DC} calculation problem. Due to the limitations among the existing literatures, the contributions of this chapter:

- Propose a V_{DC} calculation method for the TCLC-HAPF, in which the proposed V_{DC} method in [14–16, 21] for the APF or the LC-HAPF cannot be directly applied to the TCLC-HAPF because the TCLC-HAPF has controllable active TCLC part (which is changing dependent of the loading situation) and active inverter part, while the APFs and the LC-HAPFs have fixed passive L/LC part (which is independent of the loading situation);
- Propose a simplified minimum V_{DC} calculation for the TCLC-HAPF reactive power and current harmonics compensation, which can significantly reduce the number of calculation steps compared to the conventional V_{DC} calculation methods based on the FFT [15, 16, 21], thus the digital controller can reduce the required processing time and ensure the system response time and performance;
- Develop an adaptive DC-link voltage controller for the TCLC-HAPF reactive power and current harmonics compensation based on the simplified V_{DC} calculation method;
- With the proposed adaptive DC-link voltage controller, the TCLC-HAPF can operate at its required V_{DC} level, thus lowering the system switching loss and noise.

In this chapter, the background information and motivation of this chapter are introduced in Sect. 7.1. The circuit configuration and modeling of a three-phase three-wire TCLC-HAPF are described in Sect. 7.2. Based on its modeling, the simplified V_{DC} calculation method is proposed in Sect. 7.3 and the adaptive V_{DC} control block is given in Sect. 7.4. Then, representative simulation case studies (in Sect. 7.5) and experimental results (in Sect. 7.6) are provided to verify the deduced

V_{DC} calculation method and the proposed adaptive V_{DC} controller for the TCLC-HAPF. Finally, conclusion is drawn in Sect. 7.7.

7.2 Circuit Configuration of Three-Phase Three-Wire TCLC-HAPF

The circuit configuration of a three-phase three-wire TCLC-HAPF is given in Fig. 7.1. Figure 7.2 shows the single-phase TCLC-HAPF equivalent circuit models at the fundamental and harmonic frequencies. In this TCLC-HAPF topology, the TCLC part and the active inverter part can complement each other's disadvantages. As the TCLC part offers the reactive power compensation range and provides a large fundamental voltage drop between the load voltage and the active inverter voltage, the voltage rating of the active inverter part can be significantly reduced. On the other hand, the active inverter part can solve the inherent problems of using the TCLC alone, such as inrush current, resonance problem, noise of thyristors turning on/off, mistuning of firing angles and low harmonic compensation ability.

From Fig. 7.1, v_{sx}, i_{sx}, v_x, i_{Lx} and i_{cx} ('x' denotes phase a, b or c) represent the source voltage, source current, load voltage, load current and compensating current respectively, L_s is the system inductance. For the TCLC part, L_c, C_{PF} and L_{PF} are its coupling inductor, parallel capacitor and its thyristor controlled reactor (TCR); and the pair of T_{x1} and T_{x2} is the bi-directional thyristor switch of the TCLC part. For the active inverter part, v_{invx} is the output voltage of the voltage source inverter (VSI); T_x and $\bar{T}_x$ are the switching devices; C_{DC} and V_{DC} are the DC-link capacitor

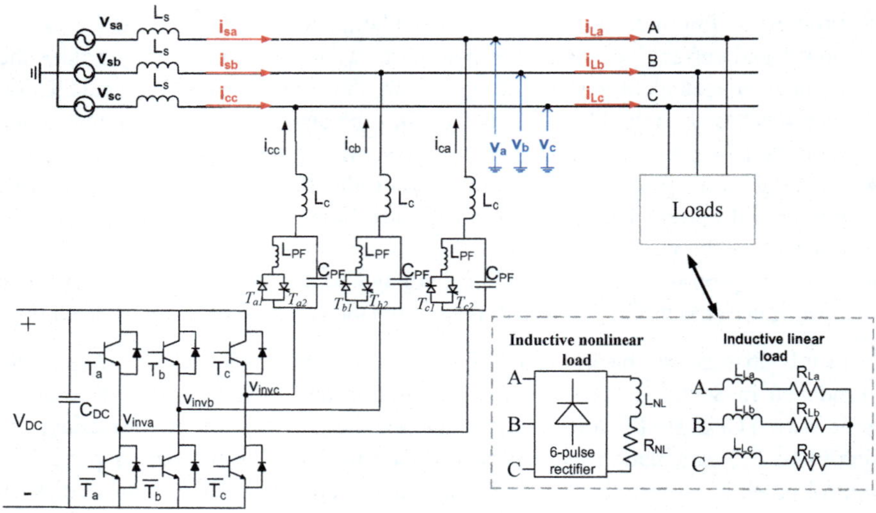

Fig. 7.1 Circuit configuration of a three-phase three-wire TCLC-HAPF

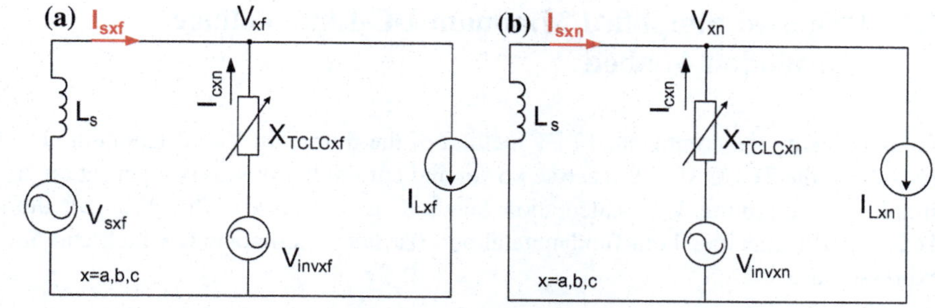

Fig. 7.2 Single-phase equivalent circuit models of the TCLC-HAPF: **a** at fundamental frequency, **b** at n^{th} order harmonic frequency

and its voltage. Figure 7.2 shows the single-phase equivalent circuit models of the TCLC-HAPF. In the following, the subscripts 'f', 'h' and 'n' represent the fundamental, total harmonic and harmonic order.

For the fundamental frequency circuit model as shown in Fig. 7.2a, V_{xf} and V_{invxf} are the fundamental load and VSI voltage; I_{sxf}, I_{cxf} and I_{Lxf} are the fundamental source, reactive compensating and load current, X_{TCLCxf} is the fundamental reactance of the TCLC part.

For the nth order harmonic frequency circuit model as shown in Fig. 7.2b, V_{invxn} is the nth order harmonic output voltage of the VSI; I_{sxn}, I_{cxn} and I_{Lxn} are the nth order harmonic system, compensating and load current, X_{TCLCxn} is the nth order harmonic reactance of the TCLC part, with the harmonic order nth = $6k \pm 1$th, $k = 1, 2 \ldots \infty$ for three-phase three-wire system [26, 27].

From Figs. 7.1 and 7.2, the phase fundamental and harmonic reactance values of the TCLC part can be calculated through (7.1) [19]:

$$
\begin{aligned}
X_{TCLCxf}(\alpha_x) &= \frac{\pi X_{LPFf} X_{CPFf}}{X_{CPFf}(2\pi - 2\alpha_x + \sin 2\alpha_x) - \pi X_{LPFf}} + X_{Lcf} \\
X_{TCLCxn}(\alpha_x) &= \frac{\pi X_{LPFn} X_{CPFn}}{X_{CPFn}(2\pi - 2\alpha_x + \sin 2\alpha_x) - \pi X_{LPFn}} + X_{Lcn}
\end{aligned}
\tag{7.1}
$$

where $X_{Lcf} = \omega L_c$, $X_{LPFf} = \omega L_{PF}$, $X_{CPFf} = 1/(\omega C_{PF})$; $X_{Lcn} = n\omega L_c$, $X_{LPFn} = n\omega L_{PF}$, $X_{CPFn} = 1/(n\omega C_{PF})$, ω ($= 2\pi f$) is the fundamental angular frequency. α_x is the phase firing angle of the thyristor switches, which can be obtained from the TCLC-HAPF controller and the detailed discussion of the TCLC-HAPF control will be presented in Sect. 7.4.

With the help of Fig. 7.2, the minimum DC-link voltage calculation for the TCLC-HAPF will be proposed and discussed in Sect. 7.3. In the following analysis, v_{sx} and v_x as shown in Fig. 7.1 are assumed to be pure sinusoidal without harmonic components, that is $V_{sx} = V_x = V_{xf}$ [14, 21] for simplification, and all the parameters are in root mean square (rms) values.

7.3 Proposed Simplified Minimum DC-Link Voltage Calculation Method

To avoid using the complicated FFT method of the minimum V_{DC} calculation [15, 16, 21] for the TCLC-HAPF reactive power and current harmonics compensation, a simplified minimum V_{DC} calculation method is proposed. The V_{DC} of the TCLC-HAPF includes both fundamental and harmonic components which can be expressed as:

$$V_{DCx} = \sqrt{V_{DCxf}^2 + V_{DCxh}^2} \tag{7.2}$$

$$V_{DC} = \max(V_{DCa},\quad V_{DCb},\quad V_{DCc}) \tag{7.3}$$

where V_{DCxf} and V_{DCxh} are the required DC-link voltage for compensating fundamental reactive power and current harmonics of each phase. The final V_{DC} is calculated as the maximum value among the three phase values. In this section, the V_{DCxf} and V_{DCxh} will be separately discussed in Sects. 7.3.1 and 7.3.2, and a comparison of the V_{DC} calculation for the TCLC-HAPF by using the conventional FFT method [15, 16, 21] and the proposed method will be discussed in Sect. 7.3.3.

7.3.1 *Deduction of DC-Link Voltage (V_{DCxf}) at Fundamental Frequency*

From Fig. 7.2a, the phase fundamental inverter output voltage (V_{invxf}) can be expressed as:

$$V_{invxf} = \left|V_x - \left|X_{TCLCxf}(\alpha_x)\right| \cdot \left|I_{cxf}\right|\right| \tag{7.4}$$

where V_x, I_{cxf} and $X_{TCLCxf}(\alpha_x)$ is the fundamental load voltage, fundamental reactive compensating current and TCLC part fundamental impedance, respectively. The reactive power ($Q_{cx\_TCLCf}(\alpha_x)$) provided by the TCLC part and the load reactive power (Q_{Lxf}) can be expressed as:

$$Q_{cx\_TCLCf}(\alpha_x) = \frac{V_x^2}{X_{TCLCxf}(\alpha_x)} \tag{7.5}$$

$$Q_{Lxf} = V_x \cdot I_{Lxfq} = V_x \cdot (-I_{cxf}) \tag{7.6}$$

where I_{Lxfq} is the phase fundamental load reactive current and is equal to $-I_{cxf}$ after the TCLC-HAPF compensation at ideal case. By combining (7.4)–(7.6), the required DC-link voltage (V_{DCxf}) at the fundamental frequency can be expressed as:

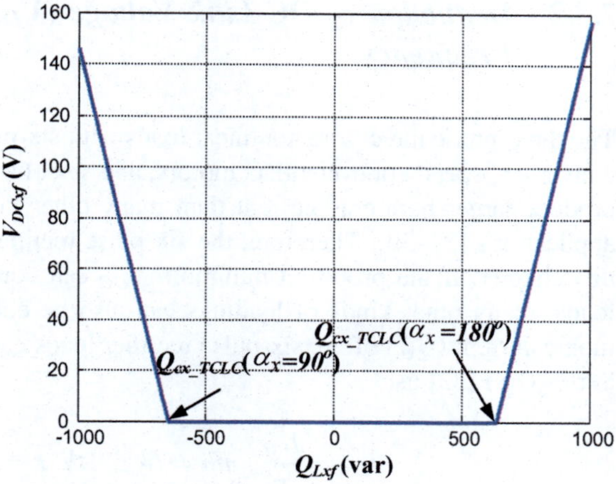

Fig. 7.3 The relationship between V_{DCxf} and Q_{Lxf}

$$V_{DCxf} = \sqrt{6} \cdot V_{invxf} = \sqrt{6} \cdot V_x \left| \frac{|Q_{Lxf}| - |Q_{cx\_TCLCf}(\alpha_x)|}{Q_{cx\_TCLCf}(\alpha_x)} \right| \tag{7.7}$$

In (7.7), the $Q_{cx\_TCLCf}(\alpha_x)$ can also be expressed as in (7.8) and it varies within two fixed boundaries according to the range ($90° < \alpha_x < 180°$) of the firing angle as in (7.9) and (7.10).

$$Q_{cx\_TCLCf}(\alpha_x) = \frac{V_x^2}{\frac{\pi X_{LPFf} X_{CPFf}}{X_{CPFf}(2\pi - 2\alpha_x + \sin 2\alpha_x) - \pi X_{LPFf}} + X_{Lcf}} \tag{7.8}$$

$$Q_{cx\_TCLCf}(\alpha_x = 90°) = \frac{V_x^2}{X_{TCLCxf}(\alpha_x = 90°)} = \frac{V_x^2}{\frac{X_{LPFf} X_{CPFf}}{X_{CPFf} - X_{LPFf}} + X_{Lcf}} \tag{7.9}$$

$$Q_{cx\_TCLCf}(\alpha_x = 180°) = \frac{V_x^2}{X_{TCLCxf}(\alpha_x = 180°)} = \frac{V_x^2}{X_{Lcf} - X_{CPFf}} \tag{7.10}$$

Moreover, if the Q_{Lxf} is within the compensation range of the TCLC part, that is $Q_{cx\_TCLCf}(\alpha_x) = -Q_{Lxf}$ and $V_{DCxf} = 0$ can be achieved; otherwise, $V_{DCxf} > 0$. Based on (7.7)–(7.10), the relationship between V_{DCxf} and Q_{Lxf} can be plotted as shown in Fig. 7.3.

7.3.2 Deduction of DC-Link Voltage (V_{DCxh}) at Harmonic Frequency

The three-phase three-wire nonlinear loads with six-pulse rectifiers such as: AC-DC converters, speed controlled DC motors, and steel hardening machines, etc. usually produce larger harmonic current than many other kinds of loads in the industrial applications [27–30]. Therefore, the six-pulse rectifier loads are mainly focused in this chapter. If the proposed minimum V_{DC} can compensate this kind of rectifier loads, many other kinds of loadings can also be compensated [31, 32]. The harmonic current (I_{Lxn}) of the six-pulse rectifier loads can be obtained through Fourier Series (FS) [33] as:

$$I_{Lxn} = \frac{I_{Lxf}}{n},\ nth = 6k \pm 1th,\ k = 1, 2\ldots\infty \tag{7.11}$$

In (7.11), the I_{Lxn} at each harmonic order can be expressed in terms of the fundamental load current (I_{Lxf}). Based on (7.11), the required DC-link voltage (V_{DCxh}) at harmonic frequency can be expressed as:

$$V_{DCxh} == \sqrt{6} \cdot I_{Lxf} \cdot \sqrt{\sum_{n=2}^{\infty}\left[\left(\frac{1}{n} \cdot X_{TCLCxn}(\alpha_x)\right)^2\right]} = \sqrt{6} \cdot I_{Lxf} \cdot A(\alpha_x) \tag{7.12}$$

$$V_{DCxh} = \sqrt{6} \times \sqrt{\sum\nolimits_{n=2}^{\infty}(V_{invxn}^2)} = \sqrt{6} \times \sqrt{\sum_{n=2}^{\infty}\left[(X_{TCLCxn}(\alpha_x) \cdot I_{Lxn})^2\right]} \tag{7.13}$$

where $A(\alpha_x) = \sqrt{\sum_{n=2}^{\infty}\left[\left(\frac{1}{n} \cdot X_{TCLCxn}(\alpha_x)\right)^2\right]}$

In (7.13), $A(\alpha_x)$ is a constant related to the $X_{TCLCxn}(\alpha_x)$ and n, where $X_{TCLCxn}(\alpha_x)$ is in terms of the firing angle (α_x) as shown in (7.1), and a look up table (LUT) (α_x vs $A(\alpha_x)$) can be built in order to simplify the $A(\alpha_x)$ calculation. Besides, the fundamental load current I_{Lxf} can be calculated by using the single-phase instantaneous p-q theory [25] as:

$$S_{Lx} = \frac{\sqrt{\bar{p}_{Lx}^2 + \bar{q}_{Lx}^2}}{2} \tag{7.14}$$

$$I_{Lxf} = \frac{S_{Lx}}{V_x} \tag{7.15}$$

where S_{Lx} is the apparent power of the loading, $\bar{p}_{Lx}$ and $\bar{q}_{Lx}$ are the DC components of the instantaneous load active and reactive power [25].

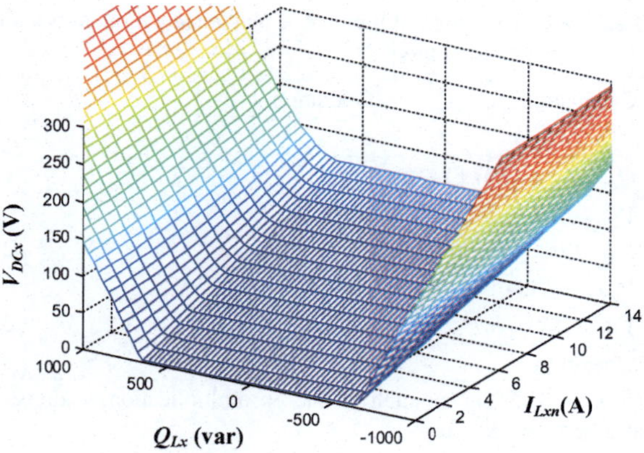

Fig. 7.4 The relationship among V_{DCx}, I_{Lxn} and Q_{Lx}

Based on the above deduction, from the TCLC-HAPF system parameters as shown in Table 7.3, the phase required DC-link voltage V_{DCx} of the TCLC-HAPF can be plotted as shown in Fig. 7.4.

When there is no harmonic current problem, the required V_{DCx} will obtain the same results as shown in Fig. 7.3. The minimum $V_{DCx} = 0$ can be achieved if Q_{Lx} is within the compensation range of the TCLC part. In addition, the larger the harmonic current contents, the larger the V_{DCx} is required.

7.3.3 Comparison Between Conventional and Proposed Minimum V_{DC} Calculation Methods

To compare the calculation steps between the conventional V_{DC} calculation method (based on FFT) [15, 16, 21] and the proposed V_{DC} calculation method for the TCLC-HAPF, Tables 7.1 and 7.2 summarize their required number of

Table 7.1 The required number of mathematical operators for the conventional minimum V_{DCx} calculation by using FFT [15, 16, 21]

Calculation	Operation							
	+	−	×	/	x^2	$\sqrt{x}$	|x|	sin x
I_{cxfq} (7.6)		1		1				
$V_{DCxf} = \sqrt{6}\, V_{invxf}$ with (7.4)	2	3	9	1		1	3	1
FFT I_{Lxn} *(complex)* [34]	13566		4360					
I_{Lxn} [33]	7				14	7		
V_{DCxh} (7.12)	20	14	57	7	7	2		7
V_{DCx} (7.2)	1				2	1		
Operations	13596	18	4426	9	23	11	3	8

Note '+' addition, '−' subtraction, '×' multiplication, '/' division, 'x^2' square, '$\sqrt{x}$' square root, '\|x\|' absolute, 'sin x' sine

Table 7.2 The required number of mathematical operators for the proposed simplified minimum V_{DCx} calculation method

Calculation	Operation							
	+	−	×	/	x^2	$\sqrt{x}$	|x|	sin x
$Q_{cx\_TCLCf}(\alpha_x)$ (7.8)	2	2	7	2	1			1
V_{DCxf} (7.7)		1	2	1		1	3	
I_{Lxf} (7.14), (7.15)	1			2	2	1		
V_{DCxh} (7.13)			2			1		
V_{DCx} (7.2)	1				2	1		
Operations	4	3	11	5	5	4	3	1

Note '+' addition, '−' subtraction, '×' multiplication, '/' division, 'x^2' square, '$\sqrt{x}$' square root, '|x|' absolute, 'sin x' sine

mathematical operators. The assumptions used in this comparison are: (1) the considered harmonic current order is up to 23rd, and (2) the sampling rate is 25 kHz and 512-point FFT algorithm are used in the conventional method.

As shown in Table 7.1, the conventional V_{DC} calculation method (based on the 512-point FFT algorithm) [34] requires a large amount of additions and multiplications. In contrast, the number of additions and multiplications used in the proposed method are 99.97 and 99.76% less than the conventional method as shown in Table 7.2. Moreover, the other mathematical operators of the proposed method are also less than the conventional method by using FFT [15, 16, 21].

7.4 Control Block of the Proposed Adaptive DC-Link Voltage Controlled TCLC-HAPF

In this section, the control block of the adaptive DC-link voltage controlled TCLC-HAPF is proposed based on the above V_{DC} calculation method. And the overall control block diagram is shown in Fig. 7.5, which consists of the following three sub-control blocks: Sect. 7.4.1 TCLC Control Block, Sect. 7.4.2 Active VSI Control Block and Sect. 7.4.3 Adaptive DC-Link Voltage Control Block.

7.4.1 TCLC Control Block

For the TCLC control block as shown in Fig. 7.5, the fundamental load reactive power (Q_{Lxf}) is calculated by the single-phase instantaneous p-q theory [18, 25], which is used to control the firing angle (α_x) of the thyristor switch (T_{1x} and T_{2x}). When the Q_{Lxf} varies within the compensation range of the TCLC part, which means $Q_{cx\_TCLCf}(\alpha_x = 180°) < -Q_{Lxf} < Q_{cx\_TCLCf}(\alpha_x = 90°)$, the corresponding α_x

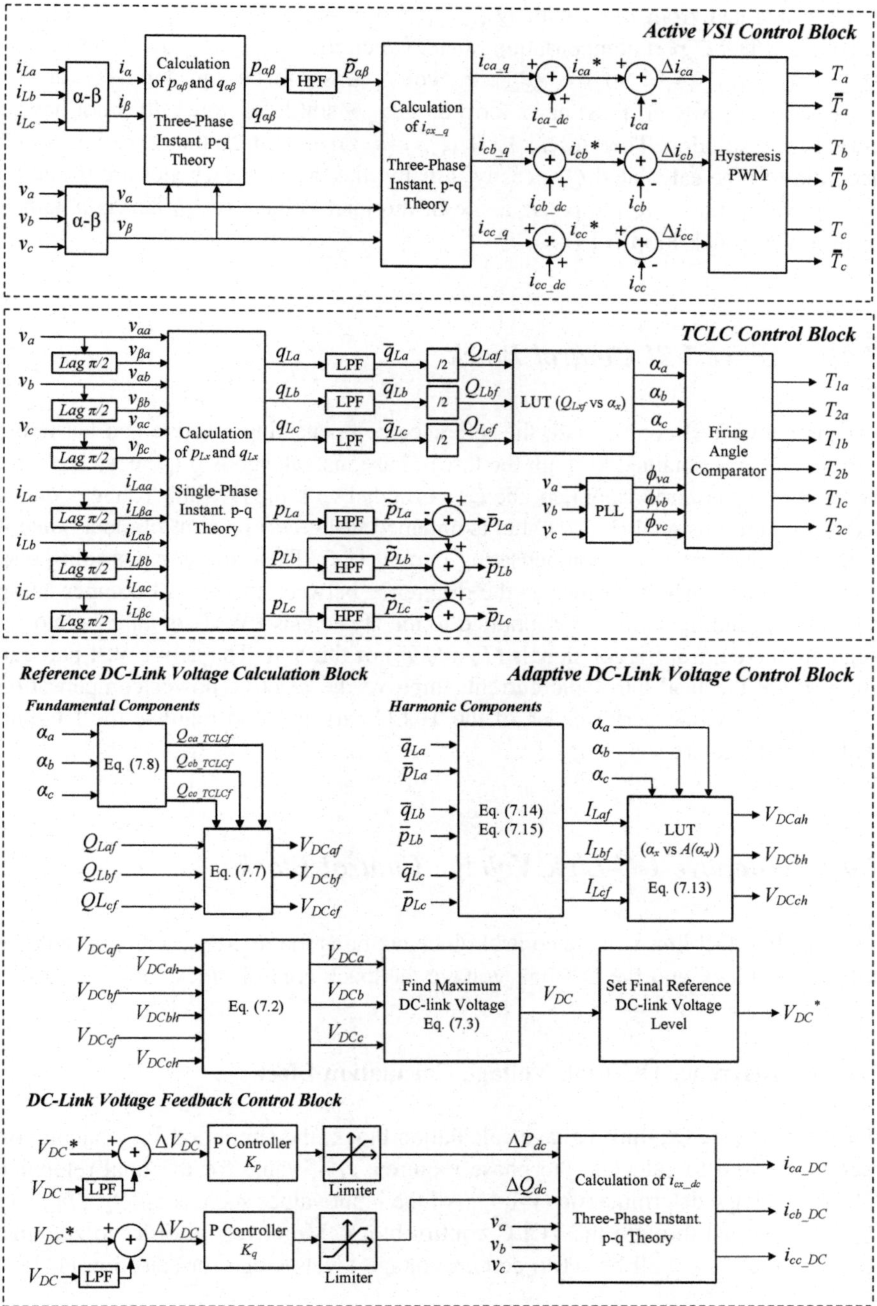

Fig. 7.5 Control block diagram of the adaptive DC-link voltage controlled TCLC-HAPF

can be obtained from (7.8) with $Q_{cx\_TCLCf}(\alpha_x) = -Q_{Lxf}$. Otherwise, if the Q_{Lxf} is outside the TCLC part compensation range, which means $Q_{cx\_TCLCf}(\alpha_x = 180°) > -Q_{Lxf}$ or $Q_{cx\_TCLCf}(\alpha_x = 90°) < -Q_{Lxf}$, α_x would be set to be equal to 180° or 90° respectively. However, (7.8) has a term of $-2\alpha_x + \sin(2\alpha_x)$, which does not have a closed-form solution. Therefore, a LUT (Q_{Lxf} vs. α_x) is built, so that α_x can be found according to the calculated Q_{Lxf} easily. Finally, the thyristor switches are triggered by comparing the α_x to the phase angle of the load voltage (ϕ_{vx}), that is obtained from a phase lock loop (PLL).

7.4.2 Active VSI Control Block

For the active VSI control part, the reference compensating reactive and harmonic current (i_{cx_q}) is obtained through the three-phase instantaneous p-q theory [18, 35]. When the V_{DC} control is applied, the i_{cx_q} is added with the DC-link voltage control feedback current signal (i_{cx_DC}), that is obtained from the V_{DC} control block, then it yields the final reference compensating current (i^*_{cx}). Then, the compensating current error (Δi_{cx}), which defines as the difference between the sensed compensating current (i_{cx}) and the i^*_{cx}, will be inputted to the Hysteresis PWM signal generator to generate the control trigger signals (T_x and T_x) of the VSI. The active VSI part can compensate the load harmonic current, improve the reactive power compensation ability and dynamic performance of the TCLC part and also regulate the DC-link voltage to its reference value.

7.4.3 Adaptive DC-Link Voltage Control Block

The adaptive DC-link voltage control block contains the reference DC-link voltage calculation block and the DC-link voltage feedback control block.

7.4.3.1 Reference DC-Link Voltage Calculation Block

For the reference DC-link voltage calculation block, the simplified V_{DC} calculation method is used to calculate the phase required V_{DCx} value for the final reference DC-link voltage determination (V_{DC}*). All the input values such as α_x, Q_{Lxf}, $\bar{p}_{Lx}$ and $\bar{q}_{Lx}$ are obtained through the TCLC control block. Moreover, in order to stabilize the V_{DC}, the V_{DC}* will be set to certain voltage levels range for selection [14].

7.4.3.2 DC-Link Voltage Feedback Control Block

For the DC-link voltage feedback control block, two proportional (P) controllers with the gains (K_p and K_q) are applied for generating the DC-link voltage feedback compensating current signals i_{cx_DC} for the adaptive V_{DC} control. The DC-link voltage error signal (ΔV_{DC}), which is the difference between the reference V_{DC}* and the sensed V_{DC}, would be input into two proportional controllers to generate the DC active and reactive feedback control signals (ΔP_{DC} and ΔQ_{DC}) [13]. ΔQ_{dc} is used to step change the DC-link voltage during the start-up process, while the ΔP_{DC} is used to maintain the DC-link voltage as its reference due to the system loss, in which the DC-link voltage control with feedback both active ΔP_{DC} and reactive ΔQ_{DC} components [13, 14, 16] can achieve both the start-up DC-link voltage self-charging function, maintaining the DC-link voltage and perform dynamic reactive power compensation simultaneously. Similar as the setting of the two controllers in [13], in order to simplify the control process, ΔP_{DC} and ΔQ_{DC} are actually calculated by the same controller, i.e. $K_p = K_q$ and $\Delta Q_{DC} = -\Delta P_{DC}$. Then ΔP_{DC} and ΔQ_{DC} would be input to the three-phase instantaneous p-q theory [35] to generate the DC-link voltage feedback compensating current signal (i_{cx_DC}), which is used to update the reference i_{cx}* in the active VSI block in order to adaptively control the DC-link voltage of the TCLC-HAPF.

7.5 Simulation Case Studies

In this section, simulation case studies of the TCLC-HAPF compensation are executed by using the PSCAD/EMTDC platform. The parameters of the TCLC-HAPF system used in the simulations are summarized in Table 7.3, and the maximum inductive and capacitive reactive power provided by the TCLC part is $Q_{cx\_TCLC}(\alpha_x = 90°) = 647var$ and $Q_{cx\_TCLC}(\alpha_x = 180°) = -633var$.

To verify the proposed V_{DC} calculation method and adaptive V_{DC} control for the TCLC-HAPF, two different balanced loading simulation case studies are performed, namely: Sect. 7.5.1 under the TCLC part reactive power compensation range and Sect. 7.5.2 over the TCLC part reactive power compensation range.

For under compensation case, the load varies from *Load* 1 to *Load* 2, the load reactive power changes from 70var to 103var, which is within the designed compensation range of the TCLC part of the TCLC-HAPF ($-633\text{var} < Q_{cx\_TCLC}(\alpha_x) < 647\text{var}$). Moreover, the simulated TCLC-HAPF compensation performances with the proposed adaptive V_{DC} control will be compared with the conventional fixed V_{DC} controlled case.

Table 7.3 System and TCLC-HAPF parameters

Parameters		Physical values
System	f, v_{sx}, L_{sx}	50 Hz, 110 V, 0.5 mH
TCLC-HAPF	L_{cx}, L_{PF}, C_{PF}, C_{DC}	2.5 mH, 30 mH, 160 μF, 3300 μF

For the over compensation case, when the *Load* 2 and *Load* 3 are connected to the system, the load reactive power rises to 936var, which is outside the TCLC part compensation range, the adaptive V_{DC} control can also help to increase the V_{DC} level to ensure the excessive load reactive power compensation.

7.5.1 Under Compensation by Adaptive V_{DC} Controlled TCLC-HAPF

Figure 7.6 shows the three-phase simulated system voltage and current before TCLC-HAPF compensation during *Load* 1 varying to *Load* 2. Table 7.4 summarizes the corresponding simulation results before the TCLC-HAPF compensation. When *Load* 1 is connected, the three-phase simulated total harmonic distortion ($THDi_{sx}$) of the source current (i_{sx}) is 27.9% and source power factor (PF_{sx}) is 0.96. When *Load* 2 is connected, the $THDi_{sx}$ is 27.5% and PF_{sx} is 0.96, in which both $THDi_{sx}$ cannot satisfy the IEEE Standard [36].

The calculated minimum V_{DC} values for compensating *Load* 1 and *Load* 2 by using the conventional FFT and the proposed methods are listed in Table 7.5, where the V_{DC} calculated by the proposed method can cover the V_{DC} value calculated by the conventional method, and the difference between them is small, which verifies the proposed minimum V_{DC} calculation method.

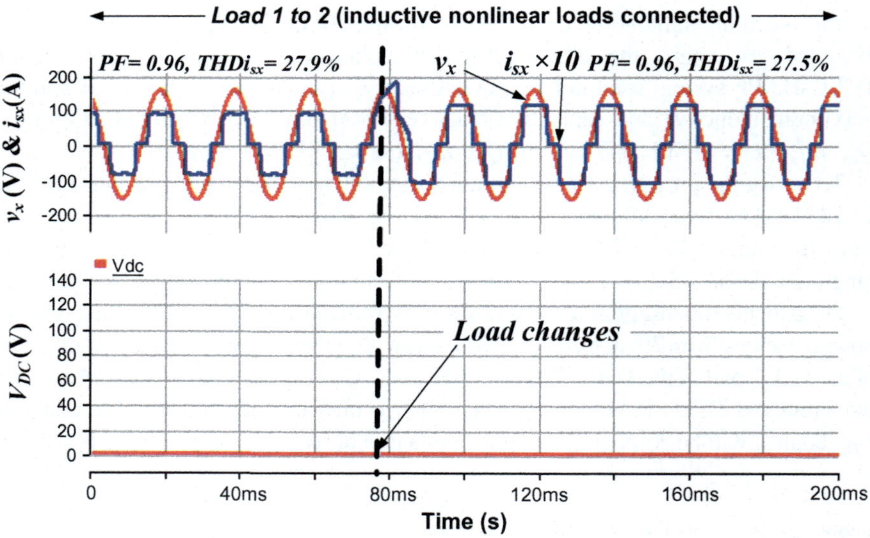

Fig. 7.6 Simulated system voltage and current before TCLC-HAPF compensation during under compensation case (Load 1 varying to Load 2)

Table 7.4 Simulation results before TCLC-HAPF compensation during under compensation case

Case	Phase	i_{sx} (A)	THD_{isx} (%)	Q_{sxf} (var)	PF_{sx}
Load 1	*a*, *b*, *c*	6.69	27.9	70	0.96
Load 2	*a*, *b*, *c*	9.03	27.5	103	0.96

Table 7.5 Simulated adaptive V_{DC} levels of the TCLC-HAPF during under compensation case

	Conventional method with FFT	Proposed method	Final reference
Case	Required V_{DC}	Required V_{DC}	V_{DC} level
Load 1	24.8 V	26.1 V	30 V
Load 2	32.2 V	33.7 V	40 V

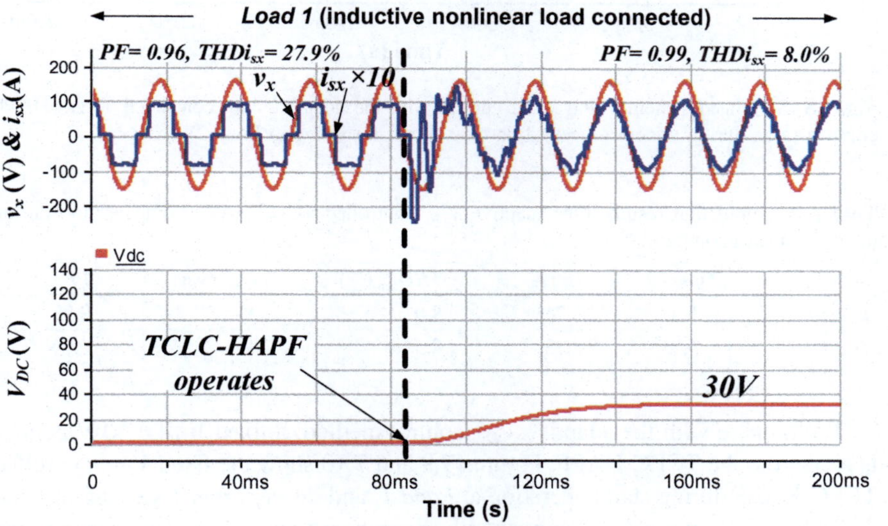

Fig. 7.7 Simulated system voltage and current with the adaptive V_{DC} controlled TCLC-HAPF compensation during starts operation at Load 1

The three-phase simulated system voltage and current with the proposed adaptive V_{DC} controlled TCLC-HAPF compensation is shown in Figs. 7.7 and 7.8, and the corresponding compensation results are summarized in Table 7.6. Figures 7.7 and 7.8 show the adaptive V_{DC} controlled TCLC-HAPF during starts operation at *Load* 1, and during *Load* 1 varying to *Load* 2. From Figs. 7.7 and 7.8, they show that V_{DC} level is changing adaptively according to the load variations. Table 7.6 shows that the THD_{isx} decreases to less than 9% and the PF_{sx} is improved to 0.99 after the adaptive V_{DC} controlled TCLC-HAPF compensation, in which $THDi_{sx}$ satisfy the IEEE Standard [36].

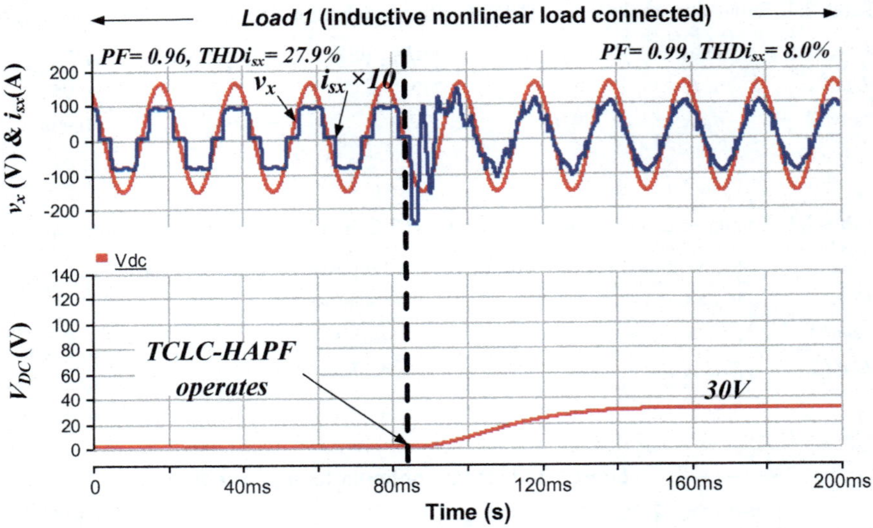

Fig. 7.8 Simulated system voltage and current with the adaptive V_{DC} controlled TCLC-HAPF compensation during under compensation case (Load 1 varying to Load 2)

Table 7.6 Simulation results after adaptive V_{DC} controlled TCLC-HAPF compensation during under compensation case

Case	Phase	i_{sx} (A)	THD_{isx} (%)	Q_{sxf} (var)	PF_{sx}
Load 1	*a*, *b*, *c*	6.71	8.0	11	0.99
Load 2	*a*, *b*, *c*	8.73	8.3	11	0.99

To compare with the adaptive V_{DC} control method, a fixed V_{DC} = 60 V control is applied to the TCLC-HAPF. Figures 7.9 and 7.10 show the fixed V_{DC} controlled TCLC-HAPF during starts operation at *Load* 1, and during *Load* 1 varying to *Load* 2. Table 7.7 summarizes the corresponding compensation results. The compensated THD_{isx} decreases to less than 8% and the PF_{sx} is improved to 0.99 after the fixed V_{DC} controlled TCLC-HAPF compensation. Compared Fig. 7.10 with Fig. 7.8, the fixed and adaptive DC voltage control can obtain similar steady-state compensation results. But the proposed control strategy solely requires lower DC voltage levels for compensation. Moreover, Fig. 7.11 shows the compensating current i_{ca} of phase *a* and its frequency spectrum with *THD* values for the fixed and proposed adaptive V_{DC} controlled TCLC-HAPF. From Fig. 7.11, i_{ca} of the adaptive V_{DC} control method obtains lower switching noise than the fixed V_{DC} case.

Based on the above simulation results, they show that the TCLC part can compensate the load reactive power, and the V_{DC} provided by the VSI of the TCLC-HAPF can deal with the harmonic current problem. Figures 7.6, 7.7, 7.8,

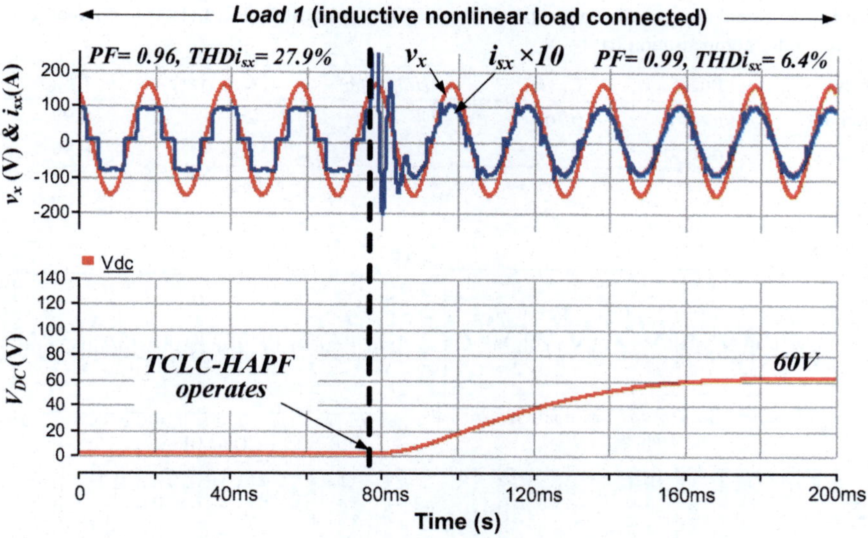

Fig. 7.9 Simulated system voltage and current with the fixed V_{DC} controlled TCLC-HAPF compensation during starts operation at Load 1

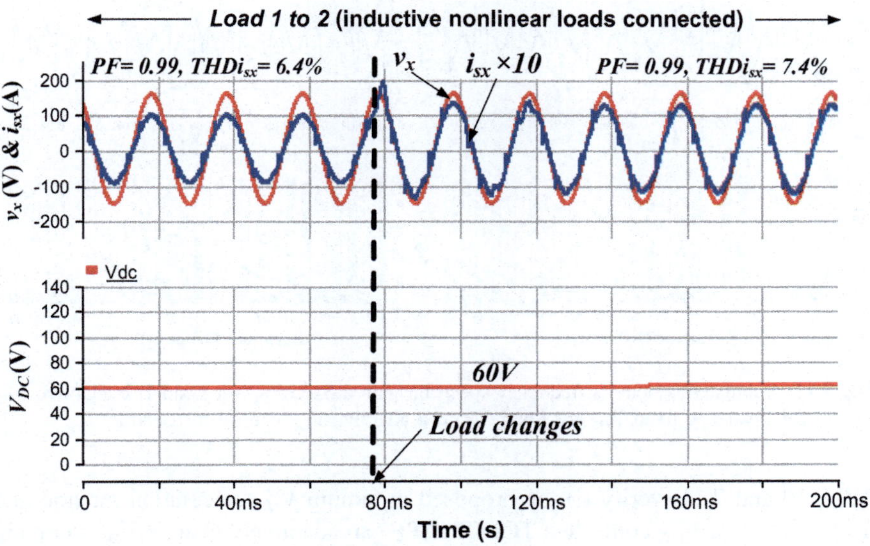

Fig. 7.10 Simulated system voltage and current with the fixed V_{DC} controlled TCLC-HAPF compensation during under compensation case (Load 1 varying to Load 2)

Table 7.7 Simulation results after fixed $V_{DC} = 60$ V controlled TCLC-HAPF compensation during under compensation case

Case	Phase	I_{sx} (A)	THD_{isx} (%)	Q_{sxf} (var)	PF_{sx}
Load 1	*a*, *b*, *c*	6.76	6.4	20	0.99
Load 2	*a*, *b*, *c*	8.77	7.4	26	0.99

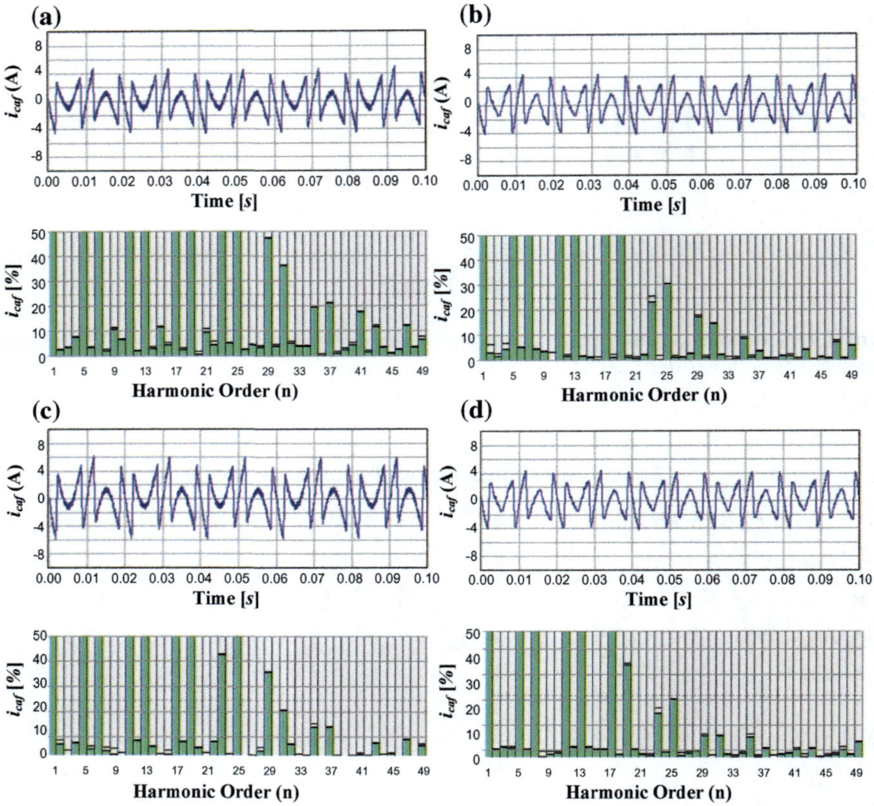

Fig. 7.11 Simulated i_{ca} and its frequency spectrum with **a** fixed V_{DC} for Load 1, **b** adaptive V_{DC} control for Load 1, **c** fixed V_{DC} for Load 2, and **d** adaptive V_{DC} control for Load 2

7.9, 7.10 and 7.11, verify (i) the proposed minimum V_{DC} calculation method and (ii) the adaptive V_{DC} controlled TCLC-HAPF can adaptively change V_{DC} to obtain lower switching noise and similar compensation performances compared with the fixed V_{DC} case.

7.5.2 Over Compensation by Adaptive V_{DC} Controlled TCLC-HAPF

Figure 7.12 shows the three-phase simulated system voltage and current before TCLC-HAPF compensation during *Load* 2 varying to *Load* 2 + *Load* 3. Table 7.8 summarizes the corresponding simulation results for *Load* 2 + *Load* 3 before the TCLC-HAPF compensation. When *Load* 2 + *Load* 3 are connected to the system, the three-phase simulated $THDi_{sx}$ becomes 16.4% and PF_{sx} becomes 0.79, in which $THDi_{sx}$ cannot satisfy the IEEE Standard [36].

The calculated required minimum V_{DC} values for compensating *Load* 2 + *Load* 3 case is listed in Table 7.9. Figure 7.13 shows the three-phase simulated system voltage and current with the proposed adaptive V_{DC} controlled TCLC-HAPF compensation and Table 7.10 summarizes the corresponding compensation results. When *Load* 2 + *Load* 3 are connected to the system, the V_{DC} level adaptively changes from 40 to 120 V according to the load condition, the $THDi_{sx}$ and PF_{sx} are compensated to 4.0 and 0.99%, respectively.

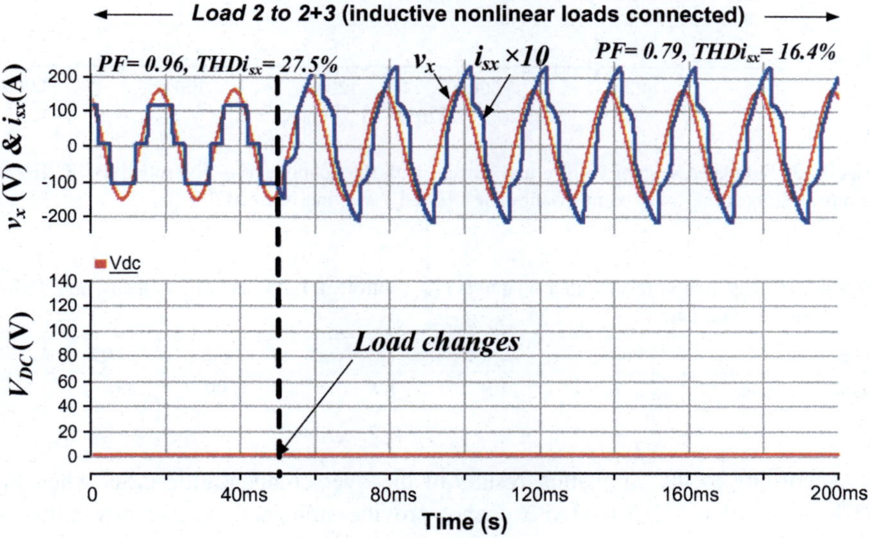

Fig. 7.12 Simulated system voltage and current before TCLC-HAPF compensation during over compensation case (Load 2 varying to Load 2 + 3)

Table 7.8 Simulation results before TCLC-HAPF compensation during over compensation case

Case	Phase	I_{sx} (A)	Q_{sxf} (var)	PF_{sx}	THD_{isx} (%)
Load 2 + *Load* 3	*a, b, c*	14.4	936	0.79	16.4

Table 7.9 Simulated adaptive V_{DC} level of the TCLC-HAPF during over compensation case

Case	Required V_{DC}	V_{DC} level
Load 2 + *Load* 3	119.5 V	120 V

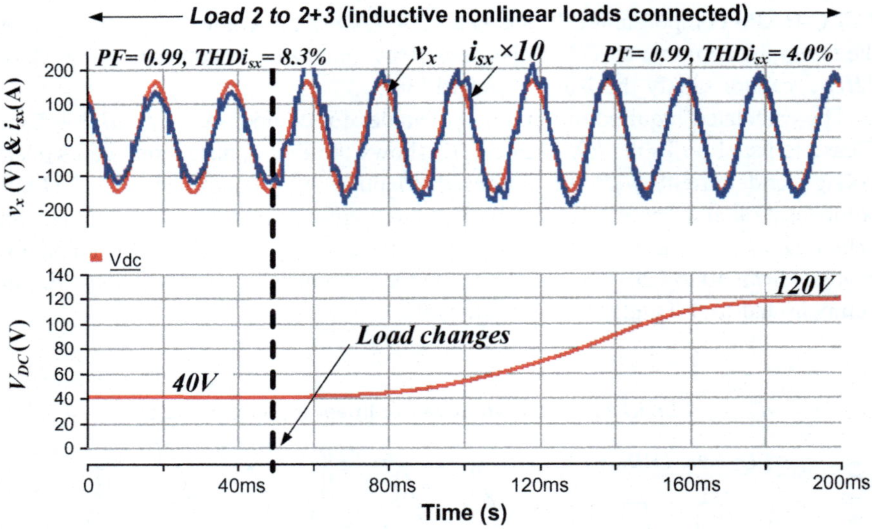

Fig. 7.13 Simulated system voltage and current with the adaptive V_{DC} controlled TCLC-HAPF compensation during over compensation case (Load 2 varying to Load 2 + 3)

Table 7.10 Simulation results after adaptive V_{DC} controlled TCLC-HAPF compensation during over compensation case

Case	Phase	I_{sx} (A)	Q_{sxf} (var)	PF_{sx}	THD_{isx} (%)
Load 2 + *Load* 3	*a*, *b*, *c*	11.8	89	0.99	4.0

According to the simulation results of the over compensation case, when the TCLC part of the TCLC-HAPF cannot provide sufficient reactive power to the loads, the proposed adaptive V_{DC} control method can also dynamically increase the V_{DC} level to increase the TCLC-HAPF reactive power compensation range.

Figures 7.12 and 7.13 and Tables 7.8, 7.9 and 7.10 verify that the adaptive V_{DC} controlled TCLC-HAPF can dynamically compensate the reactive power and suppress the current harmonics when the load reactive power falls outside the TCLC part compensation range.

7.6 Experimental Results

The following experimental results are performed on an 110 V–5 kVA three-phase three-wire TCLC-HAPF laboratory prototype. The control system of the TCLC-HAPF is composed of two paralleled DSP-TMS320F2812s to separately control the TCLC part and the active inverter part. For the TCLC part, the thyristor modules are SanRex PK110FG160. For the active inverter part, the insulated gate bipolar transistors (IGBTs) are PM300DSA60. The system and the TCLC-HAPF parameters as shown in Table 7.3 are also used for the experimental testing. However, owing to the current limitation of the laboratory prototype, only under compensation case will be tested in this section. The adaptive DC-link voltage controlled TCLC-HAPF is verified in the following two parts: Sect. 7.6.1 the dynamic performance of the TCLC-HAPF according to the load variation (*Load* 1 varies to *Load* 2); Sect. 7.6.2 the comparison of the VSI switching noise and switching loss with the fixed V_{DC} controlled TCLC-HAPF.

Figure 7.14 shows the three-phase experimental system voltage and current before the TCLC-HAPF compensation during *Load* 1 varying to *Load* 2, and Table 7.11 lists the experimental results before the TCLC-HAPF compensation. When *Load* 1 is connected, the three-phase $THDi_{sx}$ values are 24.5, 23.7 and 24.1%, and PF_{sx} values are 0.96, 0.96 and 0.96, respectively. When *Load* 2 is connected, the three-phase $THDi_{sx}$ are 23.2, 22.4, 22.8% and PF_{sx} values are 0.96, 0.96 and 0.96, respectively. And, the $THDi_{sx}$ for both loadings cannot satisfy the IEEE Standard [36].

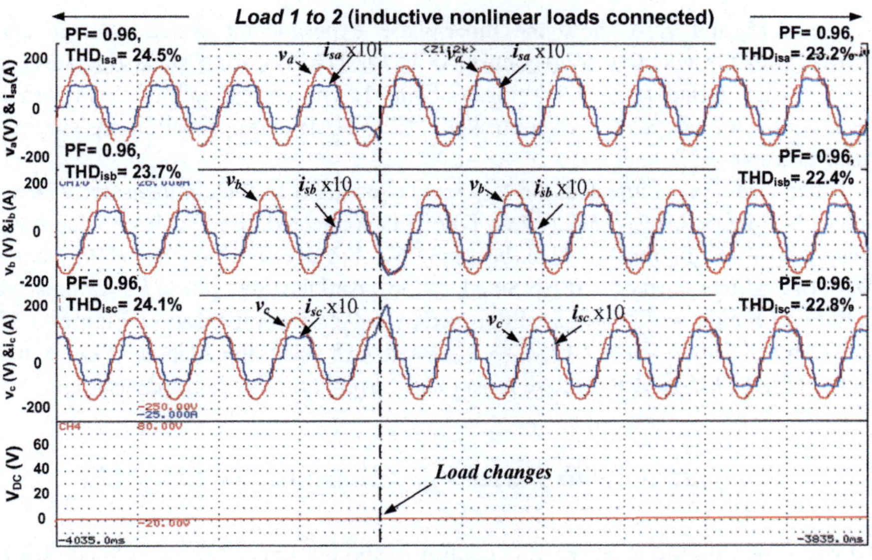

Fig. 7.14 Three-phase experimental system voltage and current before the TCLC-HAPF compensation during Load 1 varying to Load 2

Table 7.11 Experimental results before TCLC-HAPF compensation during Load 1 to Load 2

Case	Phase	i_{sx} (A)	THD_{isx} (%)	Q_{sxf} (var)	PF_{sx}
Load 1	*a*	6.6	24.5	80	0.96
	b	6.7	23.7	80	0.96
	c	6.6	24.1	80	0.96
Load 2	*a*	8.7	23.2	110	0.96
	b	8.5	22.4	110	0.96
	c	8.5	22.8	110	0.96

Table 7.12 Experimental adaptive V_{DC} levels of the TCLC-HAPF during Load 1 and Load 2 cases

Case	Required V_{DC}	V_{DC} level
Load 1	25.1 V	30 V
Load 2	35.1 V	40 V

According to the proposed simplified minimum V_{DC} calculation method, the required V_{DC} of the adaptive V_{DC} controlled TCLC-HAPF compensation for *Load* 1 and *Load* 2 are listed in Table 7.12.

7.6.1 Dynamic Performance of Adaptive V_{DC} Controlled TCLC-HAPF to Load Variation

Figures 7.15 and 7.16 show the three-phase experimental system voltage and current with the adaptive V_{DC} controlled TCLC-HAPF during starts operation at *Load* 1, and during *Load* 1 varying to *Load* 2. The corresponding experimental results after the adaptive V_{DC} controlled TCLC-HAPF compensation are summarized in Table 7.13.

When *Load* 1 is connected, the V_{DC} level rises to 30 V, the compensated $THDi_{sx}$ values become 7.9, 7.2, 7.5%. When *Load* 2 is connected, the V_{DC} level adaptively changes to 40 V, the $THDi_{sx}$ values become 6.8, 5.8, 6.4%, in which the $THDi_{sx}$ values for both loadings satisfy the IEEE Standard [36]. And the three-phase PF_{sx} values for both *Load* 1 and *Load* 2 cases are improved to 0.99 after compensation. Figures 7.15 and 7.16 and Tables 7.11, 7.12, 7.13, 7.14 and 7.15 verify the adaptive V_{DC} control method for TCLC-HAPF reactive power and current harmonics compensation.

7.6.2 Comparison with Fixed V_{DC} Controlled TCLC-HAPF

To compare with the adaptive V_{DC} control method, a fixed V_{DC} = 60 V control is applied to the TCLC-HAPF. Figures 7.17 and 7.18 show the three-phase experimental system voltage and current with the fixed V_{DC} controlled TCLC-HAPF

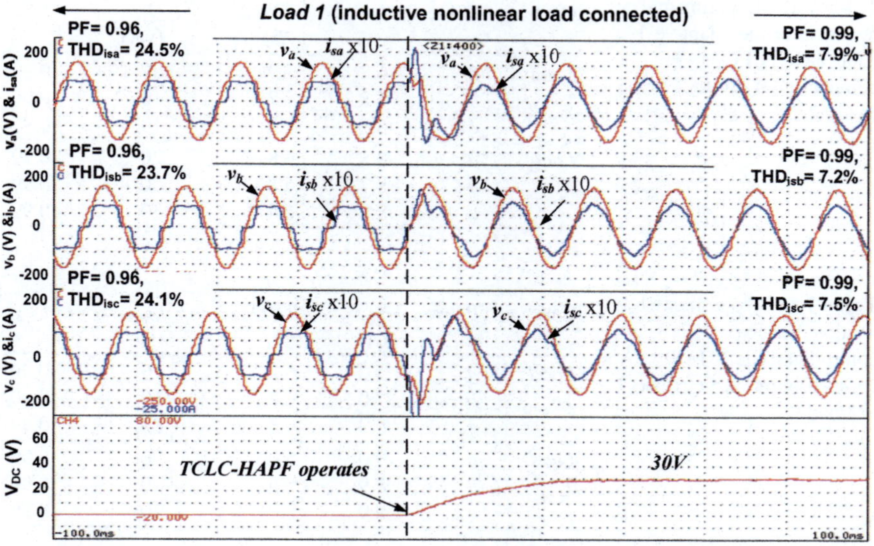

Fig. 7.15 Three-phase experimental system voltage and current with the adaptive V_{DC} controlled TCLC-HAPF compensation during starts operation at Load 1

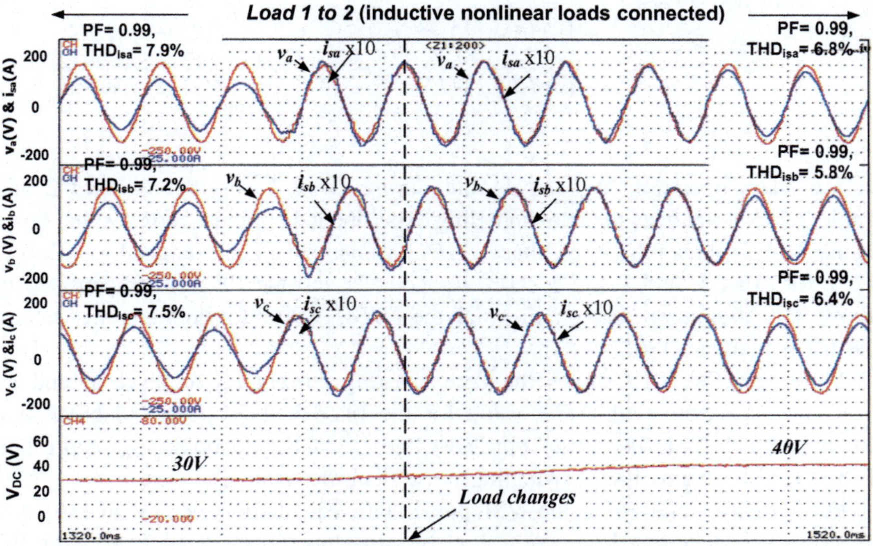

Fig. 7.16 Three-phase experimental system voltage and current with the adaptive V_{DC} controlled TCLC-HAPF compensation during Load 1 varying to Load 2

Table 7.13 Experimental results after the adaptive V_{DC} controlled TCLC-HAPF compensation during Load 1 to Load 2

Case	Phase	I_{sx} (A)	THD_{isx} (%)	Q_{sxf} (var)	PF_{sx}
Load 1	*a*	6.8	7.9	20	0.99
	b	6.8	7.2	10	0.99
	c	6.8	7.5	10	0.99
Load 2	*a*	8.8	6.8	20	0.99
	b	8.7	5.8	20	0.99
	c	8.8	6.4	30	0.99

Table 7.14 Experimental results after fixed V_{DC} controlled TCLC-HAPF compensation during Load 1 to Load 2

Case	Phase	I_{sx} (A)	THD_{isx} (%)	Q_{sxf} (var)	PF_{sx}
Load 1	*a*	7.0	6.2	10	0.99
	b	7.0	6.2	10	0.99
	c	6.9	6.3	20	0.99
Load 2	*a*	9.0	5.0	20	0.99
	b	8.9	5.5	10	0.99
	c	9.0	5.0	20	0.99

Table 7.15 Experimental VSI power loss between fixed and adaptive V_{DC} controlled TCLC-HAPF during *Load* 1 to *Load* 2

Case	Power loss (W)	
	Fixed $V_{DC} = 60$ V	Adaptive V_{DC}
Load 1	141 W	117 W ($V_{DC} = 30$ V), ↓17%
Load 2	147 W	120 W ($V_{DC} = 40$ V), ↓18%

during starts operation at *Load* 1, and during *Load* 1 varying to *Load* 2 and Table 7.14 summarizes the corresponding experimental compensation results. From Table 7.14, it shows the three-phase $THDi_{sx}$ values have been reduced to 6.2, 6.2, 6.3% for *Load* 1, and 5.0, 5.5, 5.0% for *Load* 2; and the PF_{sx} is compensated to above 0.99 after compensation, in which the $THDi_{sx}$ values for both loadings satisfy the IEEE Standard [36].

Moreover, Fig. 7.19 shows the compensating current i_{ca} of phase a and its frequency spectrum with THD value for the fixed and proposed adaptive V_{DC} controlled TCLC-HAPF. From Fig. 7.19, the i_{ca} of the adaptive V_{DC} controlled TCLC-HAPF obtains lower switching noise than the fixed V_{DC} case.

Furthermore, referred to the VSI power loss calculation method in [36], the experimental switching loss results for compensating *Load* 1 and *Load* 2 by the fixed and adaptive V_{DC} controlled TCLC-HAPF are shown in Table 7.15. From Table 7.15, the switching power loss can be reduced by 17 and 18% for *Load* 1 and *Load* 2 by the adaptive V_{DC} control. Therefore, Fig. 7.19 and Table 7.15 verify that the adaptive V_{DC} control method for TCLC-HAPF can reduce the switching noise and switching loss and obtain similar steady-state compensation results in comparison with the fixed V_{DC} control method.

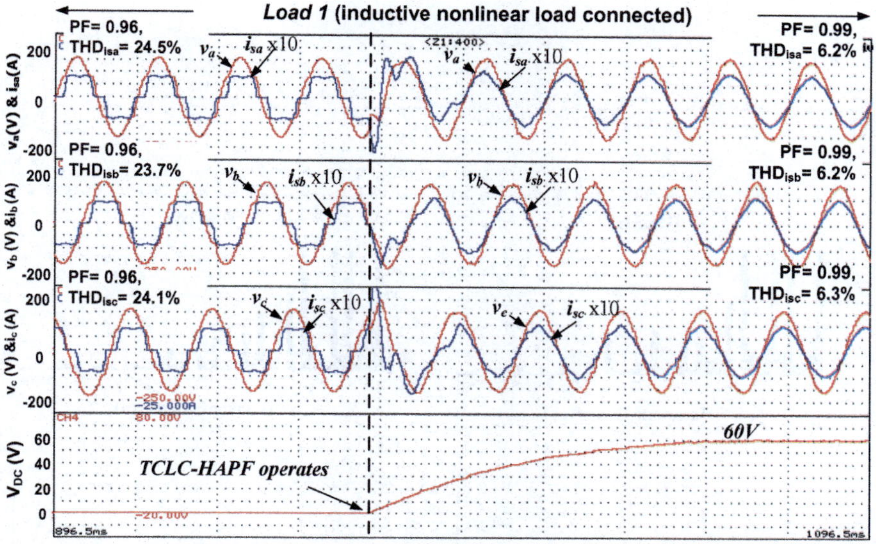

Fig. 7.17 Three-phase experimental system voltage and current with the fixed V_{DC} controlled TCLC-HAPF compensation during starts operation at Load 1

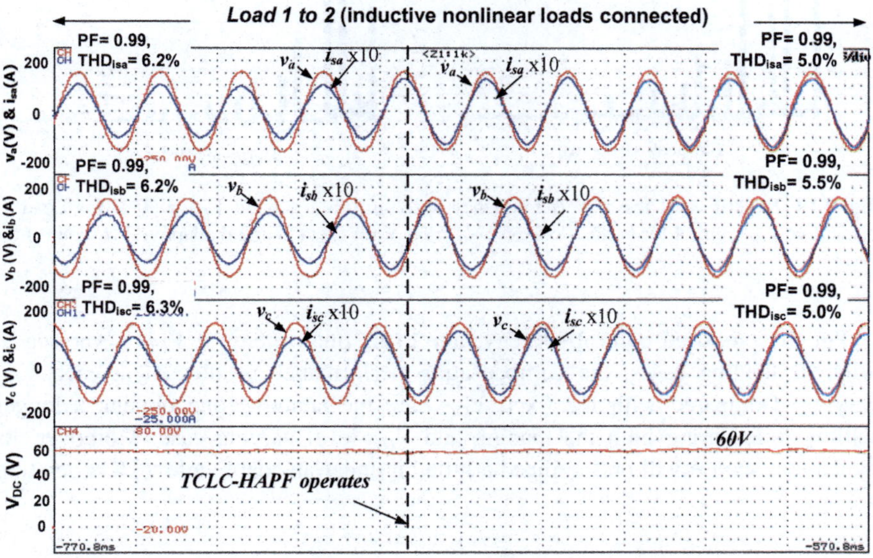

Fig. 7.18 Three-phase experimental system voltage and current with the fixed V_{DC} controlled TCLC-HAPF compensation during *Load* 1 varying to *Load* 2

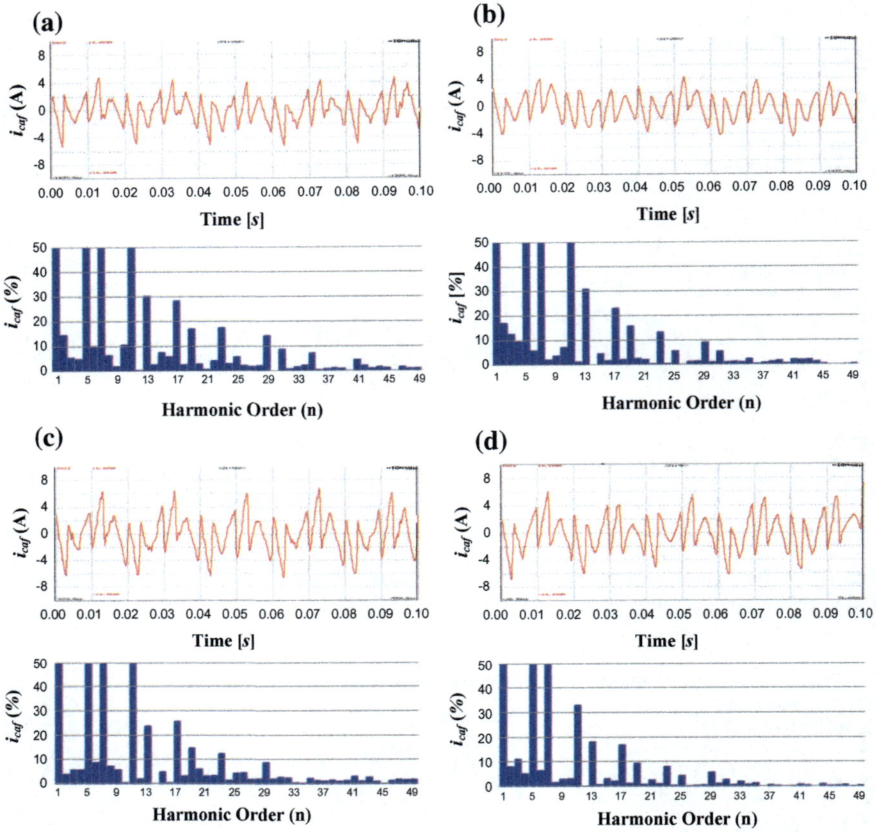

Fig. 7.19 Experimental ica and its frequency spectrum with **a** fixed $V_{DC} = 60$ V for Load 1, **b** adaptive V_{DC} control for Load 1, **c** fixed $V_{DC} = 60$ V for Load 2, and **d** adaptive V_{DC} control for Load

For the proposed control strategy, due to its final reference V_{DC}^{*} is varying at different loading cases, the compensating performance is affected at each V_{DC} varying. Compared with the fixed V_{DC} one, the adaptive one obtains a longer settling time during both the loading and V_{DC} level varying case. Moreover, its dynamic response will be sacrificed a little bit under an adaptive low DC operating voltage.

7.7 Summary

In this chapter, the adaptive DC-link voltage controller for three-phase three-wire TCLC-HAPF is proposed. Different from the conventional V_{DC} calculation method based on the complicated FFT, the proposed V_{DC} calculation can significantly reduce a large number of calculation steps, thus simplify the V_{DC} calculation. On the other hand, the proposed adaptive DC-link voltage control method for TCLC-HAPF can achieve satisfactory compensation performance, low switching loss and switching noise simultaneously. Finally, simulation and experimental results verify both adaptive DC-link voltage controller and the proposed simplified V_{DC} calculation method of TCLC-HAPF.

References

1. Q. Xu, F. Ma, A. Luo, Z. He, H. Xiao, Analysis and control of M3C based UPQC for power quality improvement in medium/high voltage power grid. IEEE Trans. Power Electron. **31** (12), 8182–8194 (2016)
2. W.R.N. Santos, E.D.M. Fernandes, E.R.C.D. Silva, C.B. Jacobina, A.C. Oliveira, P.M. Santos, Transformerless single-phase universal active filter with UPS features and reduced number of electronic power switches. IEEE Trans. Power Electron. **31**(6), 4111–4120 (2016)
3. W.R.N. Santos, E.R.C.D. Silva, C.B. Jacobina, E.D.M. Fernandes, A.C. Oliveira, R.R. Matias, D.F.G. Filho, O. Almeida, P.M. Santos, The transformerless single-phase universal active power filter for harmonic and reactive power compensation. IEEE Trans. Power Electron. **29**(7), 3563–3572 (2014)
4. M.C. Wong, Y.Z. Yang, C.S. Lam, W.H. Choi, N.Y. Dai, Y. Wu et al., Self-reconfiguration property of a mixed signal controller for improving power quality compensation during light loading. IEEE Trans. Power Electron. **30**(10), 5938–5951 (2015)
5. L. Limongi, L.D.S. Filho, L. Genu, F. Bradaschia, M. Cavalcanti, Transformerless hybrid power filter based on a six-switch two-leg inverter for improved harmonic compensation performance. IEEE Trans. Ind. Electron. **62**(1), 40–51 (2015)
6. C. Kumar, M.K. Mishra, An improved hybrid DSTATCOM topology to compensate reactive and nonlinear loads. IEEE Trans. Ind. Electron. **61**(12), 6517–6527 (2014)
7. B. Singh, K. Al-Haddad, A. Chandra, A review of active filters for power quality improvement. IEEE Trans. Ind. Electron. **45**(5), 960–971 (1999)
8. F.Z. Peng, H. Akagi, A. Nabae, A new approach to harmonic compensation in power systems—a combined system of shunt passive and series active filters. IEEE Trans. Ind. Appl. **26**(6), 983–990 (1990)
9. S.P. Litrán, P. Salmerón, Reference voltage optimization of a hybrid filter for nonlinear load compensation. IEEE Trans. Ind. Electron. **61**(6), 2648–2654 (2014)
10. J.C.A. Gil, E. Pérez, C. Ariño, H. Beltran, Optimization algorithm for selective compensation in a shunt active power filter. IEEE Trans. Ind. Electron. **62**(6), 3351–3361 (2015)
11. E.S. Sreeraj, E.K. Prejith, K. Chatterjee, S. Bandyopadhyay, An active harmonic filter based on one-cycle control. IEEE Trans. Ind. Electron. **61**(8), 3799–3809 (2015)
12. S. Srianthumrong, H. Akagi, A medium-voltage transformerless ac/dc power conversion system consisting of a diode rectifier and a shunt hybrid filter. IEEE Trans. Ind. Appl. **39**(3), 874–882 (2003)
13. W.H. Choi, C.S. Lam, M.C. Wong, Y.D. Han, Analysis of dc-link voltage controls in three-phase four-wire hybrid active power filters. IEEE Trans. Power Electron. **28**(5), 2180–2191 (2013)
14. C.S. Lam, W.H. Choi, M.C. Wong, Y.D. Han, Adaptive dc-link voltage controlled hybrid active power filters for reactive power compensation. IEEE Trans. Power Electron. **27**(4), 1758–1772 (2012)

15. C.S. Lam, X.X. Cui, W.H Choi, M.C. Wong, Y.D. Han, Minimum inverter capacity design for three-phase four-wire LC-hybrid active power filters. IET Power Electron. **5**(7), 956–968 (2012)
16. C.S. Lam, M.C. Wong, W.-H. Choi, X.-X. Cui, H.-M. Mei, J.-Z. Liu, Design and performance of an adaptive low-dc-voltage-controlled LC-hybrid active power filter with a neutral inductor in three-phase four-wire power systems. IEEE Trans. Ind. Electron. **61**(6), 2635–2647 (2014)
17. S. Rahmani, A. Hamadi, K. Al-Haddad, A combination of shunt hybrid power filter and thyristor-controlled reactor for power quality. IEEE Trans. Ind. Electron. **61**(5), 2152–2164 (2014)
18. L. Wang, C.S. Lam, M.C. Wong, An unbalanced control strategy for a thyristor controlled LC-coupling hybrid active power filter (TCLC-HAPF) in three-phase three-wire systems. IEEE Trans. Power Electron. **32**(2), 1056–1069 (2017)
19. L. Wang, C.S. Lam, M.C. Wong, A hybrid-STATCOM with wide compensation range and low dc-link voltage. IEEE Trans. Ind. Electron. **63**(6), 3333–3343 (2016)
20. M.C. Wong, J. Tang, Y.-D. Han, Cylindrical coordinate control of three-dimensional PWM technique in three-phase four-wired trilevel inverter. IEEE Trans. Power Electron. **18**(1), 208–220 (2003)
21. C.S. Lam, X.X. Cui, M.C. Wong, Y.D. Han, Minimum dc-link voltage design of three-phase four-wire active power filters. 2012 IEEE 13th Workshop on Control and Modeling for Power Electronics, June 2012, pp. 1–5
22. A.M. Al-Zamil, D.A. Torrey, A passive series, active shunt filter for high power applications. IEEE Trans. Power Electron. **16**(1), 101–109 (2001)
23. G.K. Singh, A.K. Singh, R. Mitra, A simple fuzzy logic based robust active power filter for harmonics minimization under random load variation. Electr. Power Syst. Res. **77**(8), 1101–1111 (2007)
24. M.A. Ahmed, S.A. Zaid, O.A. Mahgoub, A simplified control strategy for the shunt active power filter for harmonic and reactive power compensation. J. Electr. Eng. (JEE) **11**, 1–7 (2011)
25. V. Khadkikar, A. Chandra, B.N. Singh, Generalized single-phase p-q theory for active power filtering: simulation and DSP-based experimental investigation. IET Power Electron. **2**, 67–78 (2009)
26. J. Lundquist, On harmonic distortion in power systems. Engineering Licentiate Thesis, Chalmers University of Technology, Gothenburg, Sweden, 2001
27. F.C. De La Rosa, *Harmonics and Power Systems* (Taylor & Francis, New York, NY, USA, 2006)
28. C. Venkatesh, D. Srikanth Kumar, D.V.S.S. Siva Sarma, M. Sydulu, Modelling of nonlinear loads and estimation of harmonics in industrial distribution system. 15th National Power Systems Conference (NPSC), IIT Bombay, pp. 592–597, Dec 2008
29. C. Batard, F. Poitiers, C. Millet, N. Ginot, Simulation of power converters using Matlab-Simulink, in *MATLAB—A Fundamental Tool for Scientific Computing and Engineering Applications*, vol. 1, ed. by V. Katsikis (In Tech, 2012)
30. R.G. Ellis, P. Eng, Power System Harmonics—A Reference Guide to Causes, Effects and Corrective Measures (Rockwell International Corporation, Canada, 2001)
31. R. Visintini, Rectifiers, in *CAS CERN Accelerator School Specialized Course on Power Converters*, ed. by B. Daniel (CERN, Geneva, Switzerland, 2006), pp. 133–183

32. National Program on Technology Enhanced Learning, Operation and analysis of the three phase fully controlled bridge converter. (2015) [Online]. Available: http://www.nptel.ac.in/courses/108105066/13. Accessed 21 Aug 2015
33. P. Duhamel, M. Vetterli, Fast fourier transforms: a tutorial review and a state of the art (Invited paper). Sig. Process. **19**, 255–299 (1990)
34. H. Akagi, Y. Kanazawa, A. Nabae, Instantaneous reactive power compensators comprising switching devices without energy storage components. IEEE Trans. Ind. Appl. **IA-20**(3), 625–630 (1984)
35. IEEE Recommended Practices and Requirements for Harmonic Control in Electrical Power Systems, IEEE Standard 519-2014 (2014)
36. L. Wang, C.S. Lam, M.C. Wong, N.Y. Dai, K.W. Lao, C.K. Wong, Non-linear adaptive hysteresis band pulse width modulation control for hybrid active power filters to reduce switching loss. IET Power Electron. **8**(11), 2156–2167 (2015)

Chapter 8
Selective Compensation of Distortion, Unbalanced and Reactive Power of a Thyristor Controlled LC-Coupling Hybrid Active Power Filter (TCLC-HAPF)

Abstract When the loads required harmonic, unbalanced and reactive power is beyond the limited capacity of a thyristor controlled LC-coupling hybrid active power filter (TCLC-HAPF), the TCLC-HAPF with the conventional control methods cannot provide satisfactory compensation performance. In this chapter, a selective compensation control method is proposed for harmonic distortion, unbalanced and reactive power of TCLC-HAPF. First, the proposed control method decomposes the load power into fundamental positive-sequence reactive power, fundamental negative-sequence power (unbalanced power) and harmonic power. Then, the decomposed harmonic, unbalanced and reactive power can be selectively or fully compensated based on the capacity of the TCLC-HAPF. Finally, simulation and experimental results are provided to verify the effectiveness of the proposed selective compensation control method for the TCLC-HAPF.

Keywords Selective compensation · Positive-sequence · Negative-sequence
Active power · Reactive power · Unbalance power · Voltage dip
Voltage fault · Thyristor controlled LC-coupling hybrid active power filter (TCLC-HAPF)

8.1 Introduction

The smart grid, regarded as the next generation power grid, is considered as a promising solution for energy crisis. However, the smart grid is interconnected with different power grids, so that the different power quality problems can easily happen in weak grid areas such as harmonic distortion, lower power factor, unbalanced problem, voltage dip, voltage fault, etc. The continuous development of the different power quality compensators has favored the developing progress of the smart gird. At the early stage, the passive power filters (PPFs) and static Var compensators (SVCs) are used to solve power quality problems. The PPFs are designed for harmonic current and fixed reactive power compensation. And, the SVCs are

L. Wang et al., *Adaptive Hybrid Active Power Filters*, Power Systems,
https://doi.org/10.1007/978-981-10-8827-8_8

designed for dynamically reactive current and unbalanced power compensation [1]. However, both PPFs and SVCs are very sensitive to the voltage variation, and suffer from the resonance problem. The active power filters (APFs) can be used for solving the different power quality problems. However, their initial and operational cost is high. To reduce the power capacity of the active inverter part, the hybrid active power filters (HAPFs) have been proposed which combining PPFs or SVCs in series/parallel with APFs (PPF + APFs [2–6], PPF//APFs [7, 8], SVC//APFs [9–12] and SVC + APFs [13–17]). Among the different hybrid structures, the thyristor controlled LC-coupling hybrid active power filter (TCLC-HAPF) [13–17] has the distinctive characteristics of a much wider compensation range than the series connected structures of PPF + APFs and a lower DC-link voltage than the parallel connected structures of PPF//APFs and SVC//APFs.

The above mentioned APFs and HAPFs are normally designed as global filters to compensate all the non-efficient power that is reactive, unbalanced and harmonic power. However, due to the loads expansion and simultaneous operation of different loads, the global filters cannot provide satisfactory compensation performance if the targeted non-efficient power are beyond the designed capacity of the power filters. Therefore, it is suggested to selectively compensate then on-efficient power components (reactive, unbalanced and harmonic power) when they are excess the power filters compensation range [18–26]. In [18–20], different harmonic selective compensation techniques are proposed. However, the reactive and unbalanced power has not been taken into consideration among those works. In [21], a selective compensation technique is proposed in distributed generators (DGs) by inserting negative- and zero-sequence virtual impedances. However, the harmonic component compensation is not taken into account in [21]. Steps beyond, the selective compensation of non-efficient power components for APF have been proposed by using the current decomposition approach [22, 23], equivalent conductance approach [24] IEEE Std. 1459-2000 based approach [25] and linear matrix inequalities (LMIs) approach [26]. However, all the above selective compensation methods [18–26] are developed for APFs only. The selective compensation techniques for HAPFs especially for SVC + APFs and SVC//APFs are still lack of study. Moreover, the compensation performances under different voltage problem conditions such as voltage fault and voltage dip are also being considered [18–26]. Therefore, this chapter aims to develop a selective compensation control method for TCLC-HAPF to selectively or fully compensate harmonic distortion, unbalanced and reactive power components based on its designed capacity.

In the following, the circuit configuration of the three-phase three-wire TCLC-HAPF is provided in Sect. 8.2. And, the basic power analysis of the proposed selective compensation is illustrated in Sect. 8.3. Then, the proposed selective control compensation method for the TCLC-HAPF is presented in Sect. 8.4. To verify the proposed selective control method, simulation case studies and representative experimental results are presented in Sect. 8.5. Finally, a summary will be drawn in Sect. 8.6.

8.2 Circuit Configuration of the TCLC-HAPF

Figure 8.1 shows the circuit configuration of TCLC-HAPF, where x stands for phase a, b and c in the following analysis. v_{sx} and v_x are the source and coupling point voltage, i_{sx}, i_{Lx} and i_{cx} are the source, load and compensating current, respectively. L_s is the transmission line impedance. The TCLC-HAPF consists of a thyristor control LC (TCLC) part and an active inverter part.

The TCLC part is composed of a coupling inductor L_c, a parallel capacitor C_{PF}, and a thyristor controlled reactor (TCR) with L_{PF}. The TCLC part provides a wide and continuous fundamental reactive power and unbalanced power compensation range, which is controlled by triggering the firing angles α of the thyristors [8].

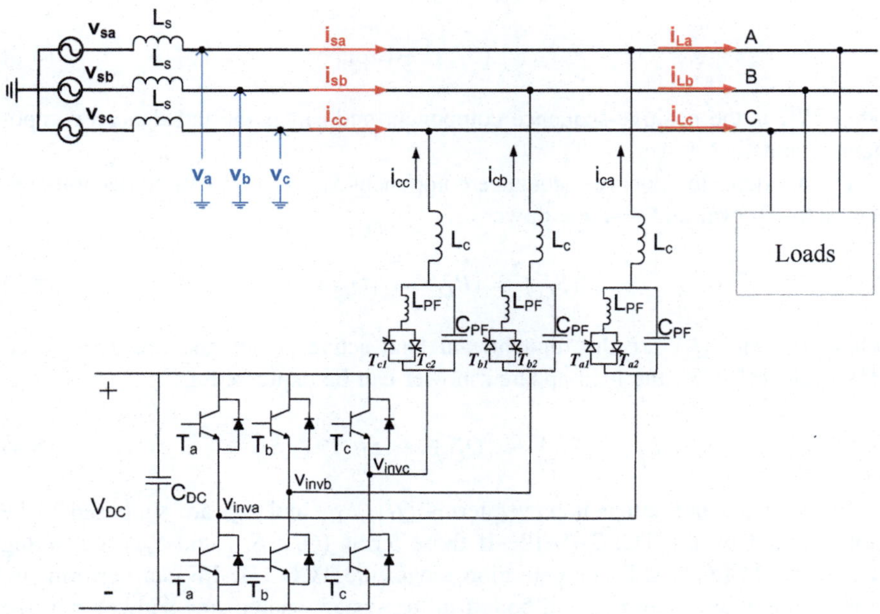

Fig. 8.1 Circuit configuration of the TCLC-HAPF

The active inverter part is a voltage source inverter (VSI) which consists of a DC-link capacitor C_{DC} and power switches IGBTs. The small capacity of the active inverter part is used to compensate the harmonic power, avoid mistuning of the firing angles and prevent the resonance problem of the TCLC part.

8.3 Power Analysis of the Proposed Selective Compensation

According to IEEE Std. 1459 [27], the definition of three-phase total effective apparent load power S_L in terms of fundamental component and harmonic component can be expressed as:

$$\begin{aligned} S_L^2 &= S_{L1}^2 + S_{Lh}^2 \\ &= (3 \cdot V_{L1} \cdot I_{L1})^2 + \left[(3 \cdot V_{L1} \cdot I_{Lh})^2 + (3 \cdot V_{Lh} \cdot I_{L1})^2 + (3 \cdot V_{Lh} \cdot I_{Lh})^2\right] \end{aligned} \tag{8.1}$$

where S_{L1} and S_{Lh} are the fundamental and harmonic components of the load apparent power. V_{L1}, I_{L1}, V_{Lh} and I_{Lh} are the fundamental and harmonic components of the load voltage and load current, respectively.

The fundamental S_{L1} load apparent power in (8.1) can be decomposed into positive-sequence component and unbalanced (negative) component [26, 27] as:

$$S_{L1}^2 = \left(S_{L1}^+\right)^2 + (S_{U1})^2 \tag{8.2}$$

where S_{L1}^+ is the positive-sequence component and S_{U1} is the unbalanced component of apparent power.

Furthermore, the positive-sequence component S_{L1}^+ in (8.2) can be decomposed into active power and reactive power as:

$$\left(S_{L1}^+\right)^2 = \left(P_{L1}^+\right)^2 + \left(Q_{L1}^+\right)^2 \tag{8.3}$$

where P_{L1}^+ and Q_{L1}^+ are the fundamental load active power and reactive power. Based on (8.1–8.3), the load apparent power can be expressed as:

$$S_L^2 = \left(P_{L1}^+\right)^2 + \left(Q_{L1}^+\right)^2 + (S_{U1})^2 + (S_{Lh})^2 \tag{8.4}$$

In (8.4), the non-efficient power terms Q_{L1}^+, S_{U1} and S_{Lh} are supposed to be compensated by the TCLC-HAPF. If those terms (Q_{L1}^+, S_{U1} and S_{Lh}) are falling within the TCLC-HAPF compensation range, the TCLC-HAPF can perform full compensation and only P_{L1}^+ will be left in the system source side. However, if the designed capacity ($S_{TCLC\text{-}HAPF}$) of the TCLC-HAPF cannot fully compensate all the non-efficient power components, the selective compensation is needed to be performed, the expression of the TCLC-HAPF capacity is given as:

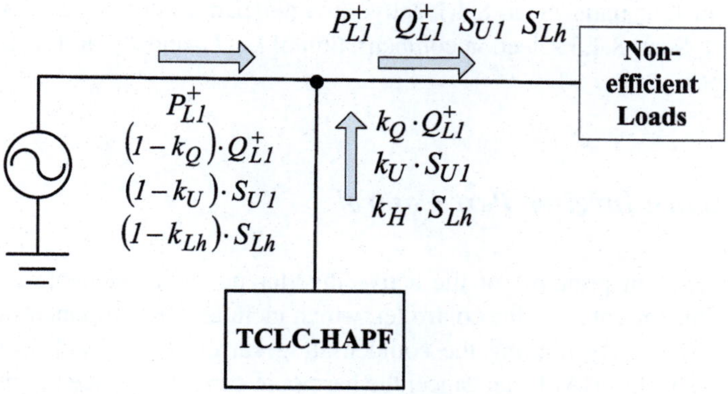

Fig. 8.2 Power flow of the proposed selective compensation

$$S_{TCLC-HAPF}^2 = \left(k_Q \cdot Q_{L1}^{+}\right)^2 + (k_U \cdot S_{U1})^2 + (k_H \cdot S_{Lh})^2 \tag{8.5}$$

where k_Q, k_U and k_H are the compensation ratio of Q_{L1}^{+}, S_{U1} and S_{Lh}, respectively. After the TCLC-HAPF compensation, the apparent power at the system source side can be expressed as:

$$S_S^2 = \left(P_{L1}^{+}\right)^2 + \left[(1-k_Q) \cdot Q_{L1}^{+}\right]^2 + [(1-k_U) \cdot S_{U1}]^2 + [(1-k_H) \cdot S_{Lh}]^2 \tag{8.6}$$

In (8.5) and (8.6), the k_Q, k_U and $k_H \in [0 \quad 1]$. However, if $S_{TCLC-HAPF}^2 \geq \left(Q_{L1}^{+}\right)^2 + (S_{U1})^2 + (S_{Lh})^2$, the TCLC-HAPF can perform full compensation with $k_Q = k_U = k_H = 1$ and $S_S = P_{L1}^{+}$. On the other hand, if $S_{TCLC-HAPF}^2 < \left(Q_{L1}^{+}\right)^2 + (S_{U1})^2 + (S_{Lh})^2$, the selective compensation is needed with the tradeoff consideration among k_Q, k_U and k_H.

Based on the above discussions, the power flow of the proposed selective compensation is illustrated in Fig. 8.2.

8.4 Proposed Selective Compensation Control Strategy of TCLC-HAPF

The TCLC-HAPF consists of the TCLC part and active inverter part. The TCLC part is controlled to provide a wide and continuous fundamental reactive power (Q_{L1}^{+}) and unbalanced power (S_{U1}) compensation range [14]. The low capacity active inverter part is mainly used to compensate the harmonic power (S_{Lh}) and also help to improve the performance of the TCLC part by enlarging its power compensation range. In the following, the control strategy of the TCLC-HAPF will be discussed in four parts: Sect. 8.4.1 active inverter part control, Sect. 8.4.2 TCLC part control, Sect. 8.4.3 selection compensation of k_Q, k_U and k_H, and D. the overall control block.

8.4.1 Active Inverter Part Control

The compensation principle of the active inverter part is to generate a reference compensating current i_{cx}^* to the controller, which includes the components of $k_Q Q_{L1}^+$, $k_U S_{U1}$ and $k_H S_{Lh}$. By limiting the compensating current i_{cx} to track its reference value i_{cx}^*, the TCLC-HAPF can cancel the harmonic current, balance the system and compensate the reactive power to certain degree.

For the unbalanced three-phase three-wire system, the instantaneous v_x and i_{Lx} can be decomposed into the positive- and negative components by using notch filter [28] as:

$$\left[v_x^+(t)\right]_{x=a,b,c} = T_2 \cdot [v_x(t)]_{x=a,b,c} - T_1 \cdot \left[v_x^D(t)\right]_{x=a,b,c} \tag{8.7}$$

$$\left[i_{Lx}^+(t)\right]_{x=a,b,c} = T_2 \cdot [i_{Lx}(t)]_{x=a,b,c} - T_1 \cdot \left[i_{Lx}^D(t)\right]_{x=a,b,c} \tag{8.8}$$

$$\left[i_{Lx}^-(t)\right]_{x=a,b,c} = T_2 \cdot [i_{Lx}(t)]_{x=a,b,c} + T_1 \cdot \left[i_{Lx}^D(t)\right]_{x=a,b,c} \tag{8.9}$$

where $v_x(t)$ and $i_{Lx}(t)$ are the three phase load voltage and load current. $v_x^D(t)$ and $i_{Lx}^D(t)$ can be obtained by delaying the $v_x(t)$ and $i_{Lx}(t)$ by 90°. $v_x^+(t)$ and $i_{Lx}^+(t)$ are individual positive sequence of the load voltage and current of each phase, $i_{Lx}^-(t)$ is negative sequence of the load current. T_1 and T_2 can be expressed as:

$$\mathrm{T}_1 = \frac{1}{2\sqrt{3}} \cdot \begin{bmatrix} 0 & 1 & -1 \\ -1 & 0 & -1 \\ 1 & -1 & 0 \end{bmatrix} \tag{8.10}$$

$$\mathrm{T}_2 = \frac{1}{3} \cdot \begin{bmatrix} 1 & -0.5 & -0.5 \\ -0.5 & 1 & -0.5 \\ -0.5 & -0.5 & 1 \end{bmatrix} \tag{8.11}$$

After obtaining the $v_x^+(t)$, $i_{Lx}^+(t)$ and $i_{Lx}^-(t)$, the three-phase positive sequence and negative sequence of the active power and reactive power can be calculated as:

$$\begin{bmatrix} p^+ \\ q^+ \end{bmatrix} = \begin{bmatrix} v_\alpha^+ & v_\beta^+ \\ -v_\beta^+ & v_\alpha^+ \end{bmatrix} \cdot \begin{bmatrix} i_{L\alpha}^+ \\ i_{L\beta}^+ \end{bmatrix} \tag{8.12}$$

$$\begin{bmatrix} p^- \\ q^- \end{bmatrix} = \begin{bmatrix} v_\beta^+ & v_\alpha^+ \\ -v_\alpha^+ & v_\beta^+ \end{bmatrix} \cdot \begin{bmatrix} i_{L\alpha}^- \\ i_{L\beta}^- \end{bmatrix} \tag{8.13}$$

In (8.12) and (8.13), the positive and negative sequences of the current and voltage in α-β plane are transformed from *a-b-c* frames by:

$$\begin{bmatrix} v_\alpha^+ \\ v_\beta^+ \end{bmatrix} = \begin{bmatrix} 1 & -1/2 & -1/2 \\ 0 & \sqrt{3}/2 & -\sqrt{3}/2 \end{bmatrix} \cdot \begin{bmatrix} v_a^+ \\ v_b^+ \\ v_c^+ \end{bmatrix} \tag{8.14}$$

$$\begin{bmatrix} i_{L\alpha}^+ \\ i_{L\beta}^+ \end{bmatrix} = \begin{bmatrix} 1 & -1/2 & -1/2 \\ 0 & \sqrt{3}/2 & -\sqrt{3}/2 \end{bmatrix} \cdot \begin{bmatrix} i_{La}^+ \\ i_{Lb}^+ \\ i_{Lc}^+ \end{bmatrix} \tag{8.15}$$

$$\begin{bmatrix} i_{L\alpha}^- \\ i_{L\beta}^- \end{bmatrix} = \begin{bmatrix} 1 & -1/2 & -1/2 \\ 0 & \sqrt{3}/2 & -\sqrt{3}/2 \end{bmatrix} \cdot \begin{bmatrix} i_{La}^- \\ i_{Lb}^- \\ i_{Lc}^- \end{bmatrix} \tag{8.16}$$

where $v_x^+(t)$ and $i_{Lx}^+(t)$ are obtained from (8.7) to (8.9).

Referring to Fig. 8.2, $k_Q Q_{L1}^+$, $k_U S_{U1}$ and $k_H S_{Lh}$ can be obtained as:

$$k_Q \cdot Q_{L1}^+ = k_Q \cdot (\bar{q})^+ \tag{8.17}$$

$$k_U \cdot \mathrm{S}_{U1} = k_U \cdot \sqrt{[(\bar{p})^-]^2 + [(\bar{q})^-]^2} \tag{8.18}$$

$$k_H \cdot S_{Lh} = k_H \cdot \left(\sqrt{[(\tilde{p})^+]^2 + [(\tilde{q})^+]^2 + [(\tilde{p})^-]^2 + [(\tilde{q})^-]^2}\right)_{\mathrm{rms}} \tag{8.19}$$

The detailed selection compensation of k_Q, k_U and k_H will be explained in Sect. 8.4.3. By adding the scale factors (k_Q, k_U and k_H) into the reference current calculation, the positive and negative compensating current in α-β plane can be given as:

$$\begin{bmatrix} \mathrm{i}_{c\alpha}^+ \\ \mathrm{i}_{c\beta}^+ \end{bmatrix} = \frac{1}{(v_\alpha^+)^2 + \left(v_\beta^+\right)^2} \cdot \begin{bmatrix} v_\alpha^+ & -v_\beta^+ \\ v_\beta^+ & v_\alpha^+ \end{bmatrix} \begin{bmatrix} k_H \cdot (\tilde{\mathrm{p}})^+ \\ k_{\mathrm{Q}} \cdot (\bar{\mathrm{q}})^+ + k_H \cdot (\tilde{\mathrm{q}})^+ \end{bmatrix} \tag{8.20}$$

$$\begin{bmatrix} \mathrm{i}_{c\alpha}^- \\ \mathrm{i}_{c\beta}^- \end{bmatrix} = \frac{1}{(v_\alpha^+)^2 + \left(v_\beta^+\right)^2} \cdot \begin{bmatrix} v_\beta^+ & -v_\alpha^+ \\ v_\alpha^+ & v_\beta^+ \end{bmatrix} \cdot \begin{bmatrix} k_U \cdot (\bar{\mathrm{p}})^- + k_H \cdot (\tilde{\mathrm{p}})^- \\ k_U \cdot (\bar{\mathrm{q}})^- + k_H \cdot (\tilde{\mathrm{q}})^- \end{bmatrix} \tag{8.21}$$

The final reference current in a-b-c plane are transformed from α-β plane by:

$$\begin{bmatrix} i_{ca}^* \\ i_{cb}^* \\ i_{cc}^* \end{bmatrix} = \sqrt{\frac{2}{3}} \cdot \begin{bmatrix} 1 & 0 \\ -1/2 & \sqrt{3}/2 \\ -1/2 & -\sqrt{3}/2 \end{bmatrix} \cdot \begin{bmatrix} i_{c\alpha}^+ + i_{c\alpha}^- \\ i_{c\beta}^+ + i_{c\beta}^- \end{bmatrix} \tag{8.22}$$

The active inverter part is used to improve the TCLC part characteristic by limiting the compensating current i_{cx} to track its reference value i_{cx}^*. And the i_{cx}^* is calculated by the above proposed method because it is valid for different voltage and current conditions.

8.4.2 TCLC Part Control

The purpose of the TCLC part is to provide the reference compensating reactive power Q_{cx}^* for each phase in order to compensate the reactive and unbalanced power ($k_Q Q_{L1}^+$ and $k_U S_{U1}$) of the loading [14]. From the previous calculated reference current i_{cx}* which consists of the load reactive power, unbalanced and harmonic current components ($k_Q Q_{L1}^+$, $k_U S_{U1}$ and $k_H S_{Lh}$), with the help of single-phase instantaneous p-q theory [29], v_x and i_{cx}*, the instantaneous reference compensating reactive power q_{cx}^* in each phase can be calculated as:

$$\begin{bmatrix} q_{ca}^* \\ q_{cb}^* \\ q_{cc}^* \end{bmatrix} = \begin{bmatrix} v_b \cdot i_{ca}^{*D} - v_a^D \cdot i_{ca}^* \\ v_b \cdot i_{cb}^{*D} - v_b^D \cdot i_{cb}^* \\ v_c \cdot i_{cc}^{*D} - v_c^D \cdot i_{cc}^* \end{bmatrix} \tag{8.23}$$

where v_x, i_{cx}*, v_x^D and $i_{cx}{}^{*D}$ are the load voltage, reference compensating current and their values delay by 90°. Then, the fundamental compensating reactive power $Q_{cx}^* \approx -\bar{q}_{cx}^*/2$ can be obtained by using q_{cx}^* in (8.23) and three low pass filters (LPFs). With the calculated reference Q_{cx}^*, the required TCLC part impedance can be calculated as:

$$\begin{bmatrix} X_{af} \\ X_{bf} \\ X_{cf} \end{bmatrix} = \begin{bmatrix} \dfrac{3 \cdot \bar{V}_x^2 \cdot (Q_{cc}^* - Q_{cb}^* - Q_{ca}^*)^{-1} \cdot (Q_{cb}^* - Q_{ca}^* - Q_{cc}^*)^{-1}}{(Q_{ca}^* - Q_{cb}^* - Q_{cc}^*)^{-1} + (Q_{cb}^* - Q_{ca}^* - Q_{cc}^*)^{-1} + (Q_{cc}^* - Q_{cb}^* - Q_{ca}^*)^{-1}} \\ \dfrac{3 \cdot \bar{V}_x^2 \cdot (Q_{ca}^* - Q_{cb}^* - Q_{cc}^*)^{-1} \cdot (Q_{cc}^* - Q_{cb}^* - Q_{ca}^*)^{-1}}{(Q_{ca}^* - Q_{cb}^* - Q_{cc}^*)^{-1} + (Q_{cb}^* - Q_{ca}^* - Q_{cc}^*)^{-1} + (Q_{cc}^* - Q_{cb}^* - Q_{ca}^*)^{-1}} \\ \dfrac{3 \cdot \bar{V}_x^2 \cdot (Q_{ca}^* - Q_{cb}^* - Q_{cc}^*)^{-1} \cdot (Q_{cb}^* - Q_{ca}^* - Q_{cc}^*)^{-1}}{(Q_{ca}^* - Q_{cb}^* - Q_{cc}^*)^{-1} + (Q_{cb}^* - Q_{ca}^* - Q_{cc}^*)^{-1} + (Q_{cc}^* - Q_{cb}^* - Q_{ca}^*)^{-1}} \end{bmatrix} \tag{8.24}$$

where $\bar{V}_x$ is the rms value of positive sequence phase voltage which can be instantaneous calculated as:

$$\bar{V}_x = \|v\| \Big/ \sqrt{3} = \sqrt{v_a^{+2} + v_b^{+2} + v_c^{+2}} \Big/ \sqrt{3} \tag{8.25}$$

Moreover, the expression of the TCLC impedances (X_{af}, X_{bf} and X_{cf}) can also be expressed in terms of TCLC passive components and firing angles (α_x) as:

$$\begin{bmatrix} X_{af}(\alpha_a) \\ X_{bf}(\alpha_b) \\ X_{cf}(\alpha_c) \end{bmatrix} = \begin{bmatrix} \frac{\pi X_{L_{PF}} X_{C_{PF}}}{X_{C_{PF}}[2\pi - 2\alpha_a + \sin(2\alpha_a)] - \pi X_{L_{PF}}} + X_{L_c} \\ \frac{\pi X_{L_{PF}} X_{C_{PF}}}{X_{C_{PF}}[2\pi - 2\alpha_b + \sin(2\alpha_b)] - \pi X_{L_{PF}}} + X_{L_c} \\ \frac{\pi X_{L_{PF}} X_{C_{PF}}}{X_{C_{PF}}[2\pi - 2\alpha_c + \sin(2\alpha_c)] - \pi X_{L_{PF}}} + X_{L_c} \end{bmatrix} \tag{8.26}$$

where X_{LPF}, X_{CPF}, X_{Lc} are the reactance of L_{PF}, C_{PF} and L_c, respectively. The required X_{xf} obtained from (8.24) is controlled by firing angle through (8.26). However, the (8.26) does not have a closed-form solution. A look up table (LUT) has been installed to directly obtain the firing angle α_x with known X_{xf}. By comparing the firing angle α_x with the phase angle $\varphi_{Vxf-Vnf}$ of voltage between TCLC part ($V_x - V_{nf}$), the trigger signals to control TCLC part thyristors can be obtained. The phase angle of voltage between TCLC part ($V_x - V_{nf}$) can be expressed as:

$$\begin{bmatrix} \varphi_{Vaf-Vnf} \\ \varphi_{Vbf-Vnf} \\ \varphi_{Vcf-Vnf} \end{bmatrix} = \begin{bmatrix} \theta_a - \tan^{-1}\left(\frac{X_{cf} - X_{bf}}{\sqrt{3}(X_{bf} + X_{cf})}\right) \\ \theta_b - \tan^{-1}\left(\frac{X_{af} - X_{cf}}{\sqrt{3}(X_{af} + X_{cf})}\right) \\ \theta_c - \tan^{-1}\left(\frac{X_{bf} - X_{af}}{\sqrt{3}(X_{bf} + X_{af})}\right) \end{bmatrix}_{\tan^{-1}\theta \in [-90^\circ, 90^\circ]} \tag{8.27}$$

where the θ_x is phase angle of load voltage V_x, which can be obtained by using phase lock loop (PLL). By comparing the α_x with the $\varphi_{Vxf-Vnf}$, the trigger signals to control the TCLC part thyristors can be obtained.

8.4.3 *Compensation Priority Selection Among* $\mathbf{k_Q}$, $\mathbf{k_U}$ *and* $\mathbf{k_H}$

In (8.17, 8.18), the $k_Q Q_{L1}^{+}$, $k_U S_{U1}$ and $k_H S_{Lh}$ can be calculated. If the load harmonic, unbalanced and reactive power within the designed capacity of the TCLC-HAPF, the below different power quality standards as shown in Table 8.1 can be satisfied simultaneously. However, if the designed capacity of the TCLC-HAPF is insufficient, the below power quality standards may not be satisfied at the same time, and their compensation priority selection can be done by changing the gains (k_Q, k_U and k_H).

The compensation priority mainly depends on (1) power quality standards [30–34], (2) their effects and (3) their corresponding penalty. Table 8.1 summarizes the

Table 8.1 Power quality standards, effects and penalty of harmonic, unbalanced and reactive power problems

		Harmonic power ($k_H S_{Lh}$)	Unbalanced power ($k_U S_{U1}$)	Reactive power ($k_Q Q^{+}_{L1}$)
Related standards	International	IEEE Std 519-2014 [30]	IEEE Std 1159-2009 [31]	N/A
	China	GB/T14595-93 [32]	GB/T 15543-2008 [33] GB755-87 [34]	Regulation of power factor adjusting charge [35]
	Requirements	$THDv < 5\%$ $THDi < 5$–20%	$UF_v < 0.5$–2% $UF_i < 10\%$	$PF > 0.8$
Effects		(1) Damage to sensitive loads, (2) Resonance problem, (3) Increase power loss, etc.	(1) Reduce the transmission efficiency, (2) Increase power loss and temperature of transformer, etc.	(1) Increase rating of transformer and generator, and increase power loss (2) Cause the voltage drop, etc.
Penalty (GB standards)		Terminate electricity supply	Terminate electricity supply	Extra charge/cost

power quality standards, effects and penalty of harmonic, unbalanced and reactive power problem [35].

Based on Table 8.1, if the harmonic current level is high, the harmonic sensitive loads can easily get damaged. And, the resonance problem can also be caused by the harmonic current. Moreover, if the harmonic current level of the customer loads cannot satisfy Chinese standard GB/T14595-93 [32], the electricity company has the right to terminate the electricity supply for that customer. In addition, the unbalanced voltage and current can reduce the transmission efficiency and increase the power loss and temperature of transformer, etc. According to Chinese standards GB/T 15543-2008 [33] and GB755-87 [34], the same penalty can be issued for unbalanced problem. On the other hand, the penalty of reactive power problem in China is the extra charge, which is not as strict as harmonic and unbalanced problem [35].

Based on the above analysis, the compensation priority has been assigned to the gain of harmonics k_H first, then to the gain of negative sequence k_U and the last is the reactive component gain k_Q in this chapter. Certainly, users can also define the compensation priority and compensation portion of each non-efficient power by themselves. The flow diagram to determine the gains (k_Q, k_U and k_H) of TCLC-HAPF are given in Fig. 8.3.

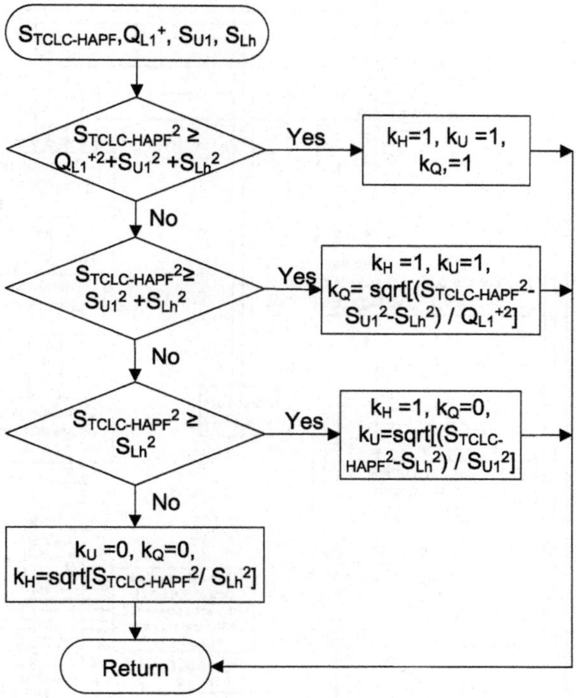

Fig. 8.3 The flowchart of compensation priority assignments of k_Q, k_U and k_H

In Fig. 8.3, if k_H, k_U and k_Q are equal to 1, the harmonic, unbalanced and reactive power are fully compensated. In contrast, if k_H, k_U and k_Q are equal to 0 (no compensation), the harmonic, unbalanced and reactive power are not compensated. As shown in Fig. 8.3, if the TCLC-HAPF capacity $S_{TCLC\text{-}HAPF}$ can compensate the loading harmonic, unbalanced and reactive power simultaneously, $k_H = 1$, $k_U = 1$ and $k_Q = 1$. If the $S_{TCLC\text{-}HAPF}$ can fully support the harmonic and unbalanced power compensation only, $k_H = 1$, $k_U = 1$ and $k_Q < 1$ (partial compensation). And if the $S_{TCLC\text{-}HAPF}$ can fully support the harmonic power compensation only, $k_H = 1$, $k_U < 1$ and $k_Q = 0$. Otherwise, if the $S_{TCLC\text{-}HAPF}$ cannot fully support the harmonic power compensation, $k_H < 1$, $k_U = 0$ and $k_Q = 0$. Based on Fig. 8.3, the gains k_Q, k_U and k_H can be obtained correspondingly.

8.4.4 Control Block of TCLC-HAPF

Figure 8.4 shows the TCLC-HAPF overall control block diagram with the proposed selective compensation control strategy. Based on Fig. 8.4, it can be seen that the proposed selective compensation control strategy consists of the active inverter part control and the TCLC part control. The major connection between these two control loops is the calculated reference compensating current i_{cx}*. The reference current

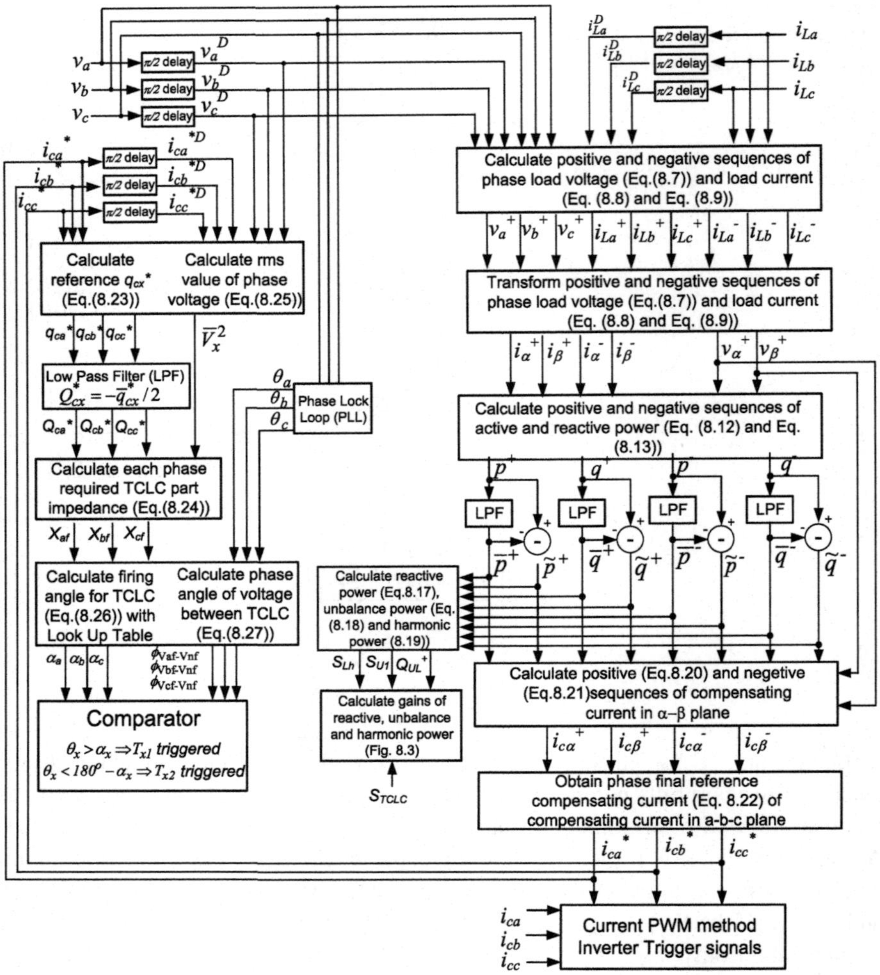

Fig. 8.4 Control block of the proposed selective compensation control strategy of TCLC-HAPF

i_{cx}* is calculated within the active inverter part control loop. And the target of the active inverter part instantaneously limits the compensating current i_{cx} to track its reference value i_{cx}*, which includes the unwanted components of the loading fundamental reactive power ($k_Q Q_{L1}^+$), unbalanced power ($k_U S_{U1}$) and harmonic power ($k_H S_{Lh}$). The trigger signals for the active inverter part are generated by comparing the i_{cx} and i_{cx}* through hysteresis current PWM method.

On the other hand, the TCLC part control loop extracts the fundamental power component from the reference current i_{cx}*. With the extracted fundamental reactive power of each phase, the required TCLC impedance can be calculated. Then, the corresponding trigger signals for the TCLC part can be generated based on the calculated TCLC impedance.

8.5 Simulation and Experimental Verifications

To verify the proposed selective compensation control method among distortion, unbalanced and reactive power of the TCLC-HAPF, PSCAD simulation verifications are carried out to investigate the system performance under different voltage and current conditions. Moreover, a laboratory-scaled hardware prototype is also constructed to obtain the experimental results. The simulation and experimental system parameters for TCLC-HAPF power quality compensation are shown in Table 8.2.

The TCLC part is used to provide the $k_Q Q_{L1}^{+}$ and $k_U S_{U1}$ compensation. Therefore, the total amount of the $k_Q Q_{L1}^{+}$ and $k_U S_{U1}$ provided by the TCLC part is given by:

$$S_{TCLCf} = 3 \cdot Q_{cx\_TCLCf}(\alpha_x) = 3 \cdot \frac{\bar{V}_x^2}{X_{TCLC}(\alpha_x)} = \frac{\bar{V}_x^2}{\frac{\pi X_{L_{PF}} X_{C_{PF}}}{X_{C_{PF}}[2\pi - 2\alpha_x + \sin(2\alpha_x)] - \pi X_{L_{PF}}} + X_{L_c}} \tag{8.28}$$

where $\bar{V}_x$ is the rms value of load voltage, X_{L_c}, $X_{L_{PF}}$ and $X_{C_{PF}}$ are the reactance of L_c, L_{PF} and C_{PF}, respectively. When both thyristors are turned off for the whole fundamental period (firing angle $\alpha_x = 180°$), the TCLC part can provide the maximum capacitive compensating reactive power $Q_{cx(MaxCap)}$. On the other hand, when one of the thyristors is alternately turned on for half of the fundamental period (firing angle $\alpha_x = 90°$), the TCLC part can provide the maximum inductive compensating reactive power $Q_{cx(MaxInd)}$ as $\alpha_x = 90°$. With the system parameters as shown in Table 8.2 and (8.28), the fundamental compensating reactive power range can be plotted in Fig. 8.5.

From Fig. 8.5, the maximum $k_Q Q_{L1+}$ and $k_U S_{U1}$ power provided by the TCLC part are $S_{TCLCf\_max} = -3 * 630$ var $= -1.89$ kvar for inductive loading compensation and $S_{TCLCf\_max} = 3 * 600$ var $= 1.80$ kvar for capacitive loading compensation.

Based on the capacity of the hardware components, the active inverter part is designed to be $S_{Act} = 0.6$ kVA, which is mainly used for harmonic compensation and enlarging the $k_Q Q_{L1}^{+}$ and $k_U S_{U1}$ compensation range. In the following simulation and experimental case studies, the maximum capacity of TCLC-HAPF is:

$$S_{TCLC-HAPF} = \sqrt{S_{TCLCf\_max}^2 + S_{Act}^2} \tag{8.29}$$

Table 8.2 Simulation and experimental parameters for power quality compensation

	Parameters	Physical values
System parameters	V_x, f	110 V, 50 Hz
TCLC-HAPF parameters	L_c, L_{PF}, C_{PF}	5 mH, 30 mH, 160 uF
	C_{DC}, V_{DC}	5 mF, 80 V

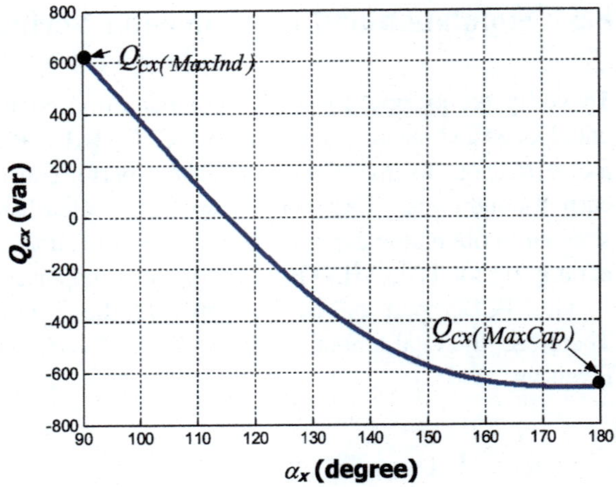

Fig. 8.5 The relationship between the firing angle and the compensating reactive power

Based on (8.29) and the above analysis, the maximum capacity of TCLC-HAPF is designed to be $S_{TCLC\text{-}HAPF}$ = 1.98 kVA for inductive loading compensation and $S_{TCLC\text{-}HAPF}$ = 1.90 kVA for capacitive loading compensation. In this chapter, with reference to the IEEE standard 519-2014 [30], the voltage THD_{vx} is required to be lower than 8% for the laboratory-scaled low voltage application in this chapter (<1 kV). For the IEEE standard 519-2014 [29], the acceptable Total Demand Distortion (*TDD*) $\leq$ 12% with I_{SC}/I_L is in 50 < 100 scale (a small rating 110 V-5 kVA experimental prototype). The nominal rated current is assumed to be equal to the fundamental load current at the worst case analysis, which results in $THD = TDD \leq 12\%$. Therefore, this chapter evaluates the TCLC-HAPF current harmonics compensating performance by setting an acceptable $THDi_{sx} \leq 12\%$.

8.5.1 PSCAD Simulations

The purpose of the simulation studies is to verify the conventional method [14] cannot provide satisfactory compensation results if the load non-effective power are beyond the limited capacity of TCLC-HAPF (in Fig. 8.6). With the same load, the proposed selective compensation method for TCLC-HAPF can selectively compensate distortion, unbalanced and reactive power components with the satisfactory results (in Fig. 8.7). Also, the proposed selective compensation method is verified under voltage dip situation (in Fig. 8.8). The detailed compensation results are given in Table 8.3.

In Fig. 8.6 and Table 8.3, the TCLC-HAPF with conventional method [14] can compensate the reactive power to 154, 64 and 44 var from 756, 809 and 346 var, and the source power factor (PF) has been improved to 0.96 from 0.57 for worst phase. However, the THD_{isx} has been compensated to 21.1, 27.2 and 13.3%, which cannot satisfy the IEEE standard 519-2014 [30]. This is due to the load

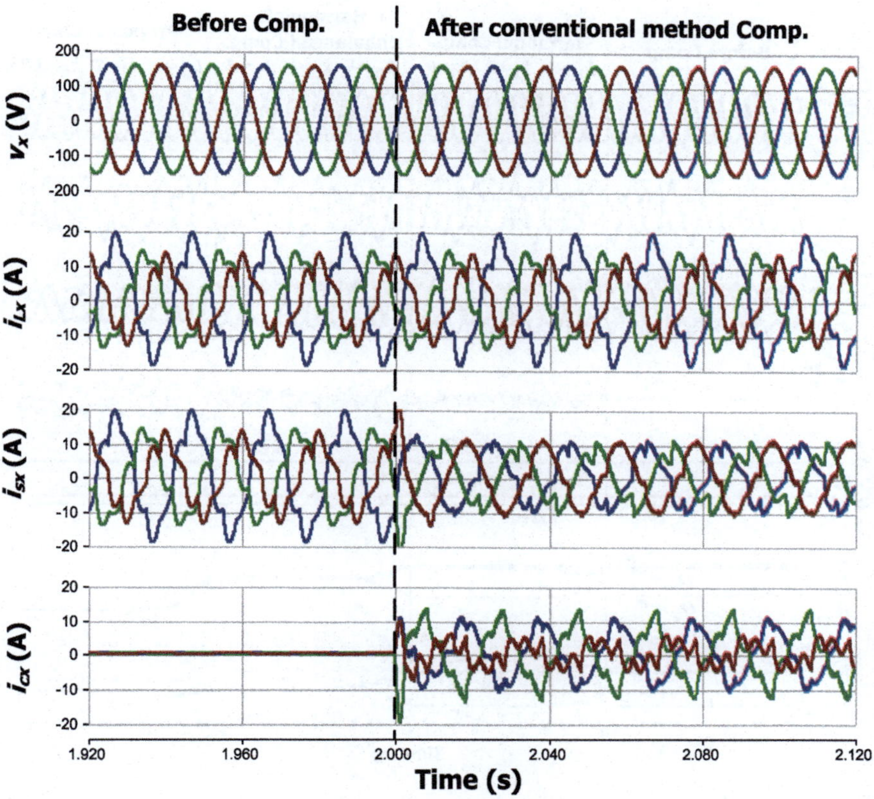

Fig. 8.6 Simulated distortion, unbalanced and reactive power compensation performances by the conventional control strategy [14] of TCLC-HAPF

non-effective power is beyond the limited capacity of TCLC-HAPF. On the other hand, from Fig. 8.7 and Table 8.3, it can be seen that the proposed selective compensation method for TCLC-HAPF can selectively compensate distortion, unbalanced and reactive power components. For harmonic compensation ($k_H = 1$, $k_U = 0$, $k_Q = 0$), the source active power and reactive power before and after the TCLC-HAPF compensation are basically keeping the same level. And the current unbalanced factor (UF_i) is still larger than >40% after compensation. However, the source current THD_{isx} has been compensated to 2.8% from the original 24.5% (worse phase among phase *a*, *b* and *c*). For harmonic and unbalanced compensation ($k_H = 1$, $k_U = 1$, $k_Q = 0$), the unbalanced factor (UF_i) is reduced to 1.2% from the original 41.6%. And the source Q_{sx} and P_{sx} of each phase become the same. Meanwhile, the source current THD_{isx} has been compensated to be lower than 5%. For the proposed selective compensative method ($k_H = 1$, $k_U = 1$, $k_Q = 0.85$), the power factor (*PF*) and UF_i are improved to 0.98 and 1.1, respectively. And, the THD_{isx} is compensated to 5.9%. In addition, there is still remaining Q_{sx} value which is due to the gain $k_Q = 0.85$ instead of $k_Q = 1$.

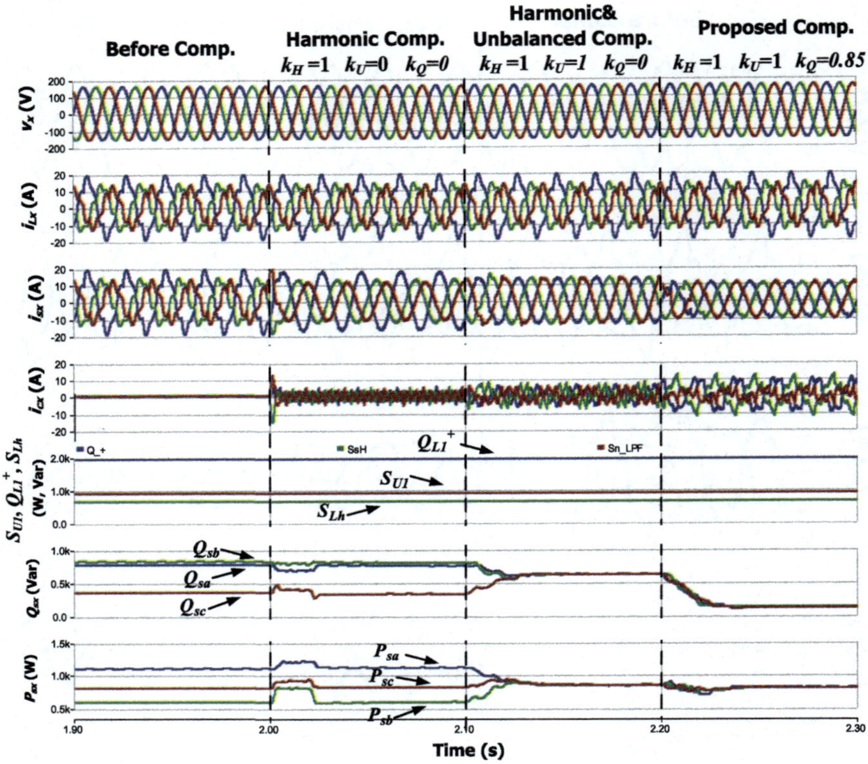

Fig. 8.7 Simulated distortion, unbalanced and reactive power compensation performances by the proposed selective compensation control strategy of TCLC-HAPF

From Fig. 8.8 and Table 8.3, it can be seen that the positive sequence of load voltage v_x^+ can be instantaneous extracted by using (8.7). The source current PF and THD_{isx} can be compensated to 0.99 and 2.7% (worse phase) even during the voltage dip condition.

8.5.2 Experimental Results

An experimental prototype of three-phase three-wire TCLC-HAPF is constructed in the laboratory. The control system has the sampling frequency of 25 kHz. The switching devices for the active inverter part are Mitsubishi IGBTs PM300DSA060. And, the switching devices for the TCLC part are thyristors SanRex PK110FG160. Moreover, the experimental parameters of the TCLC-HAPF and the test loads are approximately the same as the simulation studies in Table 8.2.

Figure 8.9 shows the experimental waveforms of the load voltage and source current before and after the TCLC-HAPF compensation with the proposed selective compensation method ($k_H = 1$, $k_U = 1$, $k_Q = 0.85$). Figures 8.10, 8.11, and 8.12 give

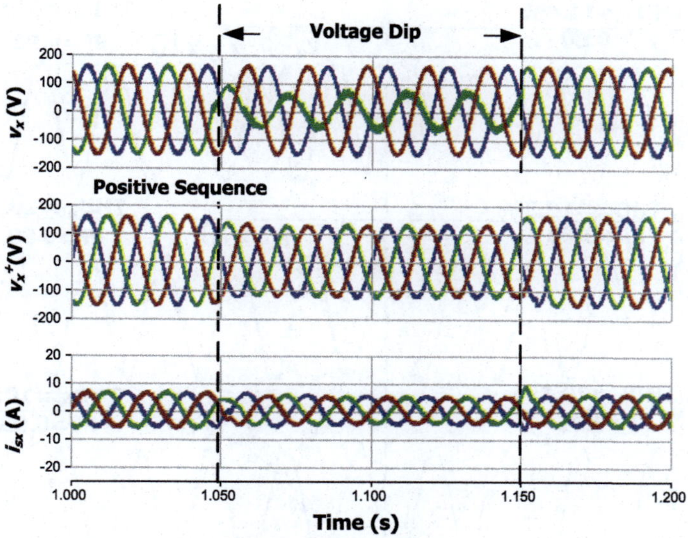

Fig. 8.8 The waveforms of the load voltage, positive sequence v_x^+ and source current by the proposed selective compensation control strategy of TCLC-HAPF during voltage dip

Table 8.3 Simulation compensation results of the proposed selective compensation method for TCLC-HAPF

	Phase	THD_{isx}(%)	Q_{sx}(var)	PF	UFi(%)
Before Compensation	*a*	15.1	756	0.89	41.6
	b	21.5	809	0.57	
	c	24.5	346	0.81	
Conventional Method [15]	*a*	21.1	154	0.99	12.0
	b	27.2	64	0.96	
	c	13.3	44	0.96	
Proposed Method - Harmonic Compensation (k_H =1, k_U=0, k_Q=0)	*a*	1.5	744	0.92	44.1
	b	2.8	800	0.59	
	c	2.6	330	0.83	
Proposed Method - Harmonic & Unbalanced Compensation (k_H =1, k_U=1, k_Q=0)	*a*	2.3	612	0.81	1.2
	b	4.6	610	0.81	
	c	4.8	620	0.81	
Proposed Selective Compensation (k_H =1, k_U=1, k_Q=0.85)	*a*	4.0	120	0.98	1.1
	b	3.9	124	0.98	
	c	5.9	118	0.98	
Proposed Selective Compensation during Voltage Dip	*a*	2.3	7	0.99	1.2
	b	2.7	9	0.99	
	c	2.5	8	0.99	

Note The shaded area means unsatisfactory results

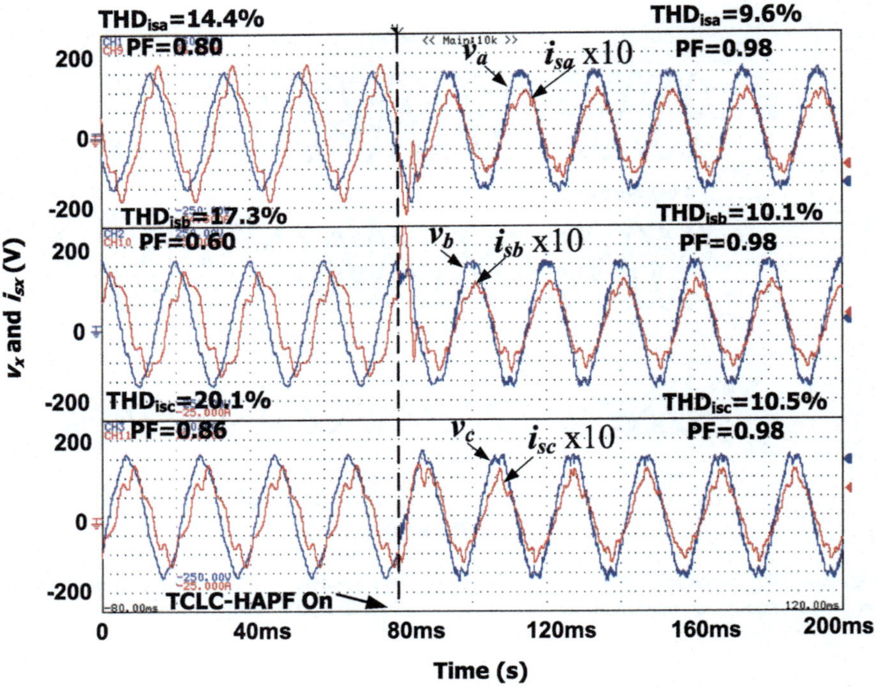

Fig. 8.9 The experimental waveforms of the load voltage and source current before and after TCLC-HAPF compensation with the proposed selective compensation method ($k_H = 1$, $k_U = 1$, $k_Q = 0.85$)

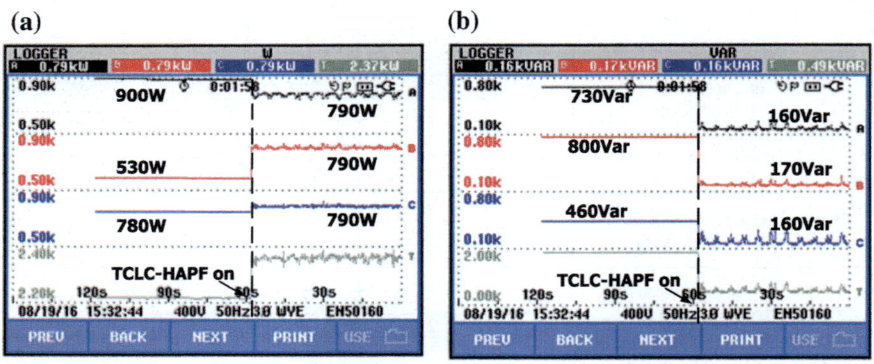

Fig. 8.10 Experimental results of dynamic performance before and after the TCLC-HAPF compensation with the proposed selective compensation method ($k_H = 1$, $k_U = 1$, $k_Q = 0.85$): **a** source active power P_{sx} and **b** source reactive power Q_{sx}

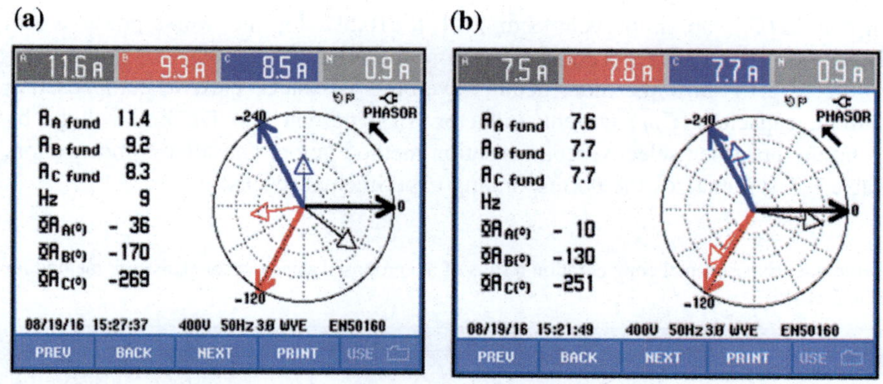

Fig. 8.11 Experimental phasor diagrams of source voltage and current **a** before compensation **b** after the TCLC-HAPF compensation with the proposed selective compensation method ($k_H = 1$, $k_U = 1$, $k_Q = 0.85$)

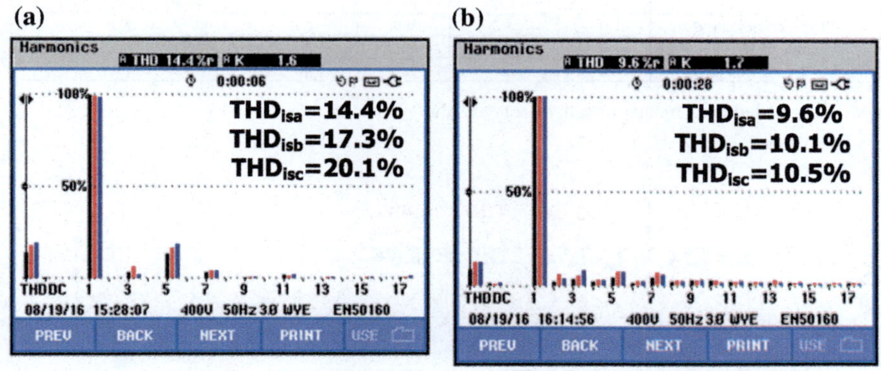

Fig. 8.12 Experimental current harmonic spectrums of source current **a** before compensation **b** after the TCLC-HAPF compensation with the proposed selective compensation method ($k_H = 1$, $k_U = 1$, $k_Q = 0.85$)

the experimental results of the source power (P_{sx} and Q_{sx}), phasor diagrams and source current harmonic spectrum, respectively.

From Fig. 8.9, it can be that after TCLC-HAPF compensation with the proposed selective compensation method, the source current i_{sx} and the coupling point voltage v_x are basically in phase with each other. The source power factors (PFs) are compensated to 0.98 from the original 0.80, 0.86 and 0.86. From Fig. 8.10, the source P_{sx} and Q_{sx} in each phase are compensated to about the same (790 W and 160 var) from the original 900, 530 and 780 W for P_{sx} and 730, 800 and 460 var for Q_{sx}. And, the remaining Q_{sx} is due to $k_Q = 0.85$ setting. Figure 8.11 shows that the magnitude of i_{sx} and phase angle between of v_x and i_{sx} of each phase become approximately the same. From Fig. 8.12, it can be seen that the source current THD_{isx} are improved from 20.1 to 10.5% for the worst phase. From Table 8.4, the UF_i is reduced to 1.3% from the original 32.2%.

Figure 8.13 show the more detail waveforms of source current, compensating current, capacitor (C_{PF}) current, inductor (L_{PF}) current and DC-link voltage by using the proposed selective compensation method before and after compensation. Table 8.4 summarizes the corresponding experimental results.

Table 8.4 Experimental compensation results of the proposed selective compensation method for TCLC-HAPF

	Phase	THD_{vx}(%)	THD_{isx}(%)	Q_{sx}(var)	PF	UFi(%)
Before Compensation (Fig. 8.9)	*a*	4.5	14.4	730	0.80	32.2
	b	4.5	17.3	800	0.60	
	c	4.6	20.1	460	0.86	
Proposed Selective Compensation (k_H=1, k_U=1, k_Q=0.85)	*a*	3.3	9.6	160	0.98	1.3
	b	3.5	10.1	170	0.98	
	c	3.4	10.5	160	0.98	
Before Compensation (Fig. 8.13)	*a*	4.5	22.7	510	0.63	29.4
	b	4.5	22.4	290	0.89	
	c	4.6	14.5	540	0.78	
Proposed Selective Compensation (k_H=1, k_U=1, k_Q=1)	*a*	3.2	9.4	50	0.99	3.6
	b	3.1	10.5	50	0.99	
	c	3.1	10.5	40	0.99	

Note The shaded area means unsatisfactory results

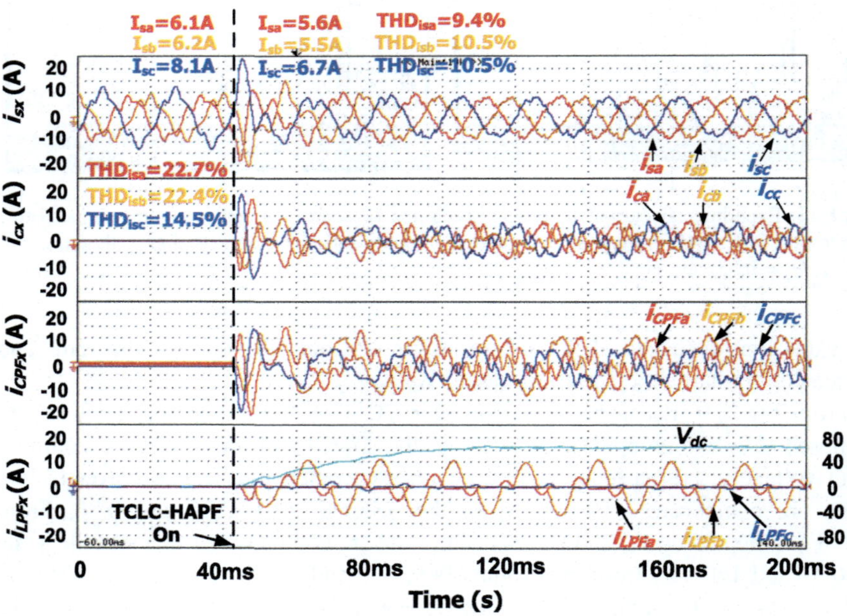

Fig. 8.13 The waveforms of source current, compensating current, capacitor (C_{PF}) current, inductor (L_{PF}) current and dc-link voltage by using the proposed selective compensation method before and after compensation

Comparing the simulation results as in Table 8.3 with the experimental results as in Table 8.4, there are differences between simulation and experimental results, which are actually due to the difference of component parameters, the resolution of the transducers, the digital computation error and the noise in the experiments. In addition, a small capacity of isolation transformer used in the laboratory also causes larger voltage distortion than the simulation case, which deteriorates the compensation results. Therefore, those factors will affect the TCLC-HAPF compensation performance during experiments.

8.6 Summary

In this chapter, a selective compensation control strategy for distortion, unbalanced and reactive power has been proposed for the TCLC-HAPF. The proposed control method can selectively or fully compensate the non-efficient power based on the limited capacity of the TCLC-HAPF. According to simulation and experimental results, it can be proved that the proposed selective control method can successfully decompose the load power into fundamental positive-sequence reactive/active power, fundamental negative-sequence power (unbalanced power) and harmonic power. And the discomposed power components can be selectively or fully compensated.

References

1. L. Wang, C.S. Lam, M.C. Wong, Design of a thyristor controlled LC compensator for dynamic reactive power compensation in smart grid. IEEE Trans. Smart Grid **8**(1), 409–417 (2017)
2. Y. Liu, W. Wu, Y. He, Z. Lin, F. Blaabjerg, H.S.H. Chung, An efficient and robust hybrid damper for LCL- or LLCL-based grid-tied inverter with strong grid-side harmonic voltage effect rejection. IEEE Trans. Ind. Electron. **63**(2), 926–936 (2016)
3. L. Wang, C.S. Lam, M.C. Wong et al., Non-linear adaptive hysteresis band pulse-width modulation control for hybrid active power filters to reduce switching loss. IET Power Electron. **8**(11), 2156–2167 (2015)
4. Z. Shuai, A. Luo, J. Shen, X. Wang, Double closed-loop control method for injection-type hybrid active power filter. IEEE Trans. Power Electron. **26**(9), 2393–2403 (2011)
5. L.R. Limongi, L.R. da Silva Filho, L.G.B. Genu, F. Bradaschia, M.C. Cavalcanti, Transformerless hybrid power filter based on a six-switch two-leg inverter for improved harmonic compensation performance. IEEE Trans. Ind. Electron. **62**(1), 40–51 (2015)
6. R.B. Gonzatti, S.C. Ferreira, C.H. da Silva, R.R. Pereira, L.E. Borges da Silva, G. Lambert-Torres, Smart impedance: a new way to look at hybrid filters. IEEE Trans. Smart Grid **7**(2), 837–846 (2016)
7. A. Luo, S. Peng, C. Wu, J. Wu, Z. Shuai, Power electronic hybrid system for load balancing compensation and frequency-selective harmonic suppression. IEEE Trans. Ind. Electron. **59** (2), 723–732 (2012)

8. E. Durna, İ. Yılmaz, M. Ermiş, Suppression of time-varying interharmonics produced by medium-frequency induction melting furnaces by a HAPF system. IEEE Trans. Power Electron. **32**(2), 1030–1043 (2017)
9. J. Dixon, Y. del Valle, M. Orchard, M. Ortuzar, L. Moran, C. Maffrand, A full compensating system for general loads, based on a combination of thyristor binary compensator, and a PWM-IGBT active power filter. IEEE Trans. Ind. Electron. **50**(5), 982–989 (2003)
10. B. Chen, C. Zhang, C. Tian, J. Wang, J. Yuan, A hybrid electrical magnetic power quality compensation system with minimum active compensation capacity for V/V cophase railway power supply system. IEEE Trans. Power Electron. **31**(6), 4159–4170 (2016)
11. A. Luo, Z. Shuai, W. Zhu, Z.J. Shen, Combined system for harmonic suppression and reactive power compensation. IEEE Trans. Ind. Electron. **56**(2), 418–428 (2009)
12. S.Y. Lee, C.J. Wu, Reactive power compensation and load balancing for unbalanced three-phase four-wire system by a combined system of an SVC and a series active filter. Proc. Inst. Elect. Eng. Elect. Power Appl., **147**(6), 563–578 (2000)
13. S. Rahmani, A. Hamadi, K. Al-Haddad, A combination of shunt hybrid power filter and thyristor-controlled reactor for power quality. IEEE Trans. Ind. Electron. **61**(5), 2152–2164 (2014)
14. L. Wang, C.S. Lam, M.C. Wong, An unbalanced control strategy for a thyristor controlled LC-coupling hybrid active power filter (TCLC-HAPF) in three-phase three-wire systems. IEEE Trans. Power Electron. **32**(2), 1056–1069 (2017)
15. L. Wang, C.S. Lam, M.C. Wong, A hybrid-STATCOM with wide compensation range and low DC-link voltage. IEEE Trans. Ind. Electron. **63**(6), 3333–3343 (2016)
16. L. Wang, C.S. Lam, M.C. Wong, Hardware and software design of a low dc-link voltage and wide compensation range thyristor controlled LC-coupling hybrid active power filter, in *TENCON 2015 IEEE Region 10 Conference Proceedings* (Nov 2015)
17. L. Wang, C.S. Lam, M.C.W. Wong, Modeling and parameter design of thyristor controlled LC-coupled hybrid active power filter (TCLC-HAPF) for unbalanced compensation. IEEE Trans. Ind. Electron. **64**(3), 1827–1840 (2017)
18. M. Aleenejad, H. Mahmoudi, P. Moamaei, R. Ahmadi, A new fault-tolerant strategy based on a modified selective harmonic technique for three-phase multilevel converters with a single faulty cell. IEEE Trans. Power Electron. **31**(4), 3141–3150 (2016)
19. J. He, Y.W. Li, R. Wang, C. Wang, A measurement method to solve a problem of using dg interfacing converters for selective load harmonic filtering. IEEE Trans. Power Electron. **31** (3), 1852–1856 (2016)
20. Y. Zhang, Y.W. Li, N.R. Zargari, Z. Cheng, Improved selective harmonics elimination scheme with online harmonic compensation for high-power PWM converters. IEEE Trans. Power Electron. **30**(7), 3508–3517 (2015)
21. X. Zhou, F. Tang, P.C. Loh, X. Jin, W. Cao, Four-leg converters with improved common current sharing and selective voltage-quality enhancement for islanded microgrids. IEEE Trans. Power Del. **31**(2), 522–531 (2016)
22. B. Singh, V. Verma, J. Solanki, Neural network-based selective compensation of current quality problems in distribution system. IEEE Trans. Ind. Electron. **54**(1), 53–60 (2007)
23. B. Singh, V. Verma, Selective compensation of power-quality problems through active power filter by current decomposition. IEEE Trans. Power Del. **23**(2), 792–799 (2008)
24. S. Orts-Grau et al., Selective compensation in four-wire electric systems based on a new equivalent conductance approach. IEEE Trans. Ind. Electron. **56**(8), 2862–2874 (2009)
25. S. Orts-Grau et al., Selective shunt active power compensator applied in four-wire electrical systems based on IEEE Std. 1459. IEEE Trans. Power Del. **23**(4), 2563–2574 (2008)
26. J.C. Alfonso-Gil, E. Pérez, C. Ariño, H. Beltran, Optimization algorithm for selective compensation in a shunt active power filter. IEEE Trans. Ind. Electron. **62**(6), 3351–3361 (2015)
27. IEEE Standard Definitions for the Measurement of Electric Power Quantities Under Sinusoidal, Nonsinusoidal, Balanced, or Unbalanced Conditions, IEEE Std 1459–2010 (Revision IEEE Std 1459–2000), pp. 1–52 (2010)

28. D. Yazdani, M. Mojiri, A. Bakhshai, G. JoÓs, A fast and accurate synchronization technique for extraction of symmetrical components. IEEE Trans. Power Electron. **24**(3), 674–684 (2009)
29. V. Khadkikar, A. Chandra, B.N. Singh, Generalized single-phase p-q theory for active power filtering: simulation and DSP-based experimental investigation. IET Power Electron. **2**, 67–78 (2009)
30. IEEE Recommended Practices and Requirements for Harmonic Control in Electrical Power Systems, IEEE Standard 519–2014, June 2014
31. IEEE Recommended Practice for Monitoring Electric Power Quality, IEEE Std. 1159-2009, June 2009
32. Standardization Administration of the People's Republic of China, *GB/T 14549-1993 Power quality — Harmonics in public supply network* (China Electric Power Press, Beijing, 1993)
33. Standardization Administration of the People's Republic of China, *GB/T 15543-2008 Power quality—Three-phase voltage unbalance* (China Electric Power Press, Beijing, 2008)
34. Standardization Administration of the People's Republic of China, *GB755-87 Rotating electrical machines—rating and performance* (China Electric Power Press, Beijing, 1987)
35. Ministry of Water Resources and Electric Power of China, *Regulation of Power Factor Adjustment Charge* (China Electric Power Press, Beijing, 1983)

Chapter 9
Implementation of a 110 V-5 kVA Three-Phase Three-Wire of Thyristor Controlled LC-Coupling Hybrid Active Power Filter (TCLC-HAPF) Experimental Prototype

Abstract This chapter presents the hardware and software design of an 110 V-5 kVA three-phase three-wire Thyristor Controlled LC-Coupling Hybrid Active Power Filter (TCLC-HAPF). The overall system configuration and experimental setups of the TCLC-HAPF are presented firstly. Then, the hardware design of TCLC-HAPF is introduced, including: the thyristor and IGBT drivers, the transducer with signal conditioning circuits. After that, the software design of digital control system of the TCLC-HAPF is presented. Finally, the performance of the designed TCLC-HAPF hardware prototype under different voltage and current conditions is verified by experimental results.

Keywords Digital signal processor (DSP) · Insulated gate bipolar translator (IGBT) · Current harmonics · Thyristor controlled LC-coupling hybrid active power filter (TCLC-HAPF)

9.1 Introduction

In this chapter, an 110V 5-kVA three-phase three-wire Thyristor Controlled LC-Coupling Hybrid Active Power Filter (TCLC-HAPF) experimental prototype is designed and constructed in the laboratory. The hardware and software design of the proposed TCLC-HAPF is introduced. The brief settings of TCLC-HAPF can be described as below and more details can be found at other Sects. 9.2–9.5.

The digital control system of the TCLC-HAPF consists of two paralleled digital signal processors (DSPs) TMS320F2812s, and the basic settings of both DSPs are the same. The sampling frequency of the control system is 25 kHz in both simulation and experiment (for both DSPs). For every 1/25 kHz(s) period, the timer will provide a signal to process analog to digital (A/D) conversion and the corresponding interrupt. After processing the proposed control strategy, the output PWMs signal will be generated and the maximum pulse width modulation (PWM) switching frequency is 12.5 kHz. The SanRex PK110FG160 thyristors are

L. Wang et al., *Adaptive Hybrid Active Power Filters*, Power Systems,
https://doi.org/10.1007/978-981-10-8827-8_9

used for the TCLC part, while the Mitsubishi IGBT intelligent power modules PM300DSA60 are employed as the switching devices of the inverter.

The layout of the chapter can described as below. In Sect. 9.2, the circuit configuration and experimental prototype of the TCLC-HAPF are presented firstly. In Sect. 9.3 and 9.4, the hardware and software design of TCLC-HAPF are provided, respectively. Then, the performances of designed TCLC-HAPF under different voltage and current conditions are carried out in Sect. 9.5. Finally, the summary is drawn in Sect. 9.6.

9.2 Circuit Configuration of the TCLC-HAPF Experimental Prototype

The experimental prototype of the 110 V-5 kVA three-phase three-wire TCLC-HAPF has been developed and built in the laboratory. The system configuration of the three-phase three-wire TCLC-HAPF experimental prototype is shown in Fig. 9.1. And the experimental setup of the TCLC-HAPF experimental prototype in the laboratory is illustrated in Fig. 9.2.

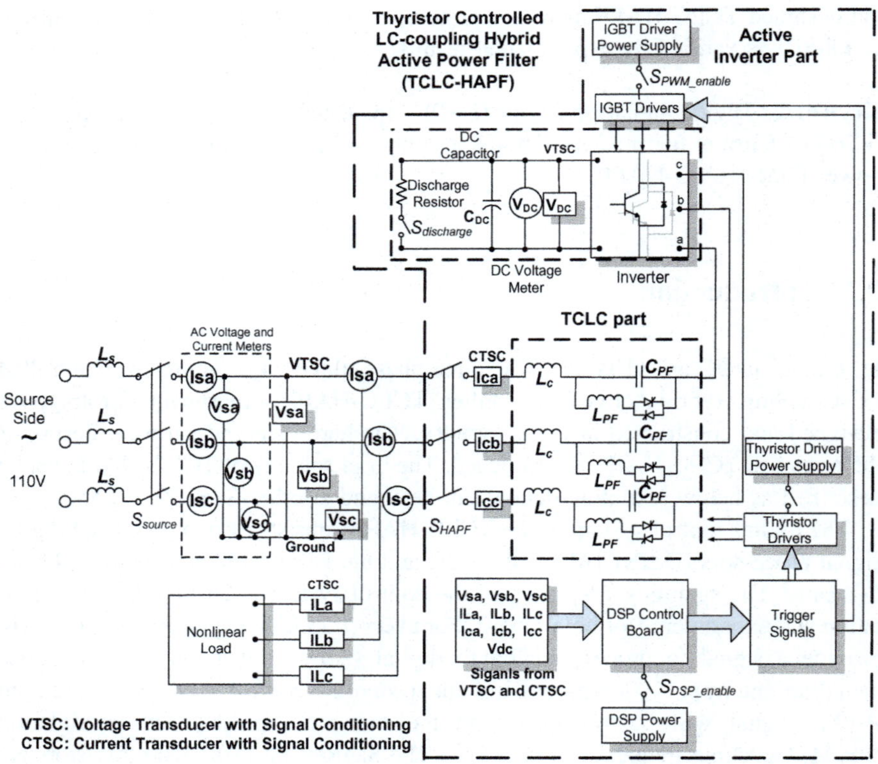

Fig. 9.1 System configuration of the 110 V-5 kVA TCLC-HAPF experimental prototype

Fig. 9.2 Experimental setup of the 110 V-5 kVA TCLC-HAPF experimental prototype

As shown in Fig. 9.1, the TCLC part consists of coupling inductor L_c, parallel inductor L_{PF}, parallel capacitor C_{PF} and thyristors with their drivers. And the active inverter part is composed of a voltage source inverter (VSI) with DC-link capacitor C_{DC}, IGBTs and their drivers. The control system includes transducer with signal conditioning circuits and DSP controllers. The outputs of four voltage transducers and six current transducers with signal conditioning circuits are considered as the inputs to the DSP-based control system for control algorithm processing and to generate control signals to the thyristor and the IGBT drivers.

Based on the circuit system configuration in Fig. 9.1, the hardware and software realization of the TCLC-HAPF experimental prototype is discussed in the following.

9.3 Hardware Design of TCLC-HAPF Experimental Prototype

In this part, the hardware design of the TCLC-HAPF experimental prototype will be presented in two parts: Sect. 9.3.1: Power semiconductor switching devices and their drivers, and Sect. 9.3.2: Design of the transducer with signal conditioning board.

9.3.1 Power Semiconductor Switching Devices and Their Drivers

The power semiconductor switching device has undergone a revolutionary change, from bipolar technology-based devices, such as bipolar junction transistors (BJT), thyristors and gate turn-off (GTO) thyristors, to insulated bipolar devices and MOS-based devices, such as insulated gate bipolar transistors (IGBTs), p metal-oxide-semiconductor (MOS)-based devices, such as MOS field effect transistors (MOSFETs)and MOS-controlled thyristors (MCTs) [1, 2]. Moving from bipolar to MOSFET has resulted in operating frequency increases and capacity slightly decreases. A summary of power switching devices' operating voltage and frequency is provided in Fig. 9.3 [1].

9.3.1.1 Thyristor and Its Drivers in TCLC Part

The TCLC part is used to provide the large fundamental voltage drop between load voltage and inverter voltage and the power semiconductor switching devices in TCLC part are design to be triggered once in a fundamental period. Therefore, the large capacity and relatively low operating frequency power semiconductor device: SanRex thyristor PK110FG160 [3] is selected. The SanRex thyristor PK110FG160 dual intelligent power module has the rated current and maximum voltage of 110 A and 1600 V respectively. To drive the thyristor, the driver POWERSEM PSHI23 [4] is selected to drive the thyristors. In the PSH123 board, the pin connections settings and hardware connection of PSH123 board are plotted in Figs. 9.4 and 9.5, respectively.

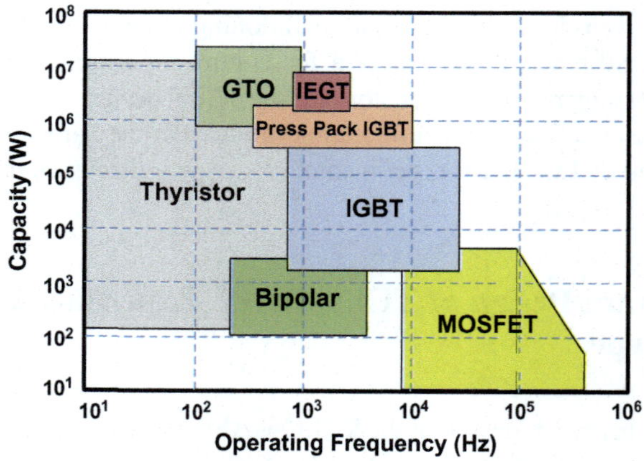

Fig. 9.3 Operating voltage and frequency of power semiconductor switching devices [1]

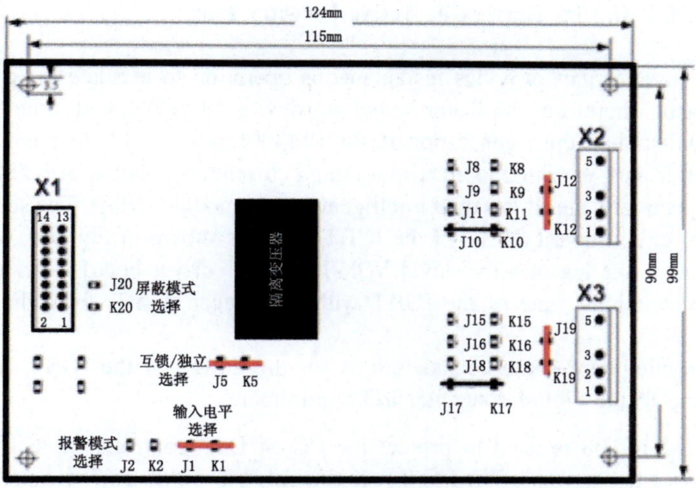

Fig. 9.4 Pin connections of the driver PSHI23 [4]

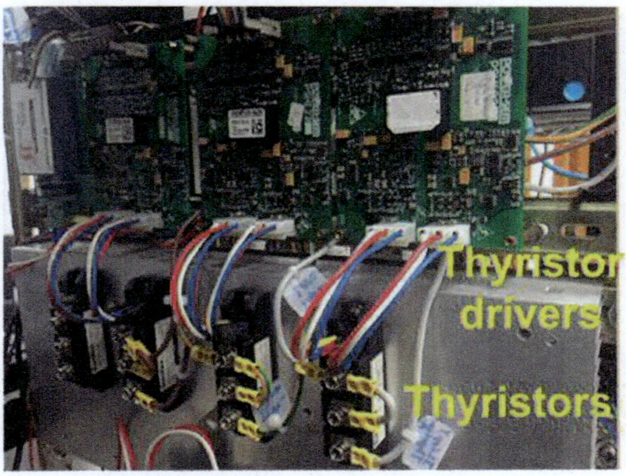

Fig. 9.5 Hardware connection between thyristor power switches and their drivers

a. Power supply of PSHI23 is set to be 15 V for typical operation;
b. Input voltage level is set to be 5 V logic with $V_{IH} = 2.4$ V, $V_{IL} = 0.5$ V in order to capable with the controller which has output voltage level in the range of 0–3 V.
c. Inter Lock Function is set to be cancelled (suitable for individual thyristor driving).
d. The outputs are set to provide individual gate ON and OFF signals (Fig. 5).

9.3.1.2 IGBT and Its Drivers in Active Inverter Part

The active inverter part provides instantaneous operation to regulate the compensating current. Therefore, the faster switching devices IGBTs are selected for the VSI. The Mitsubishi third generation IGBT PM300DSA60 is [5] selected for the active inverter part, in which its maximum rated current and voltage are 300 A and 600 V, respectively, and it is a dual intelligent power module (IPM). The schematic for the print circuit board (PCB) of the IGBT driver is shown in Fig. 9.6, which is used to drive a one leg of a two-level VSI. The IGBT driver board is designed to control the switching state of the IGBT with the trigger signals from the digital controller.

The followings are some considerations for the design of the IGBT driver in consideration of the digital controller I/O requirement.

a. The SN74HCT08 is used to protect the I/O of DSP controller. It works as a buffer since the current of the I/O is recommended to within 1.67 mA per pin for 3.3 V-tolerance; while the typical working current of HCPL4504 is 16 mA.
b. +VL, R7 and R8 are combined as resistor divider to pull up the signal "FO" and limit the voltage within 3.3 V.

The RC low-pass filter is adopted to filter the noise which could probably be on the fault signal. The selection of R1 should consider the current of controller I/O, "FO" pull up potential and RC low-pass filter's loading effect.

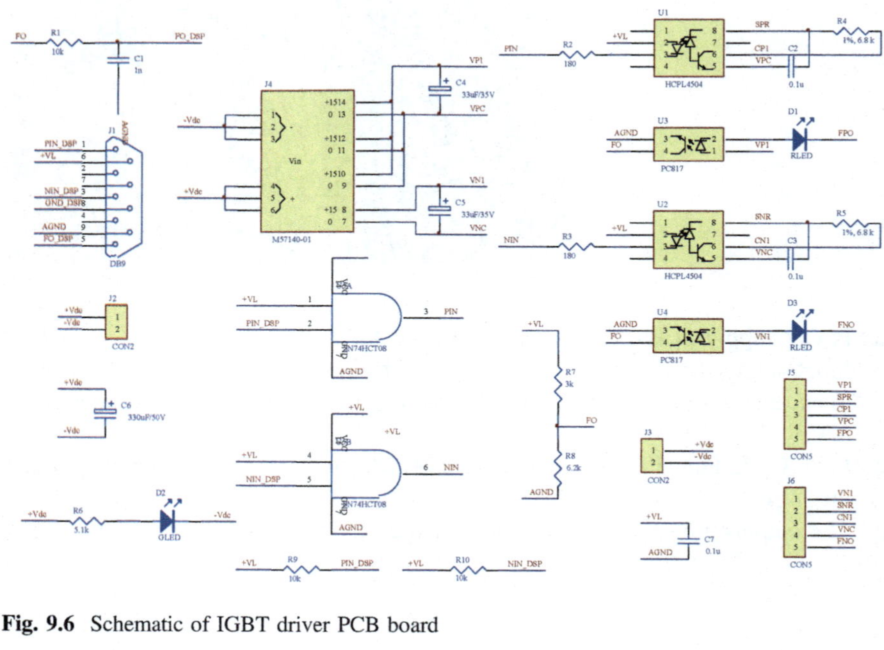

Fig. 9.6 Schematic of IGBT driver PCB board

Fig. 9.7 Hardware connection between IGBT power switches and corresponding drivers

The final hardware connection between the IGBT power switches and their corresponding drivers are shown in Fig. 9.7.

9.3.2 Transducer with Signal Conditioning Boards

The three-phase load voltage v_x, load current i_x, compensating current i_{cx} and DC-link voltage v_{DC} of the TCLC-HAPF are measured by transducer with signal conditioning boards. The purpose of the transducer with signal conditioning boards is to transfer the large voltage and current electrical signals into small analog signals. Then, the output small signals from the signal conditioning boards are sent to the analog-to-digital (A/D) converter to convert the small analog signals into digital signals for digital controller further processing. The proposed signal conditioning circuit is given in Fig. 9.8.

For the 110 V-5 kVA TCLC-HAPF experimental system, the current in each phase are normally lower than 15 A (without considering transient current). The LEM LA-55P (50 A/50 mA) current transducers with routing double-loop are selected, and the input/output becomes 25 A/50 mA. And the LEM LV25-400 V (400 V/25 mA) voltage transducers are used for the load voltage and the DC-link voltage.

As shown in Fig. 9.8, the signal conditioning circuit is constructed by two amplifiers, a zener diode and resistors. The signal conditioning circuit output can be expressed as:

$$v_{out} = \frac{R_6 + R_7}{R_6}\left(\frac{R_3 \cdot R_5}{R_3R_4 + R_3R_5 + R_4R_5} \cdot V_{ref} + \frac{R_4 \cdot R_5}{R_3R_4 + R_3R_5 + R_4R_5} \cdot v_{in}\right) \quad (9.1)$$

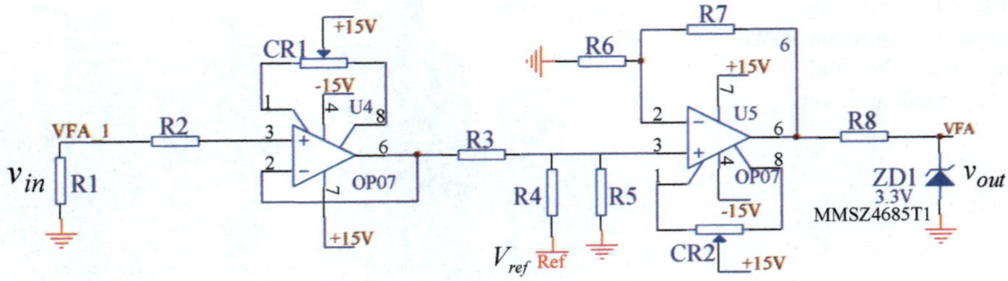

Fig. 9.8 Proposed signal conditioning circuit

Table 9.1 Parameters of signal conditioning board

Range (peak value)		v_{in} (V)	v_o (V)	R_1 (Ω)	R_3 (kΩ)
i_{Lx}	−20 to 20 A	−4 to 4	0–3.191	100	40
i_{cx}	−20 to 20 A	−4 to 4	0–3.191	100	40
v_{sx}	−160 to 160 V	−4 to 4	0–3.191	400	40
v_{dc}	0–100 V	0–5	0.412–2.988	400	4

Notes x stands of phase *a*, *b* and *c*, R_2 = 10 kΩ, R_4 = 30 kΩ, R_5 = 20 kΩ, R_6 = 40 kΩ and R_7 = 10 kΩ

where $v_{in} = R_1 \cdot i_{out}$.

In (9.1), V_{ref} is designed to be 3 V which is provided by the chip REF5030. With carefully selected resistor values $R_1 \sim R_7$, the peak of v_{out} can be designed to be within 0-3.3 V to consistent with the I/O of the digital controller. In addition, v_{out} is limited by the zener diode to avoid over-voltage problem.

According to the experimental conditions, the maximum measurement ranges for load current and compensating current are set as $\pm 20A_{peak}$. And the maximum measurement range of the load voltage and DC-link capacitor voltage are set as $\pm 160V_{peak}$ and +100 V respectively. And the parameters of signal conditioning board are provided in Table 9.1. And, the hardware connection of the transducer with signal condition boards is provided in Fig. 9.9.

9.4 Software Design of TCLC-HAPF Experimental Prototype

The control system is composed of two parallel DSP-TMS320F2812s to separately control the TCLC part and the active inverter part. And the hardware appearance of the DSP controllers and their extension board are given in Fig. 9.10. Each TMS320F2812 consists of two event managers (EVs), EVA and EVB. And each EV module contains two general purpose (GP) timers (totally 4 timers). In the control systems of TCLC-HAPF, each timer is designed for different purposes: Timer 1 and Timer 3 are responsible for generating PWM signals for IGBTs and

Fig. 9.9 Hardware connection of the transducer with signal condition boards

Fig. 9.10 The hardware appearance of the DSP controllers and their extension boards

trigger signals for thyristors, Timer 4 is used to activate the A/D conversion, and Timer 2 is used to define the sample rate of A/D and process signals.

For the 110 V-5 kVA TCLC-HAPF experimental prototype, the sampling frequency is set to 25 kHz in Timer 2. For every 1/25 kHz(s) period, the Timer 2 will provide a trigger signal to process A/D conversion and the corresponding interrupt. And there are totally 10 channel signals (3 load voltage v_x, 3 load current i_{Lx},

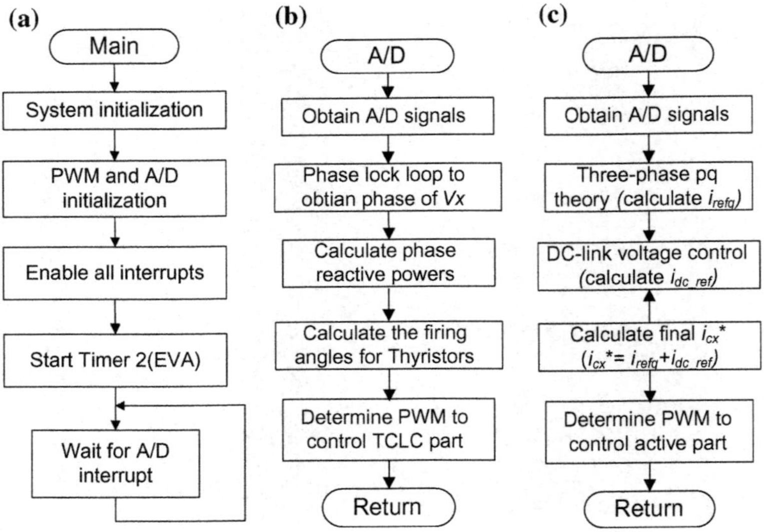

Fig. 9.11 DSP flowcharts of the TCLC-HAPF: **a** main program, **b** interrupt routine for the TCLC part and **c** interrupt routine for the active inverter part

3 compensating current i_{cx}, and 1 DC-link capacitor voltage V_{dc} are converted into digital values.

Figure 9.11 shows the program flow charts for the DSP-TMS320F2812 in performing A/D signal sampling, generating the trigger signals for controlling the TCLC part and generating the PWM signals for controlling the active inverter part. And it is noticed that the main program for control TCLC part and active inverter part are the same, while the interrupt routines for TCLC part and active inverter part are different.

9.5 Experimental Results

9.5.1 Experimental Results of TCLC-HAPF for Dynamic Inductive and Capacitive Reactive Power Compensations

Figures 9.12 and 9.13 show the dynamic compensation waveforms of load voltage v_x, source current i_{sx}, and reactive power Q_{sa} of phase *a* by applying TCLC-HAPF for inductive load and capacitive load compensation. Figure 9.14 gives the corresponding source current harmonic spectrums for inductive and capacitive reactive power compensations.

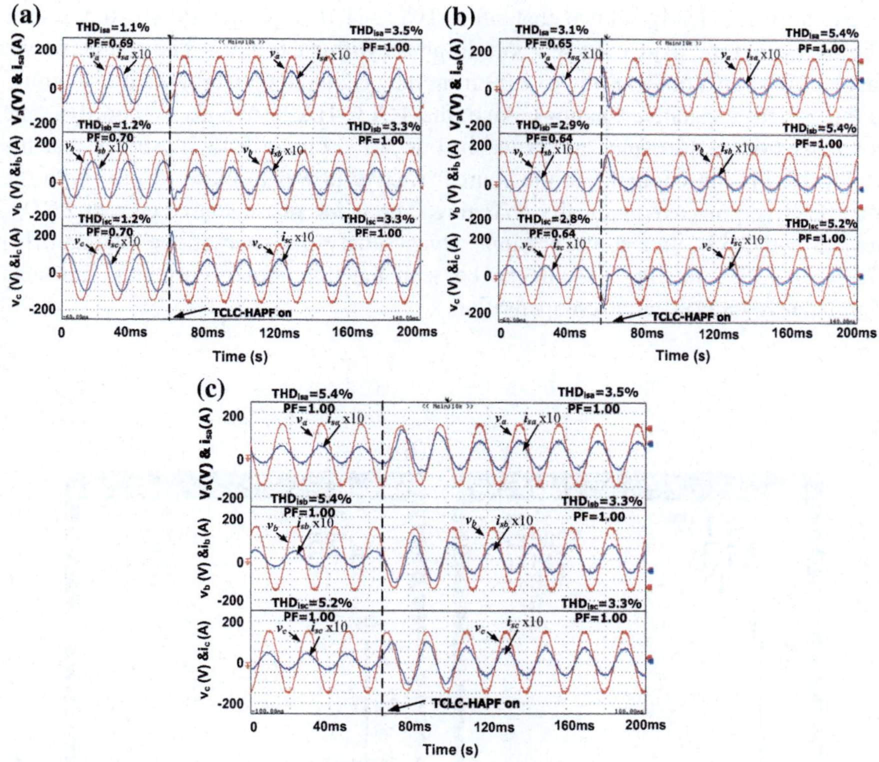

Fig. 9.12 Dynamic compensation waveforms of v_x and i_{sx} by applying TCLC-HAPF under **a** inductive load, **b** capacitive load and **c** changing from capacitive load to inductive load

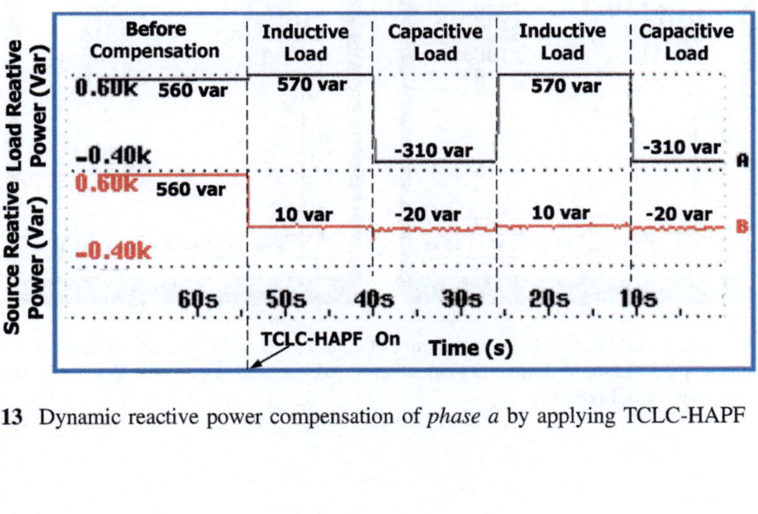

Fig. 9.13 Dynamic reactive power compensation of *phase a* by applying TCLC-HAPF

Figure 9.12 clearly shows that after TCLC-HAPF compensation, the source current i_{sx} and the load voltage v_x are in phase with each other. The source power factors (*PF*s) are compensated to 1.00 from the original 0.69 (for inductive loading) and 0.64 (for capacitive loading). Also, the TCLC-HAPF obtains a good dynamic compensation performance as shown in Fig. 9.12. Figure 9.13 illustrates that the TCLC-HAPF can compensate the source reactive power to close to zero with the load reactive power dynamically changing. The worst phase source current THD_{isx} are 3.5 and 5.4% after compensation, which satisfy the international standard [6] (THD_{isx}< 15%). In Fig. 9.14, it can be seen that the source current i_{sx} are also significantly reduced after compensation.

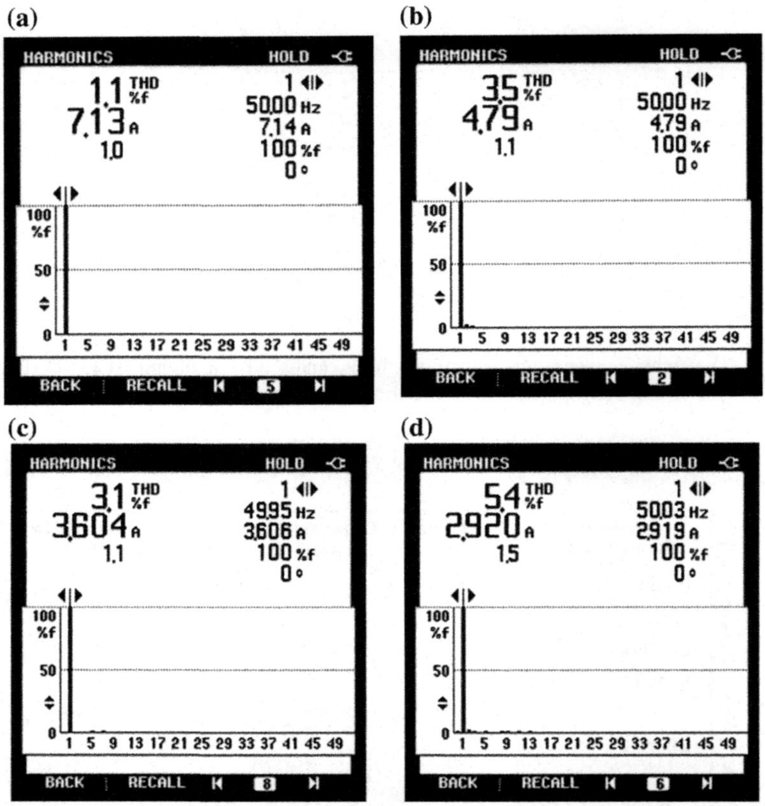

Fig. 9.14 Source current harmonic spectrums of phase a: **a** before compensation of inductive load, **b** after compensation of inductive load, **c** before compensation of capacitive load, and **d** after compensation of capacitive load

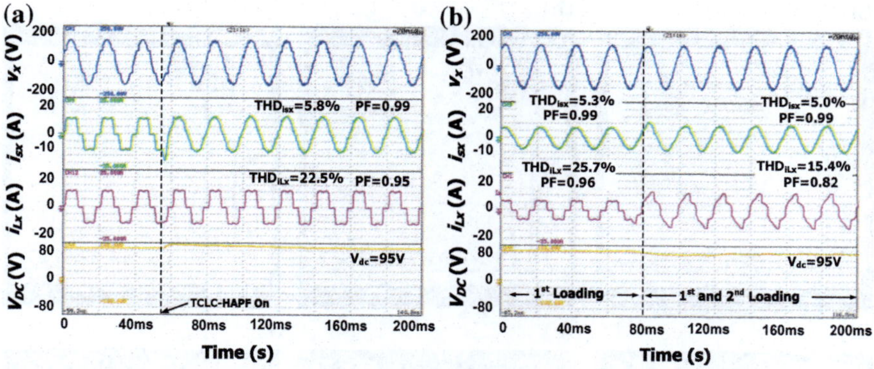

Fig. 9.15 The experimental load voltage, source current, load current and DC-link voltage waveforms with nonlinear loading dynamic changing by applying TCLC-HAPF **a** before and after TCLC-HAPF compensation for nonlinear loading compensation, **b** during TCLC-HAPF compensation before and after 2nd loading is connected

9.5.2 Experimental Results of TCLC-HAPF for Reactive Power and Harmonic Compensations

Figure 9.15 shows the dynamic waveforms of load voltage, source current, load current and DC-link voltage waveforms by applying TCLC-HAPF for different nonlinear loadings compensation. The source current harmonic spectrum before and after TCLC-HAPF compensations are given in Fig. 9.16.

From Fig. 9.15, after the TCLC-HAPF compensation, the source current $THDi_{sx}$ have been compensated to 5.8% for the nonlinear loading, 5.3% for 1st loading compensation and 5.0% for 1st and 2nd loading compensation. Meanwhile, the power factors (PF) have been improved to 0.99 for all different loadings compensation.

9.5.3 Experimental Results of TCLC-HAPF for Unbalanced Loading Compensation

Figures 9.17 and 9.18 show the experimental source voltage and current, source reactive and active power before and after TCLC-HAPF compensation with V_{DC} = 50 V. The experimental source current spectrums after compensation are provided in Fig. 9.19. And the phasor diagrams of the source voltage and current after compensation are illustrated in Fig. 9.20.

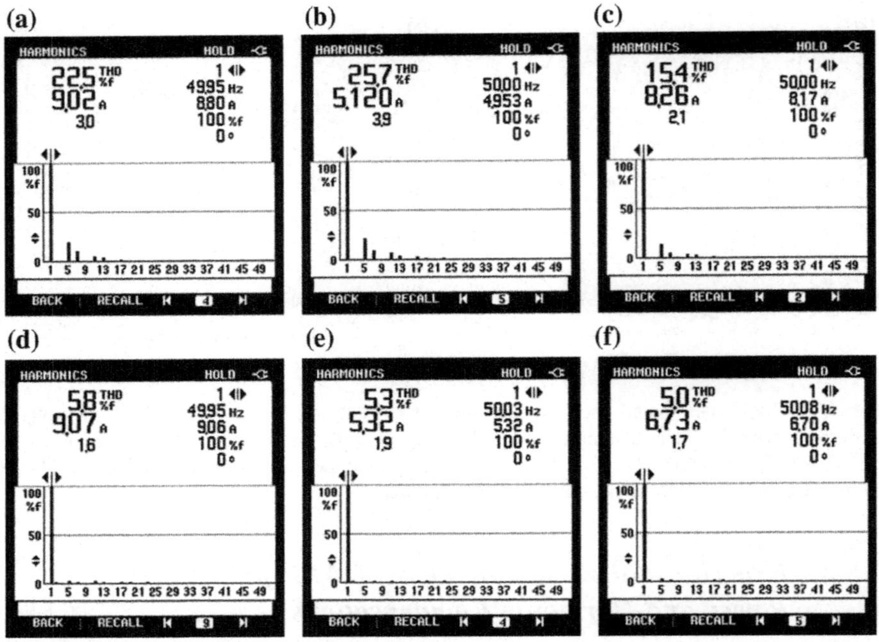

Fig. 9.16 Experimental source current harmonic spectrums **a** before compensation of the nonlinear loading, **b** before compensation of 1st loading, **c** before compensation of 1st and 2nd loadings, **d** the nonlinear loading compensation after TCLC-HAPF compensation, **e** 1st loading before compensation after TCLC-HAPF compensation, **f** 1st and 2nd loadings before compensation after TCLC-HAPF compensation

From Figs. 9.17 and 9.18, with the proposed parameter design method, the TCLC-HAPF can compensate the fundamental reactive power and balance the active power. And the PFs have been compensated to 0.99 for two sets of nonlinear loadings. In Fig. 9.19, after TCLC-HAPF compensation, the worst phase THD_{isx} become 9.5 and 10.6% from the original 22.7% for unbalanced inductive loads and 38.6% for mixed inductive and capacitive loads, which can meet the IEEE standard [6]. Moreover, the source current unbalanced factor (UBI_{fS}) becomes less than 4.5% after TCLC-HAPF compensation.

9.5.4 *Experimental Results of TCLC-HAPF Compensation During Voltage Dip*

Figure 9.21 provides the dynamic compensation waveforms of load voltage v_x and source current i_{sx} by applying TCLC-HAPF during a sudden voltage dip with V_{DC} = 50 V. During the voltage dip, the source current are 4.30, 4.98 and 4.00 A

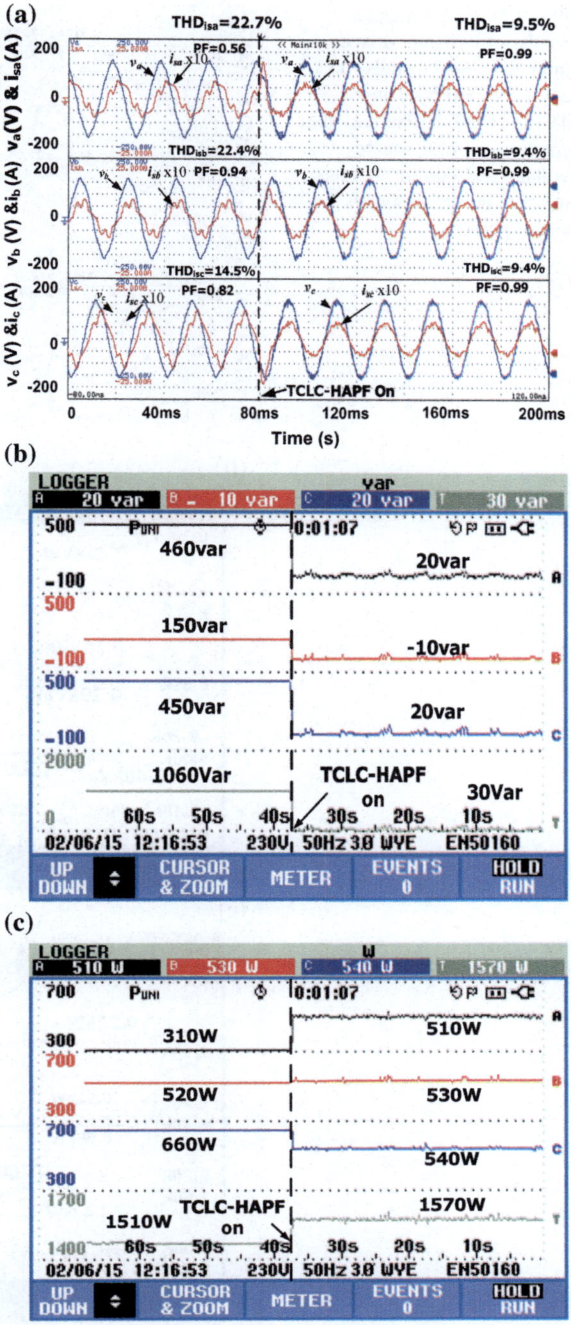

Fig. 9.17 Experimental results of dynamic unbalanced inductive loads compensation by using TCLC-HAPF before and after compensation: **a** source voltage and current, **b** source reactive power, **c** source active power

Fig. 9.18 Experimental results of dynamic mixed inductive and capacitive loads compensation by using TCLC-HAPF before and after compensation: **a** source voltage and current, **b** source reactive power, **c** source active power

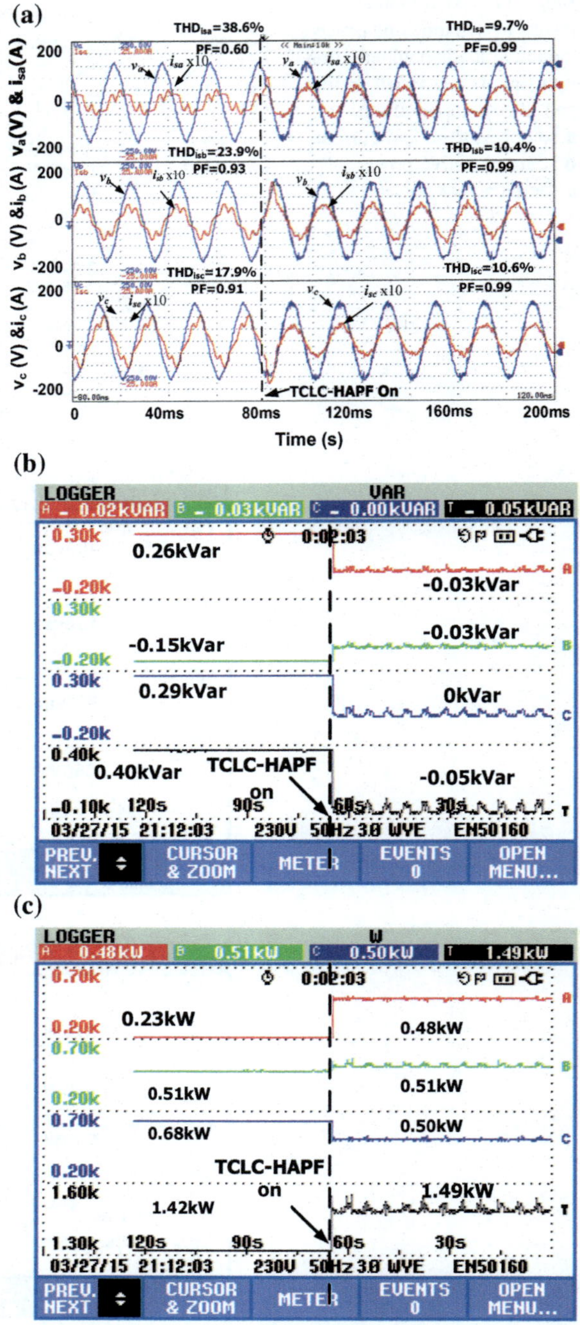

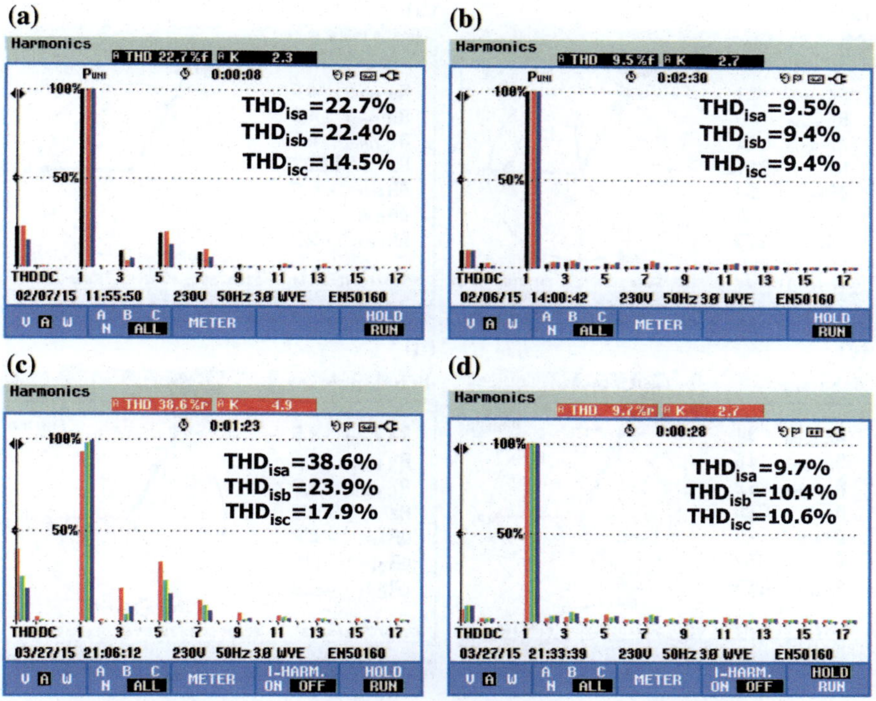

Fig. 9.19 Experimental source current spectrums: **a** before compensation of unbalanced inductive loads, **b** after TCLC-HAPF compensation of unbalanced inductive loads, **c** before compensation of unbalanced mixed inductive and capacitive loads, **d** after TCLC-HAPF compensation of mixed inductive and capacitive loads

with the THD_{isx} less than 10.0%. Therefore, it can be said that the TCLC-HAPF obtains stable performance during the voltage dip.

9.5.5 Experimental Results of TCLC-HAPF Compensation Under Voltage Fault

Figure 9.22 provides the dynamic compensation waveforms of load voltage v_x and source current i_{sx} by applying TCLC-HAPF under the voltage fault condition. Figure 9.23 shows the source current harmonic spectrums before and after TCLC-HAPF compensation. During the voltage fault, the i_{sx} can be compensated to be approximately balanced with PF ≈ 1 and $THD_{isx} < 10.0\%$. From Fig. 9.22, it can be seen that the proposed TCLC-HAPF can still obtain satisfactory performances even under asymmetric grid fault.

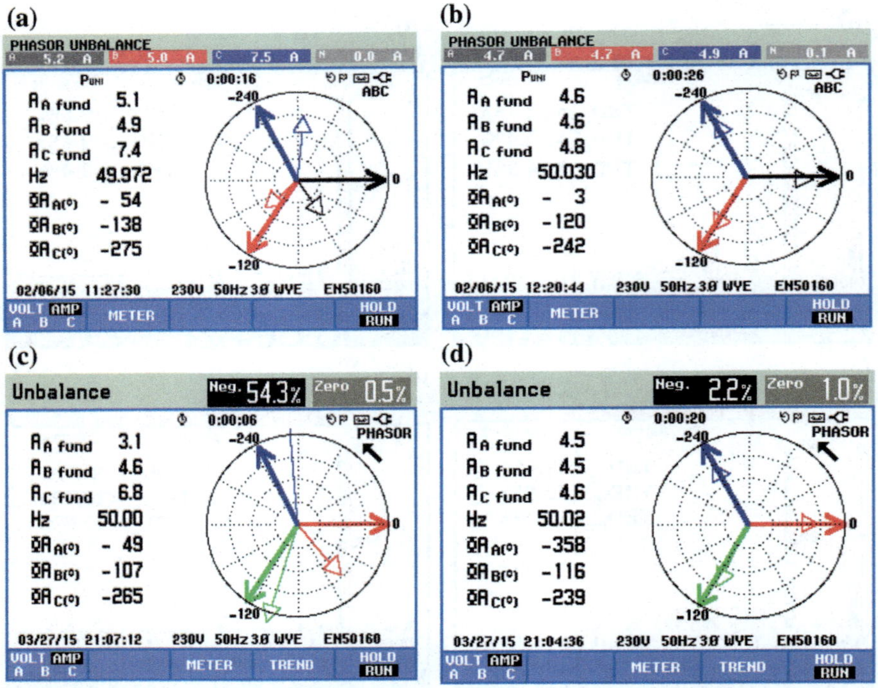

Fig. 9.20 Experimental phasor diagrams of source voltage and current: **a** before compensation of unbalanced inductive loads, **b** after TCLC-HAPF compensation of unbalanced inductive loads, **c** before compensation of unbalanced mixed inductive and capacitive loads, **d** after TCLC-HAPF compensation of mixed inductive and capacitive loads

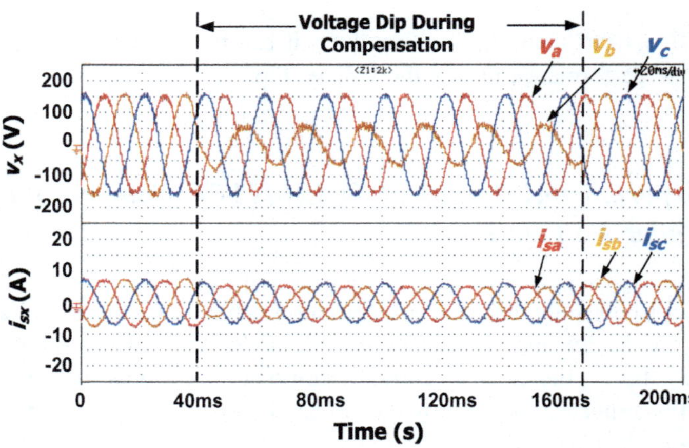

Fig. 9.21 Dynamic compensation waveforms of v_x and i_{sx} by applying TCLC-HAPF during voltage dip

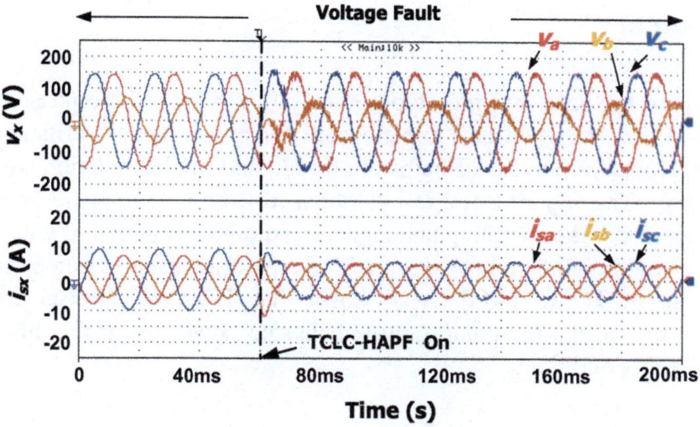

Fig. 9.22 Dynamic compensation waveforms of v_x and i_{sx} by applying TCLC-HAPF under voltage fault condition

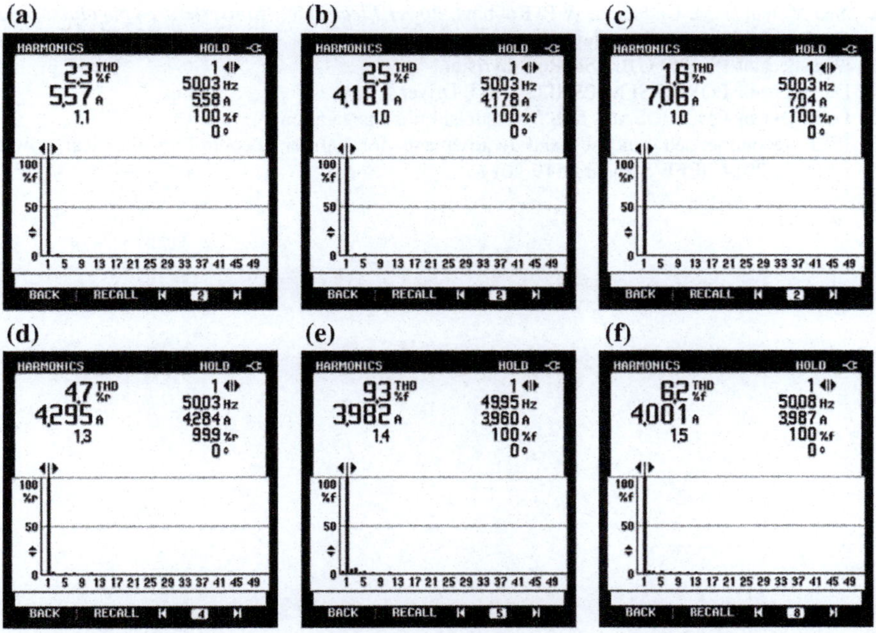

Fig. 9.23 Source current harmonic spectrum under voltage fault before compensation: **a** phase a, **b** phase b, **c** phase c, and after TCLC-HAPF compensation: **d** phase a, **e** phase b, **f** phase c

9.6 Summary

In this chapter, the hardware prototype of an 110 V-5 kVA three-phase three-wire TCLC-HAPF has designed and implemented for experimental verification of the system performances. Details of hardware design and implementation are provided in this chapter. The experimental results obtained from the hardware prototype are presented included: the sampling and signal conditioning circuits, thyristor drivers, IGBT drivers and digital control system. The experiments have been performed on the hardware of TCLC-HAPF. It is shown that TCLC-HAPF can compensate reactive power, current harmonics and unbalanced power under the different voltage and current conditions.

References

1. B.J. Baliga, Trends in power semiconductor devices. IEEE Trans. Electron. Dev. **43**(10), 1727–1731 (1996)
2. Ned Mahan, T.M. Undeland, W.P. Robbins, *Power Electronics, Converters, Applications, and Design*, 3rd edn. (Willey, New York, 2003)
3. Datasheet of PK110FG160 SanRex thyristor
4. Datasheet of POWERSEM PSHI23 IGBT Driver
5. Datasheet of PM300DSA60 IGBT Mitsubishi third generation
6. IEEE recommended practices and requirements for harmonic control in electrical power systems, 2014, IEEE Standard 519-2014

Chapter 10
Conclusions and Prospective for Further Work

Abstract In this chapter, conclusions are drawn for the study of the thyristors controlled LC-coupling hybrid active power filters (TCLC-HAPFs). Specifically, the detailed chapter summary and achievements are included. Also, some possible future research works are provided.

Keywords Thyristor-controlled LC-coupling hybrid active power filter (TCLC-HAPF) · Hybrid-coupled static synchronous compensator (hybrid-STATCOM) · Low dc-link voltage · Reactive power · Current harmonics · Wide compensation range · Unbalanced compensation

10.1 Conclusions

This book discusses the design and control of a wide compensation range and low DC-link voltage TCLC-HAPF for harmonic current, reactive power and unbalanced power compensation in three-phase three-wire distribution power system. And, an 110 V-5 kVA experimental prototype is designed and tested to show its validity. The key contributions and achievements of each chapter are summarized in the following:

(1) The overview and summary of the major power quality issues (low power factor, harmonic pollution and unbalanced problem) have been provided by using the power quality measurement data in Macau in Chap. 1. Moreover, there is a large potential market for reactive power and harmonic compensation in China. In order to solve the above major power quality issues, the historical review of different existing power quality compensators topologies and control methods are also provided in Chap. 1.
(2) The TCLC-HAPF and other power quality compensators are compared in Chap. 2 in terms of V-I characteristic, cost, power loss, reliability and tracking performance. Among them, the TCLC-HAPF is chosen for further research investigation since TCLC-HAPF has relatively lower cost, higher reliability,

L. Wang et al., *Adaptive Hybrid Active Power Filters*, Power Systems,
https://doi.org/10.1007/978-981-10-8827-8_10

lower loss than the multilevel APF and HAPF. Also, the TCLC-HAPF obtains wider compensation range and better tracking performance than the HAPF.

(3) Two techniques are proposed in Chap. 3 for TCLC part and active inverter part of TCLC-HAPF. For TCLC part, the solution to prevent harmonic current injection by the thyristor pair is proposed and proved by simulation and experimental results. For active inverter part, a nonlinear hysteresis PWM method is proposed to reduce the power loss while keeping the current harmonics level at an acceptable level, which is proved by simulation results.

(4) The unbalanced parameter design method of TCLC-HAPF is proposed in Chap. 4 in comparison with the conventional balanced parameter design method. Through the simulation and experimental results, it is proved that the proposed unbalanced parameter design method for TCLC-HAPF is capable for balancing the source active power, and compensating load reactive power and harmonic current under unbalanced loading, while the balanced parameter design method cannot achieve unbalanced compensation with satisfactory results.

(5) The control strategy for TCLC-HAPF is investigated in Chap. 5 to compensate reactive power, harmonic current and balance active power under unbalanced loading with low DC-link voltage. The validity of the proposed control strategy is proved by simulation and experimental results.

(6) The minimizing inverter capacity design and comparative performance evaluation of different SVC-HAPFs (FC-TCR-HAPF and TCLC-HAPFs) are proposed in Chap. 6. Through the minimizing inverter capacity design, simulation and experimental performance evaluations, the TCLC-HAPF structure is final selected with acceptable compensation performance and low inverter capacity.

(7) A simplified minimum DC-link voltage calculation method is proposed in Chap. 7 for TCLC-HAPF reactive power and current harmonics compensation, which can significantly reduce the large amount of the calculation steps by using the FFT method. Also, an adaptive DC-link voltage controller for the TCLC-HAPF is developed to dynamically keep its operating at its minimum DC-link voltage level to reducing its switching loss and switching noise.

(8) The selective compensation control method of harmonic distortion, unbalanced and reactive power of TCLC-HAPF is developed in Chap. 8. The proposed selective control method can provide satisfactory compensation performance, when the loads required harmonic, unbalanced and reactive power is beyond the limited capacity of the TCLC-HAPF.

(9) An 110 V-5 kVA three-phase three-wire TCLC-HAPF experimental prototype is constructed in the laboratory and presented in this chapter, to test the validity of the proposed parameter design methods and control strategies. Experimental results show that the TCLC-HAPF can effectively compensate the harmonic current and reactive power, and balance active power under the different voltage and current conditions like unbalanced current, voltage fault, voltage dip, etc.

10.2 Perspectives for Future Works

Based on the present research, the following points are suggested for future studies.

1. As hysteresis PWM control is simple and widely applied. The nonlinear hysteresis PWM control is proposed for TCLC-HAPF in this book, and other PWM control strategies can be further investigated for TCLC-HAPF system such as: space vector modulation (SVM), deadbeat PWM, etc.
2. The control systems of the TCLC-HAPF hardware prototype are using two parallel DSP-TMS320F2812s. Meanwhile, the additional extension board is required to fit for the physical connection between DSPs and the drivers of IGBT and thyristor. Therefore, the industrial control controller/system is required to further develop to improve the control efficiency of the TCLC-HAPF.
3. The TCLC-HAPF is a cost-effective solution for medium/high voltage level reactive power, harmonic current and unbalanced power compensation. Therefore, the increase of power rating of the TCLC-HAPF hardware system is necessary to show the advantages of the TCLC-HAPF, e.g. 10 kV.
4. The necessary protection is very important for high voltage application. Otherwise, the system components can be easily damaged due to the suddenly voltage and current changes. The different protection techniques and devices are necessary to be studied and considered in the future.

Biography of Authors

Lei Wang received the B.Sc. degree in Electrical and Electronics Engineering from University of Macau (UM), Macao, China, in 2011, M.Sc. degree in Electronics Engineering from Hong Kong University of Science and Technology (HKUST), Hong Kong, China, in 2012, and Ph.D. degree in Electrical and Computer Engineering from University of Macau (UM), Macao, China, in 2017.

Currently, he is a Postdoctoral Fellow in the Power Electronics Laboratory of University of Macau. His research interests included power electronics, power quality and distribution flexible AC transmission system (DFACTS), power quality compensation, and renewable energy.

He received the Champion Award in the "Schneider Electric Energy Efficiency Cup" competition held in Hong Kong, in 2011, and the Macao Science and Technology R&D Award for Postgraduates (Ph.D.) in 2018.

L. Wang et al., *Adaptive Hybrid Active Power Filters*, Power Systems,
https://doi.org/10.1007/978-981-10-8827-8

Man-Chung Wong, received his B.Sc. and M.Sc. degrees in Electrical and Electronics Engineering from University of Macau (UM), Macao, China in 1993 and 1997, respectively and Ph.D. degree in Electrical Engineering from Tsinghua University, Beijing, China in 2003. He was a Visiting Fellow in Cambridge University, UK in 2014. He is currently an Associate Professor in Department of Electrical and Computer Engineering, UM. His research interests are power electronics converters, pulse with modulation, active power filters, hybrid active power filters, and hybrid power quality compensator for high-speed railway power supply system. Most of the research results are published in world top journals such as IEEE transactions and IET journals. He has co-authored 4 Springer books, over 100 journal and conference papers in the area of power electronics, and 8 patents were granted in China and USA. Recently, an industrial power filter platform was developed and installed in a practical power system based on his research results.

He received several awards based on his research results such as Macao Young Scientific Award from Macau International Research Institute in 2000, Young Scholar Award of UM in 2001, Second Prize for Tsinghua University Excellent Ph.D. Thesis Award in 2003, and Macao Science and Technology Invention Award (Third-Class) in 2012 and 2014, respectively. He supervised several students to receive merit paper awards in conferences and champions in student project competitions.

He was several conference committee members and General Chair of IEEE TENCON 2015 Macau. He is with IEEE Macau Section and IEEE Macau Power Joint Chapter for many years. In years 2014–2015, he was the Chair of IEEE Macau Section. Recently, he is North Representative of IEEE Region 10 Power and Energy Society and Chair of IEEE Macau PES/PELS Joint Chapter.

Chi-Seng Lam received the B.Sc., M.Sc., and Ph.D. degrees in Electrical and Electronics Engineering from the University of Macau (UM), Macao, China, in 2003, 2006, and 2012, respectively. From 2006 to 2009, he was an Electrical and Mechanical Engineer in the Campus Development and Engineering Section, UM. In 2009, he returned to the Power Electronics Laboratory of UM to work as a Technician and at the same time started to pursue his Ph.D. degree in part-time, and completed his Ph.D. in less than 3 years. In 2013, he was a Postdoctoral Fellow in The Hong Kong Polytechnic University, Hong Kong, China. He is currently an Assistant Professor in the State Key Laboratory of Analog and Mixed-Signal VLSI, UM. He has co-authored/co-edited 4 books: *Design and Control of Hybrid Active Power Filters* (Springer, 2014), *Parallel Power Electronics Filters in Three-phase Four-wire Systems—Principle, Control and Design* (Springer, 2016), *Tutorials in Circuits and Systems Selected Topics in Power, RF, and Mixed-Signal ICs* (River Publishers, 2017), and *Adaptive Hybrid Active Power Filters* (Springer, in press), 4 US patents, 2 Chinese patents, and over 70 technical journals (IEEE-TIE, IEEE-TPEL, IEEE-TSG, IEEE-TIAS, IEEE-PWRD, IET-PEL, IET-EL, etc.) and conference papers. His research interests include integrated power electronics controllers, power management integrated circuits, power quality compensators, smart grid technology, renewable energy, etc.

He received the IEEE PES Chapter Outstanding Engineer Award in 2016, the Macao Science and Technology Invention Award (Third-Class) and R&D Award for Postgraduates (Ph.D.) in 2014 and 2012, respectively. He also received Macao Government Ph. D. Research Scholarship in 2009–2012, Macao Foundation Postgraduate Research Scholarship in 2003–2005, the 3rd RIUPEEEC Merit Paper Award in 2005, BNU Affinity Card Scholarship in 2001, as well as Award of Dean Honor List of UM, in 1999–2000, 2000–2001 2001–2002, and 2002–2003, respectively.

He is an IEEE Senior Member. In 2007, 2008, and 2015, he was the GOLD Officer, Student Branch Officer, and Secretary of IEEE Macau Section. He is currently the Vice-Chair of IEEE Macau Section, Chair of IEEE CAS Macau Chapter, and Secretary of IEEE

Macau PES/PELS Joint Chapter. He served as a member of Organizing Committee or Technical Program Committee of several international conferences. Since 2009, he is also a certified FMP (with Highest Score Prize awarded by Labour Affairs Bureau, Macao SAR, and Macau Institute of Management) of International Facility Management Association (IFMA).

Printed by Printforce, the Netherlands